Absorbable Metals for Biomedical Applications

Absorbable Metals for Biomedical Applications

Editors

Hendra Hermawan
Mehdi Razavi

MDPI • Basel • Beijing • Wuhan • Barcelona • Belgrade • Manchester • Tokyo • Cluj • Tianjin

Editors

Hendra Hermawan	Mehdi Razavi
Department of Mining, Metallurgical and Materials Engineering	Department of Internal Medicine, College of Medicine
Laval University	University of Central Florida
Quebec City	Orlando
Canada	United States

Editorial Office
MDPI
St. Alban-Anlage 66
4052 Basel, Switzerland

This is a reprint of articles from the Special Issue published online in the open access journal *Materials* (ISSN 1996-1944) (available at: www.mdpi.com/journal/materials/special_issues/absorbable_met_biomed_appl).

For citation purposes, cite each article independently as indicated on the article page online and as indicated below:

LastName, A.A.; LastName, B.B.; LastName, C.C. Article Title. *Journal Name* **Year**, *Volume Number*, Page Range.

ISBN 978-3-0365-1764-3 (Hbk)
ISBN 978-3-0365-1763-6 (PDF)

Contents

About the Editors . . . . **vii**

Preface to "Absorbable Metals for Biomedical Applications" . . . . **ix**

Hendra Hermawan and Mehdi Razavi
Special Issue "Absorbable Metals for Biomedical Applications"
Reprinted from: *Materials* **2021**, *14*, 3835, doi:10.3390/ma14143835 . . . . **1**

Gabriela Gasior, Jonasz Szczepański and Aleksandra Radtke
Biodegradable Iron-Based Materials—What Was Done and What More Can Be Done?
Reprinted from: *Materials* **2021**, *14*, 3381, doi:10.3390/ma14123381 . . . . **3**

Karel Klíma, Dan Ulmann, Martin Bartoš, Michal Španko, Jaroslava Dušková, Radka Vrbová, Jan Pinc, Jiří Kubásek, Tereza Ulmannová, René Foltán, Eitan Brizman, Milan Drahoš, Michal Beňo and Jaroslav Čapek
Zn–0.8Mg–0.2Sr (wt.
Reprinted from: *Materials* **2021**, *14*, 3271, doi:10.3390/ma14123271 . . . . **29**

Giorgia Fedele, Sara Castiglioni, Jeanette A. Maier and Laura Locatelli
High Magnesium and Sirolimus on Rabbit Vascular Cells—An In Vitro Proof of Concept
Reprinted from: *Materials* **2021**, *14*, 1970, doi:10.3390/ma14081970 . . . . **51**

Tycho Zimmermann, Norbert Hort, Yuqiuhan Zhang, Wolf-Dieter Müller and Andreas Schwitalla
The Video Microscopy-Linked Electrochemical Cell: An Innovative Method to Improve Electrochemical Investigations of Biodegradable Metals
Reprinted from: *Materials* **2021**, *14*, 1601, doi:10.3390/ma14071601 . . . . **61**

Yafeng Wen, Qingshan Liu, Weikang Zhao, Qiming Yang, Jingfeng Wang and Dianming Jiang
In Vitro Studies on Mg-Zn-Sn-Based Alloys Developed as a New Kind of Biodegradable Metal
Reprinted from: *Materials* **2021**, *14*, 1606, doi:10.3390/ma14071606 . . . . **77**

Cortino Sukotjo, Tiburtino J. Lima-Neto, Joel Fereira Santiago Júnior, Leonardo P. Faverani and Michael Miloro
Is There a Role for Absorbable Metals in Surgery? A Systematic Review and Meta-Analysis of Mg/Mg Alloy Based Implants
Reprinted from: *Materials* **2020**, *13*, 3914, doi:10.3390/ma13183914 . . . . **95**

Mehdi Razavi, Mohammadhossein Fathi, Omid Savabi, Lobat Tayebi and Daryoosh Vashaee
Biodegradable Magnesium Bone Implants Coated with a Novel Bioceramic Nanocomposite
Reprinted from: *Materials* **2020**, *13*, 1315, doi:10.3390/ma13061315 . . . . **115**

Radka Gorejová, Renáta Oriňaková, Zuzana Orságová Králová, Matej Baláž, Miriam Kupková, Monika Hrubovčáková, Lucia Haverová, Miroslav Džupon, Andrej Oriňak, František Kaľavský and Karol Kovaľ
In Vitro Corrosion Behavior of Biodegradable Iron Foams with Polymeric Coating
Reprinted from: *Materials* **2020**, *13*, 184, doi:10.3390/ma13010184 . . . . **131**

Shintaro Sukegawa, Hotaka Kawai, Keisuke Nakano, Kiyofumi Takabatake, Takahiro Kanno, Hitoshi Nagatsuka and Yoshihiko Furuki
Advantage of Alveolar Ridge Augmentation with Bioactive/Bioresorbable Screws Made of Composites of Unsintered Hydroxyapatite and Poly-L-lactide
Reprinted from: *Materials* **2019**, *12*, 3681, doi:10.3390/ma12223681 . . . . **149**

Devi Paramitha, Stéphane Chabaud, Stéphane Bolduc and Hendra Hermawan
Biological Assessment of Zn–Based Absorbable Metals for Ureteral Stent Applications
Reprinted from: *Materials* **2019**, *12*, 3325, doi:10.3390/ma12203325 . . . . **159**

Genghua Cao, Lu Zhang, Datong Zhang, Yixiong Liu, Jixiang Gao, Weihua Li and Zhenxing Zheng
Microstructure and Properties of Nano-Hydroxyapatite Reinforced WE43 Alloy Fabricated by Friction Stir Processing
Reprinted from: *Materials* **2019**, *12*, 2994, doi:10.3390/ma12182994 . . . . **175**

Shintaro Sukegawa, Takahiro Kanno, Norio Yamamoto, Keisuke Nakano, Kiyofumi Takabatake, Hotaka Kawai, Hitoshi Nagatsuka and Yoshihiko Furuki
Biomechanical Loading Comparison between Titanium and Unsintered Hydroxyapatite/Poly-L-Lactide Plate System for Fixation of Mandibular Subcondylar Fractures
Reprinted from: *Materials* **2019**, *12*, 1557, doi:10.3390/ma12091557 . . . . **189**

Sébastien Champagne, Ehsan Mostaed, Fariba Safizadeh, Edward Ghali, Maurizio Vedani and Hendra Hermawan
In Vitro Degradation of Absorbable Zinc Alloys in Artificial Urine
Reprinted from: *Materials* **2019**, *12*, 295, doi:10.3390/ma12020295 . . . . **201**

Marcjanna Maria Gawlik, Björn Wiese, Valérie Desharnais, Thomas Ebel and Regine Willumeit-Römer
The Effect of Surface Treatments on the Degradation of Biomedical Mg Alloys—A Review Paper
Reprinted from: *Materials* **2018**, *11*, 2561, doi:10.3390/ma11122561 . . . . **215**

About the Editors

Hendra Hermawan

Dr. Hendra Hermawan is an associate professor at the Department of Mining, Metallurgical and Materials Engineering, Universite Laval, Canada. He obtained his BS and MS from Institut Teknologi Bandung in 2002 and his PhD from Universite Laval in 2009, all in materials engineering. He is a professional engineer registered with the Professional Engineers Ontario, Canada. His research focuses on metallurgy and degradation aspects of metals for biomedical applications mostly funded by public funding agencies. He serves the scientific community as a member of editorial board of several journals and as a reviewer for papers and grants related to materials science and engineering.

Mehdi Razavi

Dr. Mehdi Razavi has been an assistant professor of Medicine, Materials Science and Engineering, and BiionixTM (Bionic Materials, Implants & Interfaces) Cluster at the University of Central Florida College of Medicine (UCF COM) since September 2019. He is also Director of the Regenerative Medicine and Nanoengineering Laboratory (RegeN) at UCF Burnett School of Biomedical Sciences. His research interests mainly encompass biomaterials, nanomedicine, tissue engineering and regenerative medicine, stem cells, orthopedic implants and bone scaffolds, drug/gene delivery systems, ultrasound therapy, cancer nanotechnology, and pancreatic islet transplantation.

Preface to "Absorbable Metals for Biomedical Applications"

Absorbable metals, as the ASTM and ISO standards named them, also known as biodegradable metals, are metals and alloys that are intended for use in biomedical applications, mainly as materials for temporary implants, such as endovascular stents, bone plates and screws, and porous scaffolds. They are expected to be completely degraded and absorbed in the body after providing a needed function, thus eliminating the harmful potential effects of permanent implants. The introduction of these metals has shifted the established paradigm of metal implants from preventing corrosion to taking advantage of it. Interest in these metals has been growing in the past decade, as indicated by the rapid increase in scientific publications, the progressive development of standards, and the launching of commercial products. The families of absorbable metals can be grouped into iron, magnesium, zinc, and their alloys. The magnesium group is considered as the most studied family, both in basic and translational research. The iron group has been subjected to vigourous research, with significant efforts to overcome its major challenge of low in vivo corrosion rates, while the zinc group is the newly added member of absorbable metals. ASTM has officially published the F3160 standard guide for metallurgical characterization of absorbable metals and the ASTM F3268 standard guide for in vitro degradation testing of absorbable metals, while the ISO is coordinating an "umbrella"document with standards that are related to biological evaluation. This book aims to present some of the latest works in the research and development of absorbable metals to solicit the most important findings, to highlight the remaining challenges, and to provide the perspectives on future directions.

Hendra Hermawan, Mehdi Razavi

Editors

ditorial

Special Issue "Absorbable Metals for Biomedical Applications"

Hendra Hermawan [1,*] and Mehdi Razavi [2]

1 Department of Mining, Metallurgical and Materials Engineering, Université Laval, Quebec City, QC G1V 0A6, Canada
2 Biionix^{TM} (Bionic Materials, Implants & Interfaces) Cluster, Department of Internal Medicine, College of Medicine, University of Central Florida, Orlando, FL 32827, USA; Mehdi.Razavi@ucf.edu
* Correspondence: hendra.hermawan@gmn.ulaval.ca

Current temporary metal implants made from titanium or stainless steel are not absorbable. Thus, a second surgery is needed to remove the implants after the tissue heals, or (for children) after they outgrow their implants. This second surgery is costly and involves additional risks of infections and pain for the patient. Therefore, absorbable metals are currently generally preferred and being investigated. Absorbable metals have shown significant clinical potential as an alternative to polymers in implant applications, where the material is eventually replaced by healthy, functioning tissue. However, several challenges remain before these metals can be translated to humans. First, the metal alloys with sufficient strength contain aluminum, yttrium, or lithium, all of which pose serious concerns for long-term toxicity. Second, in some cases such as magnesium alloys, they degrade too rapidly, and as a result, also generate possibly harmful hydrogen gas pockets. Consequently, these implants lose their mechanical integrity before the host tissue heals.

Innovations and further improvements are required, especially for load-bearing implants. The main focus of this Special Issue is to therefore collect scientific contributions dealing with the development of absorbable metals with improved and unique corrosion and mechanical properties for applications in highly loaded implants, or cardiovascular and urethral stents. This Special Issue assembles a group of highly original manuscripts that present a range of exciting innovations in alloying [1–3] and compositing [4], along with their testing and assessments [5–7] to introduce novel medical implants based on magnesium [2–4,8,9], zinc [1,7], or iron [10,11]. As the biointerface plays an important role in implant–tissue interactions, contributions to implant coating and surface engineering strategies and their effects on the implant properties and corrosion are also discussed [8,10,12]. Two studies on mechanical testing of implants made of biodegradable polymer composites provide a complimentary benchmark toward clinical application of absorbable metal implants [13,14].

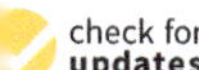

tation: Hermawan, H.; Razavi, M. ecial Issue "Absorbable Metals for omedical Applications". *Materials* 21, *14*, 3835. https://doi.org/ .3390/ma14143835

ceived: 5 July 2021
cepted: 8 July 2021
blished: 9 July 2021

blisher's Note: MDPI stays neutral th regard to jurisdictional claims in blished maps and institutional affil- ions.

Finally, the concept of absorbable metal implants might result in a significant impact on the future work of standardization agencies. However, standardization must be balanced with the main challenge in the field, which remains the successful translation of this innovative concept into medical products that guarantee the growth of the absorbable metal field. Genetic-based methods and biomarkers should be studied with more depth to evaluate their biocompatibility and bioactivity.

Contributions have been solicited from scientists working in the fields of biomaterials, tissue engineering, bioengineering, and medicine. Finally, the Editors give special thanks to the authors, and to the editorial team of *Materials*, for collaborative and peer-review process.

We hope you will enjoy reading this issue as much as we had the pleasure of assembling it.

Funding: This research received no external funding.

Conflicts of Interest: The authors declare no conflict of interest.

References

1. Klíma, K.; Ulmann, D.; Bartoš, M.; Španko, M.; Dušková, J.; Vrbová, R.; Pinc, J.; Kubásek, J.; Ulmannová, T.; Foltán, R.; et a Zn–0.8 Mg–0.2 Sr (wt.%) Absorbable Screws—An In-Vivo Biocompatibility and Degradation Pilot Study on a Rabbit Mode *Materials* **2021**, *14*, 3271. [CrossRef] [PubMed]
2. Fedele, G.; Castiglioni, S.; Maier, J.A.; Locatelli, L. High Magnesium and Sirolimus on Rabbit Vascular Cells—An In Vitro Proof o Concept. *Materials* **2021**, *14*, 1970. [CrossRef] [PubMed]
3. Wen, Y.; Liu, Q.; Zhao, W.; Yang, Q.; Wang, J.; Jiang, D. In Vitro Studies on Mg-Zn-Sn-based Alloys Developed as a New Kind o Biodegradable Metal. *Materials* **2021**, *14*, 1606. [CrossRef] [PubMed]
4. Cao, G.; Zhang, L.; Zhang, D.; Liu, Y.; Gao, J.; Li, W.; Zheng, Z. Microstructure and Properties of Nano-hydroxyapatite Reinforce WE43 Alloy Fabricated by Friction Stir Processing. *Materials* **2019**, *12*, 2994. [CrossRef] [PubMed]
5. Zimmermann, T.; Hort, N.; Zhang, Y.; Müller, W.-D.; Schwitalla, A. The Video Microscopy-linked Electrochemical Cell: A Innovative Method to Improve Electrochemical Investigations of Biodegradable Metals. *Materials* **2021**, *14*, 1601. [CrossRef [PubMed]
6. Champagne, S.; Mostaed, E.; Safizadeh, F.; Ghali, E.; Vedani, M.; Hermawan, H. In Vitro Degradation of Absorbable Zinc Alloy in Artificial Urine. *Materials* **2019**, *12*, 295. [CrossRef] [PubMed]
7. Paramitha, D.; Chabaud, S.; Bolduc, S.; Hermawan, H. Biological Assessment of Zn–based Absorbable Metals for Ureteral Sten Applications. *Materials* **2019**, *12*, 3325. [CrossRef] [PubMed]
8. Razavi, M.; Fathi, M.; Savabi, O.; Tayebi, L.; Vashaee, D. Biodegradable Magnesium Bone Implants Coated with a Nove Bioceramic Nanocomposite. *Materials* **2020**, *13*, 1315. [CrossRef] [PubMed]
9. Sukotjo, C.; Lima-Neto, T.J.; Santiago Júnior, J.F.; Faverani, L.P.; Miloro, M. Is There a Role for Absorbable Metals in Surgery? A Systematic Review and Meta-Analysis of Mg/Mg Alloy Based Implants. *Materials* **2020**, *13*, 3914. [CrossRef] [PubMed]
10. Gorejová, R.; Oriňaková, R.; Orságová Králová, Z.; Baláž, M.; Kupková, M.; Hrubovčáková, M.; Haverová, L.; Džupon, M Oriňak, A.; Kaľavský, F.; et al. In Vitro Corrosion Behavior of Biodegradable Iron Foams with Polymeric Coating. *Materials* **202** *13*, 184. [CrossRef] [PubMed]
11. Gąsior, G.; Szczepański, J.; Radtke, A. Biodegradable Iron-Based Materials—What Was Done and What More Can Be Done *Materials* **2021**, *14*, 3381. [CrossRef] [PubMed]
12. Gawlik, M.M.; Wiese, B.; Desharnais, V.; Ebel, T.; Willumeit-Römer, R. The Effect of Surface Treatments on the Degradation c Biomedical Mg Alloys—A Review Paper. *Materials* **2018**, *11*, 2561. [CrossRef] [PubMed]
13. Sukegawa, S.; Kawai, H.; Nakano, K.; Takabatake, K.; Kanno, T.; Nagatsuka, H.; Furuki, Y. Advantage of Alveolar Ridg Augmentation with Bioactive/Bioresorbable Screws Made of Composites of Unsintered Hydroxyapatite and Poly-L-lactid *Materials* **2019**, *12*, 3681. [CrossRef] [PubMed]
14. Sukegawa, S.; Kanno, T.; Yamamoto, N.; Nakano, K.; Takabatake, K.; Kawai, H.; Nagatsuka, H.; Furuki, Y. Biomechanica Loading Comparison between Titanium and Unsintered Hydroxyapatite/Poly-L-Lactide Plate System for Fixation of Mandibula Subcondylar Fractures. *Materials* **2019**, *12*, 1557. [CrossRef] [PubMed]

 materials

MDPI

Review

Biodegradable Iron-Based Materials—What Was Done and What More Can Be Done?

Gabriela Gąsior, Jonasz Szczepański and Aleksandra Radtke *

Faculty of Chemistry, Nicolaus Copernicus University in Toruń, Gagarina 7, 87-100 Toruń, Poland; ggasior@doktorant.umk.pl (G.G.); jszcze@gmail.com (J.S.)
* Correspondence: aradtke@umk.pl; Tel.: +48-60032129

Abstract: Iron, while attracting less attention than magnesium and zinc, is still one of the best candidates for biodegradable metal stents thanks its biocompatibility, great elastic moduli and high strength. Due to the low corrosion rate, and thus slow biodegradation, iron stents have still not been put into use. While these problems have still not been fully resolved, many studies have been published that propose different approaches to the issues. This brief overview report summarises the latest developments in the field of biodegradable iron-based stents and presents some techniques that can accelerate their biocorrosion rate. Basic data related to iron metabolism and its biocompatibility, the mechanism of the corrosion process, as well as a critical look at the rate of degradation of iron-based systems obtained by several different methods are included. All this illustrates as the title says, what was done within the topic of biodegradable iron-based materials and what more can be done.

Keywords: iron; corrosion rate; biodegradable material; biocompatibility; stent

check for updates

Citation: Gąsior, G.; Szczepański, J.; Radtke, A. Biodegradable Iron-Based Materials—What Was Done and What More Can Be Done? *Materials* 2021, *14*, 3381. https://doi.org/10.3390/ma14123381

Academic Editors: Hendra Hermawan and Mehdi Razavi

Received: 13 May 2021
Accepted: 17 June 2021
Published: 18 June 2021

1. Introduction

Coronary artery disease (CAD) is characterised by narrowing of the blood vessels that supply oxygenated blood to cardiac muscles: it is responsible for around 20% of all deaths in developed countries [1]. In 1977, for the first time, an angioplasty was performed. This procedure, using a balloon inserted into a narrowed blood vessel and then inflated, enabled the vessel to be restored and prolonged the life of the first 38-year-old patient by 37 years [2] Balloon angioplasty, however, was limited by unpredictable vessel dissection and recoil and by the high rate of restenosis [3]. Therefore, the next revolution in the treatment of cardiovascular medicine was the introduction of stents, which resulted in both better early results and lower rates of restenosis. At the same time, there were limitations due to stent thrombosis and neointimal hyperplasia resulting in vasoconstriction [4].

The first coronary stent made from stainless steel used in surgery was introduced in 1987 by Siegward [5]. Currently, the clinical uses of coronary stents are made from either 316 L stainless steel, Co–Cr or TiNi alloys [3]. There are two main types of metal stents: self-expanding stents, confined by a sheath that can be removed after delivery of the device, and balloon-expandable stents that are mounted on a balloon catheter that is inflated to deploy the device. Self-expanding stents have technical limitations and a tendency to induce greater neointimal hyperplasia. Therefore, balloon-expandable stents are used in most coronary stent procedures [2].

In 2007, Mani et al. formulated nine features that an ideal stent should have: (1) good expandability ratio; (2) ability to be crimped on the balloon catheter; (3) sufficient flexibility; (4) sufficient radial hoop strength and negligible recoil; (5) non-toxicity for tissues and all organisms; (6) high thromboresistivity; (7) absence of restenosis after implantation; (8) drug delivery capacity; and (9) adequate radiopacity/magnetic resonance imaging (MRI) compatibility [6].

In the meantime, the idea of biodegradable (or bioresorbable) stents arose. The biggest advantage of biodegradable stents (BDS) is that they disappear when they are no longer needed, which is about six months after the implantation. In this way, all late stent complications, like permanently diminished flow of covered side branches, bleeding problems associated with long term anticoagulation, permanent late fracture abnormal vasomotion and CT/MRI imaging artefacts are omitted [7–11]. At the same time, BDS provides mechanical support analogous to bare-metal stents. It is also a better solution for still-growing children because it helps avoids a second intervention to remove the implant [12]. This is why the list of features of an ideal stent should include a tenth feature: fully biodegradable.

Two types of materials are used to create bioresorbable stents or scaffolds (BRS): polymers and metals. Initially, more attention was paid to polymers, and already in 1998, so over 20 years ago, a scaffold composed of high-molecular-weight poly-L-lactic acid (PLLA monofilaments) was implanted per Igaki-Tamai into a human coronary artery [13]. The first report, where a total of 25 scaffolds were successfully implanted into 19 lesions of 15 patients, were described and published in 2000 [13]. Long-term (>10 years) studies in 50 patients showed that, after three years, no traces of the stent scaffold in the blood vessel could be detected [14]. The results were great, but the device failed to progress as it required a larger guide catheter for implantation than a metal stent, it needed a heated contrast, and it had the lack of a drug coating [14]. However, the proposal to use PLLA in stents was not forgotten, and research is still ongoing. In 2016, the Absorb drug-eluting device, based on PLLA, was the first bioresorbable cardiovascular scaffold approved for use in the United States by the U.S. Food and Drug Administration [2]. Other biodegradable polymers and copolymers used for research include: poly(ε-caprolactone) (PCL), poly(L-lactide-co-ε-caprolactone) (PLCL), phosphoryl choline (ChoP), poly(desaminotyrosyl-tyrosine ethyl ester) carbonate (PTD-PC), poly(anhydride ester) (PAE), poly(D,L-lactide-co-glycolide) (PLGA), and poly(vinylidenefluoride)-hexafluoropropylene (PVDF-HFP) [6,13,15–18]. Despite very promising results, the polymers also have several disadvantages that limit their use. Compared to metals, polymers have lower values of Young's modulus (0.2–7.0 GPa) than those of metals (54–200 GPa), and generally have poorer mechanical properties [19]. This makes the spacers in polymer stents thicker than in metal stents, which results in the impossibility of complete expansion as the balloon expands [2]. Therefore, more and more research is being done to create a biodegradable metal stent. This would allow the advantage of polymeric and metallic devices to be combined.

Metals degrade in the body through corrosion. Therefore, metals used in first generation stents, such as stainless steel, nitinol or titanium, which have a high corrosion resistance factor, cannot be used as resorbable materials. From research conducted over the past 20 years, three main metals have emerged that could potentially form biodegradable cardiovascular implants: Fe, Mg and Zn [20–22]. This review summarises the achievements in developing fully biodegradable iron-based stents, but the most important features of the next two metals should be mentioned. Magnesium BDS are completely biocompatible and have good mechanical properties. However, magnesium has a high corrosion rate, which means it loses integrity when it is still needed. Moreover, it releases hydrogen during degradation that is harmful to cells. Despite their drawbacks, magnesium-based devices were the first to be approved for clinical trials and are now commercially available, for example, in Biotronik's Magmaris [23]. Zinc is characterised by good biocompatibility and a corrosion factor adequate to the desired lifetime of the stent. However, its mechanical properties are too weak (Table 1). It is necessary to introduce modifications to improve the mechanical parameters and, at the same time, not affect the corrosion time [24].

Iron has a high strength, ductility, and formability (Table 1), allowing stents with thinner constructions and struts or fabrication of special shapes, like foils or foams [25]. Unfortunately, in comparison to Zn and Mg, iron has a corrosion rate so low that pure iron can hardly be called "biodegradable". But due to its biocompatibility and excellent mechanical properties, it is worth thinking about modifications that could accelerate corrosion.

Table 1. The mechanical properties of biodegradable metals.

	Yield Strength (MPa)	Young's Modulus (GPa]	Tensile Strength (MPa)	Shear Modulus (GPa)	Elastic Modulus (GPa)	Hardness (HV)	Ref
Mg	51	44–45.5	175–235	16–18	44–48	38	[26,27]
Zn	285–325	90–110	90–200	35–45	14–32	42	[28]
Fe	108–122	204–212	230–345	78–84	195–235	157	[29,30]

In the following sections of the article, readers will find the information about the iron properties, its biodegradability and its corrosion test, which can be carried out in immersion mode and during the electrochemical testing. Biological properties of iron-based materials, in terms of tissue biocompatibility, cellular biocompatibility, hemocompatibility, and clinical biocompatibility are presented and discussed also. A critical look at the rate of degradation of systems obtained by several different synthesis methods, including: spark plasma sintering, vacuum induction melting, vacuum arc melting, electroforming, powder metallurgy and template-based synthesis of porous materials, as well as by the addition of another phase to the iron, will allow the reader to select methods which are still worth optimizing because they give hope for their use in biomedical applications, and those that they do not provide any chance of obtaining iron-based material as an optimally biodegradable system.

2. Biodegradability of Pure Iron

In 2001, the first results of an in vivo study by Peuster et al. showed that pure iron degrades too slowly in the body to be considered a bioabsorbable material [31]. There was a need to modify the iron to increase its corrosion rate without losing its biocompatibility and good mechanical properties. Since then, several studies have been carried out to understand how the working environment of an implant affects the rate of corrosion and the products that result from this process.

In a physiological environment, the corrosion rate of Fe highly depends on the rate of the cathodic reaction and the amount of dissolved oxygen. An increase in the oxygen levels in the corrosive environment will also increase the iron corrosion rate. In contrast, a decrease in the amount of oxygen in contact with Fe causes a reduction in the corrosion rate [32]:

$$Fe \rightarrow Fe^{2+} + 2e^{-} \text{ (anodic reaction)} \quad (1)$$

Electrochemical equations for iron corrosion work as follows [33]:

$$O_2 + 2H_2O + 4e^{-} \rightarrow 4OH^{-} \text{ (cathodic reaction)} \quad (2)$$

As predicted in the Pourbaix diagram, ferrous hydroxide is the most common corrosion product due to the reaction:

$$Fe^{2+} + 2OH^{-} \rightarrow Fe(OH)_2 \text{ (products of the reaction)} \quad (3)$$

Other common degradation products of Fe corrosion at physiological pH are the coordination compounds of iron oxides ($FeO \cdot Fe_2O_3$). Since the human body environment is full of sodium, chloride, carbon and phosphorus, other products of iron degradation, like iron carbonates, iron phosphates, and iron chlorides, were observed [34]. That's why Qui et al. modified the Pourbaix diagram for Fe in an aqueous environment to consider the environmental conditions that the implant encounters in the body (T = 37 °C., $[Fe^{2+}] = 1 \times 10^{-6}$ mol/dm^3, $[Fe^{3+}] = 3 \times 10^{-5}$ mol/dm^3, p (O_2), = 13.3 kPa [35]. This version of the diagram can be very useful when planning iron degradation experiments in a physiological environment.

Kraus et al. presented studies on the corrosion of pure iron after a 52-week in vi test [36]. It was found that the corrosion product initially consisted of a single oxide lay but became thicker over time to form two layers of a corrosion film. The first layer adjace to the Fe surface consisted of iron and oxygen. The second layer would form initially aft the oxidation of the reaction product from Equation (3) to form stable $Fe(OH)_3$:

$$Fe(OH)_2 + 1/2H_2O + 1/4O_2 \rightarrow Fe(OH)_3$$

It is worth noting here that very often, the simulated corrosion time based on in vit tests does not coincide with the results from in vivo tests. This is due to the fact that, und the conditions of the pseudo-physiological environment, not all factors influencing t course of corrosion are taken into account:

(1) Proteins: All pseudo-physiological fluids are designed to simulate the pH and ion concentrations of blood plasma. However, they do not contain proteins, which a the most important component of blood plasma. Proteins can bind metal ions a then transport them from the implant surface, but also they could create a layer adsorbed proteins, which then works as a barrier between the metal surface of t stent and the environment, thus inhibiting biodegradation. This same layer in slight different conditions could also limit the diffusion of oxygen to certain regions of t surface and, in this way, cause a breakdown of the passive layer and preferenti corrosion of oxygen-deficient regions [37].

(2) Atmosphere: most degradation tests are carried out in air. However, in blood vesse there is an atmosphere containing around 5% of CO_2, whose influence could chan the type of degradation products [38].

(3) Cells: During healing of the wound caused by the implantation operation, the surfa of the implant is covered with a layer of cells after a few weeks. When the barri formation is complete, the implant functions in a different environment than befo This may be the main reason for the differences in long-term in vivo and in vit corrosion [25].

(4) Dissolved oxygen: Blood has about 3 cm^3/dm^3 of dissolved oxygen, (4.3 mg/dm while, for example, most solutions used in cell-based tests (Hank's solution) ha almost twice the level of dissolved oxygen—about 8 mg/dm^3. This factor is probab the most important of all, and nevertheless it is very often omitted when simulati corrosion [35].

Corrosion Tests In Vitro

To fully understand the corrosive properties of an iron-based material for applicatio in a physiological environment, two different types of in vitro testing can be performe immersion testing and electrochemical testing.

Immersion tests are divided into static and dynamic types. In a static immersio test, regulated by the ASTM G31 standard, the sample is suspended in a suitable pseud physiological fluid (mimicking the ionic composition of the physiological environme to which the test material is to be targeted) for a specified period, usually between 1 ar 4 weeks [39]. The corrosion obtained from this test is measured in terms of mass loss. Th mass loss is related to a time-related unit through the following equation:

$$DR = kW/At\rho$$

where: W = mass loss [g]; A = specimen surface [cm^2]; t = exposure time [h]; k = conversio constant for computing the degradation rate whose actual value depends on the units the variables. Usually, DR is measured in mm/year—in this situation k = 87,600.

Unfortunately, the static immersion test doesn't consider the influence of the bloo flow inside the artery on the stent, which is really important when cardiovascular devic are tested. This is the reason to use a dynamic immersion test by using (mainly polymeri tubing for circulation, and a volume for the tested device, where the flow can be controlled in order to simulate the shear stress that the stent is subjected to under real conditions.

The most common system for electrochemical tests uses potentiodynamic polarisation (PDP), which enables the determination of the polarisation resistance of an iron-based material by performing a scan in a small potential range. The obtained data make it possible to calculate the current density, and then, by using this information under the ASTM G59 standard, the rate of material degradation can be estimated [40]. Alternatively, electrochemical impedance spectroscopy (EIS), according to ASTM G106, can also be used to model material degradation using equivalent electrical circuits [41].

3. Iron Metabolism

Iron is indispensable for life because it plays a crucial role in a wide range of vital biochemical activities. These activities include oxygen sensing, transport, short-term oxygen storage, catalysis, electron transfer and energy generation [42]. These functions are based on the chemical properties of iron, which can form a variety of coordination complexes, including those with organic ligands. These complexes are both dynamic and flexible. The redox potential between ferrous Fe(II) and ferric Fe(III) cations corresponds to the energies required to drive many biological reactions. The bioavailability of iron is limited because soluble Fe(II) (heme) is readily oxidised to Fe(III) (non-heme iron) under aerobic conditions, and Fe(III) is virtually insoluble under physiological pH conditions. The ability of Fe(II) to donate electrons and Fe(III) to accept electrons in the cellular environment is not only the basis of many biochemical reactions needed by the body but also poses a biological hazard [42]. Iron can be the initiator of reactions in which products are injurious radicals [43]. Catalytic quantities of iron produce hydroxyl radicals (OH^-) from superoxide ($O_2{}^-$) and hydrogen peroxide (H_2O_2). Redox-active iron also catalyses the generation of reactive organic species, including alkoxyl (ROS), thiyl (RS), thiyl-peroxyl (RSOOS) and peroxyl (ROOS) radicals [43]. Free radicals are highly reactive species that can promote oxidation, and subsequent modification, of proteins and nucleic acids. An increase in the steady state levels of reactive oxygen species, termed oxidative stress, can lead to a variety of inflammatory, neurodegenerative, or ischemic processes. It can also accelerate ageing of the body by accelerating tissue degeneration [43]. A deficiency of iron can lead to organism disorders and diseases [44,45]. In young children, iron deficiency causes disturbances in neurological and psychosomatically development, reduces learning efficiency, and may increase the likelihood of autism [46]. In adults, it causes fatigue, increases the risk of depression and impairs the thyroid gland. Iron deficiency is associated with adverse pregnancy outcomes, like increased maternal illness, low birth weight, prematurity, and intrauterine growth restriction [47].

As iron metabolism is similar in all mammals, animal studies give reliable results and predict long term implant behaviour in the human body [42]. The total body weight of a normal adult person contains 3–5 g of iron [42]. Most of this iron is assigned to erythroid precursors for the production of haemoglobin and mature erythrocytes [48]. The haemoglobin found in each erythrocyte accounts for over 1 billion atoms of iron. The total erythroid contains 30 mg of iron/kg body weight. The liver stores most of the rest of the body's iron. Iron enters the body with food by absorption by enterocytes in the duodenum. Enterocytes contain ferric reductase, which is an enzyme that ensures that iron is available in its ferrous state. The low pH in the gastric efflux facilitates absorption, which then reduces Fe(III) to Fe(II) and delivers it to a proton-rich milieu where it can be sent for further processing or storage [42].

Most of the absorbed iron will reach the plasma [48], while the protein ferritin provides safe storage of the iron fraction retained by the cell. When the lifespan of the erythrocyte ends, it is shed through the gastrointestinal tract together with any remaining stored iron. This represents a significant mode of iron release and loss from the body. Other mechanisms of iron release involve sweating, bleeding and excretion through the epidermis [42]. These alternative mechanisms are important as there are no regulated processes for iron excretion through the liver or kidney in humans. Therefore, there are limited excretion pathways for the iron ions released from a bioabsorbable implant. However, when the body's optimal iron levels are exceeded, the body can self-regulate, e.g., by lowering iron intake from food [48]. This brief overview of human iron uptake and excretion leads to the question: Which pathway(s) will be taken by the iron originating from implants to exit the body?

4. Biological Properties of Iron-Based Materials

Biocompatibility is the property of a material to fulfil its function in the host organism by interacting with living systems without any mechanical injury, toxicity to adjacent or distant tissues, or rejection by the immune system [49,50]. In the case of biodegradable materials, there is a need to consider the biocompatibility not only of iron implants but also their corrosion products. This biodegradability should be considered at several levels.

4.1. Cellular Biocompatibility

Biocompatibility on the cellular level requires that the presence of the biomaterial is not toxic for building the artery wall with endothelial cells and cells in the blood [49]. For this reason, endothelial cells, blood cells, or smooth muscle cells (SMC) are used as the basis for cytotoxicity stent material studies [25].

For example, Huang et al., in an in vitro cytotoxicity study on silver implanted pure iron, used murine fibroblast cells (L-929), human vascular smooth muscle cells (VSMC) and human umbilical vein endothelial cells (EA. hy-926) [51]. Chang et al. used the same set of cell lines in their research [52]. Also, Čapek et al. used the murine fibroblast L929 cell line in their study [53]. In an analogical test, Liu et al. used human umbilical vein endothelial cells (ECV304) and rodent vascular smooth muscle cells [54]. Another method was chosen by Paim et al. using extracted adipose-derived stem cells from the abdominal adipose tissue of a healthy adult donor undergoing tumescent liposuction [55]. The cells were next subjected to washing and centrifuge procedures.

Cytotoxicity tests can be introduced both: in direct contact of the implant with cells as well as in indirect contact. Since the cell colony environment usually has a small, closed system with a pseudo-physiological fluid, this may affect the corrosion of the implant by the high concentration of corrosion products. There is a constant washout into the bloodstream, which is why this problem does not occur in in vivo tests. Therefore, usually in vitro research is carried out without direct contact [56].

The IC_{50} index is widely utilised to indicate. Here, the IC_{50} is the concentration of metal required to kill 50% of cells, which gives a measure of the toxicity of the element at the cellular level [57]. Using a 50 μg/mL concentration of Fe, Zhu et al. observed no cytotoxic effect after one day on human endothelial cells [33]. Assuming that the stent completely degrades in an adult, 50 kg human with a circulating blood volume of about 2800 mL, the blood iron concentration will not exceed 7 μg/mL, hence no short-term risk of cytotoxicity. The more stringent standard is ISO 10993-5, which sets the cytotoxicity limit at the level of 30%. If the cell viability is reduced by more than 30%, the material is considered cytotoxic. Based on this recommendation, Paim et al. proved that iron is not cytotoxic and can be used as a material in biodegradable stents [55].

4.2. Tissue Biocompatibility

When the stent is implanted into the body during a surgical operation, a cascade of the following successive responses takes place: at first—injury, then—blood-material interactions, provisional matrix formation, acute and chronic inflammation, granulation tissue development, foreign-body reaction (FBR) and fibrosis. It is also accompanied by thrombus formation involving activation of the extrinsic and intrinsic coagulation systems, the complement, the fibrinolytic, and the kinin-generating systems, and platelets. A temporary "scaffold" is created around the implant with a sustained release of bioactive substances from the stent to accelerate the various stages of wound healing. This location is followed by short-term acute inflammation that then becomes chronic. Chronic inflammation around the implant usually lasts from one to two weeks. If it lasts longer than three weeks, it usually means infection [58]. It is essential to know these basics when performing in vivo testing to see when the body's response is unnatural and needs to be recorded. At this point, nobody reported problems with infections or rejection of the implant due to the presence of iron [55]. There is no indication of problems with the compatibility of tissues with iron. Simultaneously, during long-term in vivo tests in animals, several teams reported that the degradation of iron causes staining of surrounding tissues [31,37,59]. Degradation products may build up in the vessel walls. However, the coloured tissues did not show any infectious responses or other damage. Long-term animal studies have shown that tissue colour returns to normal after 53 months. Liu et al. suggested that macrophages clean the corrosion product before entering the lymphatic system and finally travelling to the lymph nodes [60].

4.3. Hemocompatibility

Since stents are used in the cardiovascular system, it is crucial to determine whether they have a negative effect on the blood. That is why platelet adhesion and hemolysis are investigated. Platelet adhesion is regulated by the ASTM F756-13 standard [61]. It determines how many platelets adhere to the material. If the number of adherent platelets is too high, the implant may be thrombogenic and may cause the formation of blood clots. The hemolysis test checks that the implant does not accelerate blood hemolysis by more than 5% with reference to the negative test. 5% is the value specified by the standard [61]. Usually, for the hemocompatibility tests, either human blood is taken from volunteers or animal blood, e.g., rabbit or sheep blood, is used [51,62–66].

Iron is hemocompatible since few platelets adhere to it, and the hemolysis rate is well below 5% [55]. The problem with increased platelet adhesion may arise if Zn or Ag are added to the alloy [51,67]. This fact is fascinating, especially because pure zinc does not adhere to platelets, and silver is added to many other biocompatible materials that does not affect their thrombogenic properties. Perhaps different types of processing, or an investigation of the percentage compositions, would help avoid this effect.

4.4. Clinical Biocompatibility

We find that in vivo studies of iron-based materials are performed much less frequently than in vitro ones. This is due to ethical regulations, the high cost of in vivo research, the need for additional equipment and skills, and their long-term nature. Additionally, not all of the developed tools and materials have the potential to provide clear-cut data from in vivo testing. Considering the process of metabolising iron in the body described earlier and the fact that the U.S. Food and Drug Administration does not include iron in its regulations regarding the doses of elements in drugs and medical products due to its low inherent toxicity, we can assume that the presence of Fe does not adversely affect the organs and functioning of the human body [60].

The research carried out so far proves that iron does not affect the host in any way, does not cause enlargement of glands or accumulation of metal in organs [31]. However, more detailed animal studies are needed, especially lasting a few years, in order to understand the long-term effects of iron and its corrosion products in the body. Table 2 summarizes and compares the current results of in vivo tests, taking into account the duration of the tests and the animal models for which they were conducted.

Table 2. Summary of published in vivo test.

Material	Shape	Animal	Implantation Place	Duration	Results	Ref
Pure iron	Stent	White rabbit	Native descending aorta	6, 12, 18 months	Toxicity wasn't observed, there was no neointimal proliferation and no excess inflammatory reaction	[31]
Pure iron	Zig-zag stent	Domestic swine	Left coronary ostium	28 days	Stents started degradation without signs of thrombosis or immoderate inflammation	[59]
Pure iron	Discs	Rats	Dorsal area	1 week, 3, 6 months	Lower degradation rate of iron samples in vivo than in vitro	[55]
Fe28.5Mn28.5Si	Rectangular	Wistar rats	Bone (tibiae) and subcutaneously (back)	14, 28 days	Concentration of iron at implant surface after 28 days significantly decreased in bone and subcutaneous implants, mass of bone implant decreased after implantation whilst mass of subcutaneous implant increased	[68]
Fe	Pin	Rats	Femoral bone	4, 12, 24, 52 weeks	Fe ions from degradation were transported 1mm into tissue surrounding implant, but no local toxicity	[36]
Fe10Mn1Pd	Pin	Rats	Femoral bone	4, 12, 24, 52 weeks	No significant change to degradation rate of implant, no local toxicity	[36]
Fe21Mn0.7C1Pd	Pin	Rats	Femoral bone	4, 12, 24, 52 weeks	Slower degradation rate than previous samples, but change was not significant, no local toxicity	[36]
Fe30Mn	Wire	Rats	Femoral bone	6 months	Bone on growth was observed for bone in contact with implant, alloy did not cause adverse effects, and an iron oxide layer was observed on the implant surface	[69]
Fe-5%HA composite	Plate	Sheep	Tibiae	3, 9, 14, 35, 50, and 70 days	Degradation rate was slower than degradation rate of pure iron implant	[34]
Fe-5%TCP composite	Plate	Sheep	Tibiae	3, 9, 14, 35, 50, and 70 days	Higher degradation rate in contrast with pure iron implant	[34]

Table 2. *Cont.*

Material	Shape	Animal	Implantation Place	Duration	Results	Ref
Fe-3%HA-2%TCP composite	Plate	Sheep	Tibiae	3, 9, 14, 35, 50, and 70 days	Increase of radiopacity of implant on day 35, probably caused by accelerated bone growth and healing of tissues, degradation rate higher than pure iron implant	[34]
Fe0,074N	Stent	Rabbit, Minipigs	abdominal aorta, left anterior descending, right coronary artery	12, 36 months	Nitride modified scaffold showed non-uniform corrosion, higher corrosion rate after implantation in contrast with pure iron stent, no unusual reactions, no pathological changes to tissues	[70]
FeXMn (X = 0.5, 2.7, 6.9)	Cylindrical plate	Mice	Back (subcutaneous)	3, 6, 9	There was no significant corrosion after implantation. Implants degraded slowly probably, because of phosphate layers on surface of corroding implant	[71]
Fe35Mn1Ag	Rods	Rats	Subcutaneous	4, 12 weeks	The addition on silver increases corrosion rate in vivo, almost two times higher rate than Fe35Mn alloy	[72]
FeMn10Cu	Rods	Rabbit	Femur	30, 90 days	Addition of copper increased corrosion rate in vivo in comparison to base FeMn alloy	[73]

5. Production of Iron Biodegradable Devices

Various methods of producing materials for biodegradable devices allow not only different shapes and formats but can also significantly affect the mechanical properties and corrosion rate of the finished implants.

Metals are melted under a vacuum with the resulting product then subjected to various forming and thermomechanical processes in order to achieve the final device. The vacuum applied during the melting procedure minimises inclusion formation and removes occluded or dissolved gases from the material bulk. Casting of the medical device allows a precise shape, and it is a well-established manufacturing procedure [74,75]. Many different techniques are used to obtain new kinds of iron-based biodegradable materials, but only a few have been chosen and described in this article. After a brief description of each of the methods of obtaining biodegradable iron-based systems, the data on the corrosion rate of the described systems are presented in a tabular form.

5.1. Spark Plasma Sintering (SPS)

The SPS method is a method of rapid sintering of powdered materials. Unlike other sintering methods, in which the powder is heated using an alternating current, SPS heating the consolidated powder uses periodically repeated direct current pulses lasting from a few to several hundred milliseconds, with low voltage but high intensity (from several to tens of thousands of amperes). SPS process is characterised by a high rate of efficiency, short duration and a lower temperature regime compared to traditional sintering. With the use of the SPS method, it is possible to consolidate metal powders, ceramics, composites or intermetallic compounds [76]. Spark plasma sintering was used, for example, in the research of Huang et al., to prepare two composite specimens of Fe–5% Pd and Fe–5% Pt [77]. The sintering of the pure starting powders was carried out under vacuum at 1000 °C for a holding time of 5 min. Also, other studies used SPS to produce iron composites with palladium [53]. They sintered powders for 10 min at a temperature of 1000 °C. Cheng and Zheng obtained iron-based materials with W as the second phases [52]. The sintering process was carried out at 950 °C, 40 MPa pressure for 5 min. In Table 3 the data on corrosion rates for Fe-based materials obtained by SPS is presented.

Table 3. Corrosion rates for Fe-based materials obtained by spark plasma sintering.

Method	Material	Corrosion Rate (mm/Year)	Ref.
SPS	Fe	0.016	[52]
	Fe2W	0.075	[52]
	Fe5W	0.138	[52]
	Fe35Mn5Si	0.025	[77]
	Fe25Mn10Cu	0.258	[73]
	Fe34Mn1Cu	0.032	[73]
	Fe	0.105	[78]
	Fe3Mn	0.105	[78]

5.2. *Vacuum Induction Melting (VIM)*

The VIM process uses modern induction furnaces to melt metals. The high power installed in the device allows the full charge of the crucible to be melted quickly, minimizing the exposure of the liquid metal to oxygen from the air. Thus, it provides better control over the chemical composition of the final product. VIM is relatively flexible, featuring the independent control of time, temperature, pressure, and mass transport through melt stirring. Its advantages include (1) flexibility due to small batch sizes (2) removal of dissolved gases, for example, hydrogen and nitrogen (3) precise temperature control and (4) low losses of alloying elements by oxidation [79].

The VIM method was used many times. For example, to prepare six binary iron alloys (Fe-X, where X-Mn, Co, Al, W, B, C) ingots for use as coronary stents. The pure elements were mixed in the ratio 97:3 at.%, melted and cast under an argon atmosphere in a VIM furnace [78]. As for the corrosion rate of the systems obtained by the above method, it is comparable to the corrosion rate of the systems obtained by the STS method (Table 4).

Table 4. Corrosion rates for Fe-based materials obtained by vacuum induction melting.

Method	Material	Corrosion Rate (mm/Year)	Ref.
VIM	Fe3Co	0.128	[78]
	Fe3Al	0.112	[78]
	Fe3W	0.151	[78]
	Fe3B	0.175	[78]
	Fe3C	0.187	[78]
	Fe3S	0.145	[78]
	FeMnC	0.13	[80]
	TWIP1Pd	0.21	[80]
	Fe	0.11	[80]

5.3. Vacuum Arc Melting (VAM)

Vacuum arc melting (VAM) is a process for melting metals developed in 1839. VAM offer quick melting of the cathodic electrode under the high temperature of the direct-current arc. Then the liquid metal is rapidly solidified into an ingot in the water-cooled mold. Thanks to the dual action of vacuum and an electromagnetic field, the inclusion and gas content of the molten metal can be effectively reduced. Therefore, while VIM establishes close control of alloy formation and chemistry, VAM exhibits the desired microstructure due to greater control of the solidification rate [81]. That's why this method was used to make three alloy ingots of Fe-based bulk metallic glasses (BMG) prepared by melting the appropriate atomic percentages of Fe, Co, Cr, Mo, C, B, Mn and Y [82]. The homogenisation of the sample compositions was performed by remelting each ingot at least five times. The obtained materials were biologically compatible and showed high resistance to corrosion, as it is presented in Table 5. Therefore, their use as biodegradable material is not possible.

Table 5. Corrosion rates for Fe-based materials obtained by vacuum arc melting.

Method	Material	Corrosion Rat (mm/Year)	Ref.
VAM	Fe	0.027	[51]
	Fe implanted Ag	0.046	[51]
	Fe	0.1	[83]
	Fe30Mn	0.24	[83]
	F330Mn5Si	0.76	[83]
	F3e26Mn5Si	0.56	[83]
	Fe23Mn5Si	0.44	[83]

5.4. Electroforming

Electroforming is a different form of electroplating. In electroplating, metal is dissolved electrolytically at an anode. The resulting metal ions are transported by an electrolyte solution to be placed at a cathode. The difference between electroforming and electroplating lies in the purpose of use for the deposited metal. Electroplating is concerned with taking an existing article and applying a metallic coating. However, electroforming is obtaining metal object by utilising the electroplating process to deposit a metal on or against a master or mandrel.

Electroforming can produce shapes to tight tolerances, good surface finishes and excellent metallurgical properties because the structure of the metallic parts is formed atom by atom and layer by layer. In their research, Moravej et al. found that pure iron fabricated by electroforming technology possessed a high yield strength (YS) (360 MPa) and ultimate tensile strength (423 MPa) [84]. The ductility could be further improved 18% by annealing. Both static and dynamic in vitro degradation tests indicated a faster corrosion rate of electroformed pure iron than that of pure iron produced by casting (Table 6).

Table 6. Corrosion rates for Fe-based materials obtained by electroforming.

Method	Material	Corrosion Rate (mm/Year)	Ref.
Electroforming	Fe	0.4	[83]
	Fe	0.85	[84]

5.5. Powder Metallurgy (PM)

In short, PM consists of four basic steps: (1) powder preparation, (2) mixing and blending, (3) compacting and (4) sintering. Sometimes, this process accomplished with some secondary operations like coining, impregnation, hot forging, etc. The advantages of PM include ease of carrying out the process, high purity of the obtained products and the fact that metals and non-metals can be mixed in any proportions. That's why th method was used many times to produce iron-based biodegradable materials [53,55,8 PM materials usually show higher degradation rates than casted or wrought ones becau PM materials contain some porosity that enhances the exposed area of the sample the corrosion medium, and obtained products may contain oxides whose can chan the corrosion rate [53]. For example, Hermawan et al. used PM to prepare four F Mn alloys, namely Fe20Mn, Fe25Mn, Fe30Mn, and Fe35Mn, from high purity element powder [85]. The numbers specified the nominal weight percentage of manganese. Th sintering temperature of the compacted powder mixtures was fixed at 1200 °C for 3 h in flowing Ar-5% H_2 gas mixture. The addition of manganese powder significantly modifi the corrosion properties of the systems, increasing their the corrosion rate (Table 7). should be noted, however, that the optimal manganese addition was determined—and th at the level of 20% turned out to be the most advantageous from the point of view of th planned biodegradability. Higher percentages of manganese powder resulted in a decrea in the corrosion rate of Fe-Mn systems.

Table 7. Corrosion rates for Fe-based materials obtained by powder metallurgy.

Method	Material	Corrosion Rate (mm/Year)	Ref.
PM	Fe20Mn	1.3	[84]
	Fe25Mn	1.1	[84]
	Fe30Mn	0.7	[84]
	Fe35Mn	0.4	[84]
	Fe	0.2	[84]

5.6. Mechanical Treatments, Inducing Plastic Deformation

One of the best methods to obtain pure iron devices with improved mechanic properties is to perform mechanical treatments that induce plastic deformation, such cold rolling or cold drawing [86]. These methods create a finer grain structure as define by the Hall–Petch equation. However, these treatments may decrease the ductility of th material [81]. At the same time, they lead to a reduction in the corrosion rate, as compare to unmodified systems (Table 8) [83].

Table 8. Corrosion rates for Fe-based materials obtained by powder metallurgy followed by cold rollir

Method	Material	Corrosion Rate (mm/Year)	Ref.
PM cold rolled	Fe20Mn	0.5	[84]
	Fe25Mn	0.5	[84]
	Fe30Mn	0.7	[84]
	Fe35Mn	0,7	[84]

5.7. Template-Based Formation of Porous Material

The structure of the material has a great influence on its properties. In the eve of a need to accelerate corrosion, one of the most common methods is the creation of porous material. The mechanical properties of iron allow it to create metallic foams wi a high pore per inch (ppi) content while maintaining shape and strength [22]. The mo common method of creating porous iron foams is to apply powdered iron with option metallic additives to a polyurethane (PU) foam base, followed by heating in a furnace high temperatures until the template is destroyed [87,88].

Porous, iron-based materials degrade faster than their equivalent, fully dense ir counterparts based on size/weight. This is due to the larger contact surface with th physiological environment and more locations where corrosion can be initiated. Poro materials are also more prone to crevice corrosion, which accelerates the material adsorption [89,90]. Studies confirmed the influence of pore sizes on corrosion rate. Heiden et al. noticed that the addition of controlled porosity and hydroxyapatite as the second phase in Fe-30Mn increased the corrosion rate by almost 40 times compared to the non-porous Fe-30Mn (Table 9) [91]. Zhang et al. used as porogen ammonium bicarbonate (NH_4HCO_3) in biodegradable Fe-35Mn [92]. The degradation rate was significantly accelerated when compared to samples without porogen. However, the porous, iron-based material did not have satisfactory mechanical properties. Čapek et al. studied a Fe-Pd porosity material prepared by SPS [53]. The porous material showed faster corrosion compared to the solid condition but also scored worse in cytotoxicity tests. This does not preclude it from being used as a biocompatible material but requires additional optimization.

Table 9. Corrosion rates for porous Fe-based materials.

Method	Material	Corrosion Rate (mm/Year)	Ref.
Template-based formation of porous material	Fe	1.18	[90]
	Fe30Mn10HA	0.82	[91]

5.8. 3D Printing of Iron Materials

Three-dimensional printing has been explored as a fabrication method for patient-specific devices utilizing existing biomaterials. 3D printing of biodegradable metals, such as iron, has the potential to deliver patient-specific biodegradable scaffolds and, depending on the required properties of the scaffold at different implantation sites, a porous structure of the scaffold can be designed and fabricated by 3D printing, for example, to enhance the corrosion rate. The design can also be engineered to specifically load any drugs or biologically active substances.

Currently, electron beam melting and selective laser sintering is widely used for the fabrication of metal parts. Although, the major limitations, such as the use of an expensive 3D printer and availability of materials, may restrict its further application for metal printing [93]. In earlier studies, it was found that micro-extrusion-based 3D printing covered a broad range of available materials and was flexible enough to fabricate metal 3D parts at a low cost [93,94]. Micro-extrusion-based 3D printing is an ink-based dispensing system with a robotic arm that extrudes a continuous stream of ink through a micro-size nozzle [95]. However, Taylor et al. investigated the approach to fabricate porous iron by extrusion-based 3D printing and conventional sintering [96]. The PLGA-based copolymer was used as the binder for preparing the metallic-based polymer ink. After heating the green body in a conventional sintering furnace, a porous metallic structure was formed, although conventional sintering required additional processing time and energy to heat the sample [97]. A six-step approach to fabricate the porous iron by the amalgamation of 3D printing and pressureless microwave sintering has been developed by Sharma and Pandey [98]. This six-step process enclosed the design of a porous structure template by 3D printing, which was further used to fabricate a porous Fe scaffold using microwave sintering. Microwave sintering requires a lower energy consumption to sinter the metal part in less processing time and also results in a better mechanical properties in the sintered part when compared to the conventional sintering process [97].

6. Iron Materials Modifications

6.1. Additions

One of the most common procedures to increase the degradation rate and modify the mechanical properties of pure iron is through alloying. An additional element can be soluble in iron or dispersed as metal matrix composites (MMC). The electrochemical properties of the additive have the highest influence on the device degradation rate. There are two main ways to reduce the corrosion resistance of iron alloys: (1) the addition of elements with lower electrochemical potential than Fe (−0.44 V versus SHE) and soluble in iron to decrease the corrosion resistance of the matrix; (2) addition of elements noble than iron to create a second phase with higher electrochemical potential and cause galvani corrosion, where iron is the anode, and the more noble metal is the cathode [99].

Liu and Zheng tested the influence of the addition of low contents of common alloyin elements (Mn, Co, Al, W, Sn, B, C and S) on the corrosion rate and biocompatibility of pur iron [78]. They found that adding Co, W, C, or S slightly accelerated the corrosion of pur iron. Mn, Al, and B slowed the corrosion insignificantly. The results also showed that th addition of all alloying elements, except for Sn, improved the mechanical properties of iro after rolling. It should be emphasised that this study applies to small percentages of th additive in the alloy; larger amounts may give different results. In the afore-mentione study, there was no influence found for the added elements on the cytotoxicity of the allo Cheng and Zheng also conducted research on adding tungsten to pure iron [62]. The confirmed the acceleration of corrosion and no visible effect on the adhesion of platelet under the influence of W added by SPS. In the same study, they suggested the use of non-metal carbon nanotubes (CNT) as an alloy additive, which increased the corrosion rat and improved the strength.

Due to its higher nobility (+0.951 V vs. SHE), the addition of palladium is fairl often used in research on increasing the biodegradability of iron. Pd can also stabilise iro in the austenite form [100]. The important fact is that palladium is completely miscibl in iron. After solubilisation, it improves the mechanical properties of the iron matri However, Pd can also be precipitated as the second phase (following the associated phas diagram) or produced as MMC from metal powders under controlled temperatures [101 For example, Čapek et al. showed that alloying iron with palladium led to an increase i the mechanical properties and corrosion rates [53]. Their porous implant, with a diamete of 5 mm obtained by SPS, should fully degrade after approximately 5 years. Gao et a used palladium as the second phase in the iron matrix with selective laser melting use to prepare the samples [102]. The obtained biocomposite degraded faster than pure iro through microgalvanic corrosion between the noble Pd-rich intermetallic phase and th Fe matrix.

Platinum exhibits many properties similar to palladium since they belong to th same group in the periodic table, but the electrochemical potential of Pt is higher than P (+1.18V vs. SHE). Huang et al. compared Fe-5wt.%Pd and Fe-5wt.%Pt composites [77 The addition of either elements sped up the corrosion process of iron, and improved th mechanical properties compared to pure iron. However, better results were given b modification with platinum Pt added as the second phase to the iron as it improved th yield strength, rigidity and microhardness nearly three times that of pure iron [77].

Gold is one of the noblest metals (+1.69 V vs. SHE) and has a long history used as dental material [101]. Hulander et al. shown that the ground state cytotoxic activity of A is minimal and can be ignored, but Au salts are toxic to cells [103] Gold is partially solubl in iron, so this makes the second phase possible but under controlled conditions [104 Huang et al. fabricated Fe-Au composites by SPS with a variable percentage of gold [77 They noticed that the iron-based composites with 5 wt% of Au (or Ag) became the material with the fastest corrosion rate after 30 days of immersion. All composites degraded muc faster than pure iron. At the same time, no increase in cytotoxicity or platelet adhesio was found.

Silver is a popular choice as an addition to manufacturing traditional medical device due to its good biocompatibility and antibacterial properties [105,106]. Ag has a highe electrochemical potential (+0.800 vs. SHE) than iron, thanks to which it can cause galvani corrosion. In contrast to gold, silver is not soluble in iron [104]. Huang et al. showe that the addition of Ag to pure iron resulted in faster corrosion than the addition of Au a the second phase [107]. In these tests, silver did not show any cytotoxicity or thrombosi effect. Loffredo et al. tested the difference between Fe-16Mn-0.7C alloy and Fe-16Mn-0.7C 0.4Ag [108]. As a result, it was shown that the presence of silver decreased the ductilit caused by the development of mechanical ε-martensite. More about martensite is foun below in the Section 6.2 "Iron-Manganese alloys". In contrast to other results in the study, Čapek et al., who added 2 wt% silver to pure iron, found that Ag deteriorated the corrosion rate of Fe in both [109]: (1) the long-term immersion test and (2) potentiodynamic test.

Wang et al. noted that the addition of Ga appears to significantly improve the degradation rate of pure Fe by 5 times [110]. The primary mode of corrosion in the alloy was pitting corrosion. They suggested that, due to the more negative electrochemical potential of Ga (−0.56 V vs. SHE) compared to Fe, a more negative corrosion potential was obtained in the alloy. Also, the heterogeneous structure of the alloy would have been conducive to the occurrence of microgalvanic corrosion.

Cheng et al. researched the influence of the addition Fe_2O_3 (in variable contents) to pure iron obtained by SPS [52]. As a result, they found no significant difference in the corrosion process, cytotoxicity or platelet adhesion between composites of Fe_2O_3-Fe and pure iron.

Promising results were achieved by adding nitrogen to iron [111]. W. Lin et al. published a series of studies about nitriding iron stents [70,112,113]. Nitrided iron turns out to corrode faster than pure iron and has better mechanical properties. Long-term in vivo studies have shown the alloys to be fully biocompatible.

6.2. Iron-Manganese Alloys

As mentioned before, Mn is often an addition to iron biodegradable materials to improve the mechanical properties and biodegradability. This was started in 2007 by Hermewan et al. through the publication of research on the iron-35 wt% manganese (Fe-35Mn) alloy and suggested this alloy as a promising option due to its low magnetic susceptibility and suitable corrosion behaviour [114].

Manganese is known as an important element in the human body. Its highest concentration is in the bones, liver, kidney, pancreas and adrenal gland. In normal, healthy humans, blood Mn concentrations range from 4 to 15 µg/L, which means that the human body has a good reaction to the presence of manganese in implants. Notwithstanding, a long exposure to too high a dose of Mn is highly toxic to the body [115]. Manganese has a function as a neurotransmitter in the brain, that's why intoxication by this element induces the first signs and symptoms in the nervous system. There is evidence that long-term exposure to manganese could induce Parkinson's disease [116,117]. This intoxication has also a significant effect on cardiac function: it inhibits myocardial contraction, dilates blood vessels, and induces hypotension [115]. Therefore, in the design of Fe-Mn biodegradable cardiovascular devices, particular attention should be taken to prevent the local accumulation of manganese ions during material corrosion. Moreover, deficiency of iron ions increases the likelihood of absorption of excessive amounts of manganese and thus its intoxication, which suggests that the presence of iron in Fe-Mn alloy could prevent Mn corrosion toxicity [115,118]. On the other hand, manganese deficiency, even if rare, disrupts the healthy functioning of the central nervous system and reproductive system, as well as impedes growth.

Manganese physical properties are similar to iron. Its electrochemical potential is lower than iron (−1.18 V vs. SHE) and both these elements are soluble in each other [104]. That's why the Fe-Mn alloy has a more active corrosion potential than pure Fe.

The degradation process of manganese is similar to that described before for iron degradation [80]:

$$Mn \rightarrow Mn_2^{+} + 2e^{-} \text{ (anodic reaction)} \quad (6)$$

$$O_2 + 2H_2O + 4e^{-} \rightarrow 4OH^{-} \text{ (cathodic reaction)} \quad (7)$$

$$Mn^{2+} + 2OH^{-} \rightarrow Mn(OH)_2 \text{ (product reaction)} \quad (8)$$

However, the human artery atmosphere has around 5% more CO_2 than in the ambient atmosphere. This is an important factor in Mn degradation because Mouzou et al. showed additional reactions disturbing the corrosion process of the alloy in an atmosphere rich in CO_2 [38]:

$$CO_2 + H_2O \rightarrow H_2CO_3 \quad (1$$

$$2H_2CO_3 \rightarrow 2H^+ + 2HCO_3^- \quad (1$$

$$2HCO_3^- \rightarrow 2CO_3^{2-} + 2H^+ \quad (1$$

$$Mn \rightarrow Mn^{2+} + 2e^- \quad (1$$

$$Mn^{2+} + 2HCO_3 \rightarrow Mn(HCO_3)_2 \rightarrow MnCO_3 + CO_2 + H_2O \quad (1$$

$$Mn^{2+} + CO_3^{2-} \rightarrow MnCO_3 \quad (1$$

The same type of reaction also describes oxidation of iron present in the alloy:

$$Fe^{2+} + 2\ HCO_3 \rightarrow Fe(HCO_3)_2 \rightarrow FeCO_3 + CO_2 + H_2O \quad (1$$

$$Fe^{2+} + CO_3^{2-} \rightarrow FeCO_3 \quad (1$$

Due to the fact that $FeCO_3$ and $MnCO_3$ have a low solubility in an aqueous enviro ment, and they can form a stable passivation layer at the surface of the device, carbonat can block the degradation process [38].

Another interesting feature of manganese is that it stabilises the austenitic allotro in a large range of compositions, which helps to avoid ferromagnetism in devices. If in composition that is more than 15% Mn then antiferromagnetic behaviour will be observ in all alloys, since, as the second phase occurs as ε-martensite. However, a contributio can still be observed in the presence of α-martensite, which shows a ferromagnetic b haviour [119–121]. The accelerated degradation of Fe-Mn alloys with different Mn conte in relation to pure iron has been confirmed in many studies [38,69,89,114,122–125].

To improve the Fe-Mn combination properties, a third component can be introduc to produce a Fe-Mn-X alloy. The literature gives many different examples of these modi cations. Generally, there are several basic types of alloy additives:

(1) Noble metals as a second phase composite, like Pd [36,126,127], Cu [73,128], Ag [12 129,130] cause local galvanic corrosion, because of their high electrochemical potenti
(2) The addition of silicon drew the attention of scientists because devices made of th Fe-Mn-Si alloy have shape memory and a higher corrosion rate [54,68,72,83,131,13
(3) Elements from the second group, Mg and Ca, have strong negative electrochemic potentials (Mg = −2.38 V vs. SHE, Ca = −2.76 V vs. SHE). In addition to a Fe-M matrix creates alloys possessing active corrosion potentials. After analysing Ta curves, Hong et al. suggested that the Fe-Mn-Ca alloys have a greater tendency corrode than the Fe-Mn-Mg alloys [133].
(4) The addition of carbon to Fe-Mn has an alloy composition that aligns with that of th twinning-induced plasticity (TWIP) steels. TWIP steels have excellent strength ar ductility. The studies confirmed that TWIP (Fe-Mn-C) alloys degrade in a physiolo cal environment [134–137].

Venezuela and Dargush lately wrote a more detailed review only about Fe-M biodegradable alloys, to which I refer to those interested [138].

6.3. *Surface Modification*

Concerning the number of studies carried out to change the composition or th method of obtaining an iron-based biodegradable device, the number of studies on surfa modifications is small. However, the existing results indicate that this may be the mo appropriate method for implants with improved performances. Zhu et al. covered pu iron with a thin film of Fe-O by plasma immersion ion implantation and deposition improve the mechanical integrity and help with biocompatibility [139]. As a result, th prepared films reduced the number of adhered platelets and restrained the aggregation b also improved the surface corrosion resistance. Sandblasting the iron surface was tested

Zhou et al. which changed in an early state the surface composition, increased the density of dislocations and roughness [140]. They all investigated biocompatibility and corrosion rate. Huang and Zheng applied photolithography and electron beam evaporation of platinum discs on an iron surface, and in this way, they decreased the corrosion potential [141]. Chen et al. in their research, suggested a calcium zinc phosphate coating to delay the start of iron corrosion in the first stage, during wound healing, and to improve biocompatibility [142].

Huang et al. reported that Zn ions, implanted by a metal vapour vacuum arc in Fe, accelerated the degradation rate of the iron [67]. Also, Wang et al. used the ion implantation technique to inject Zn onto the surface of biodegradable, pure Fe [143]. In both examples, corrosion resistance decreased in comparison with pure iron and caused pitting corrosion. However, additional zinc on the surface of implantation devices enhanced the adhesion of platelets compared to pure Fe in both static and dynamic immersion tests in Hank's solution [67].

Cheng et al. created Au patterns on pure Fe using vacuum sputtering to deposit Au on the substrate [63]. They observed that the Fe with Au exhibited a higher corrosion rate than pure Fe. Furthermore, the Au pattern seemed to induce more uniform corrosion of the substrate. As expected, the corrosion film loosened the gold's adhesion to the surface and caused the disk to be eventually displaced.

Weng et al. attempted to modify the iron surface by tantalum ion implantation [144]. The results showed that Ta/Fe oxide mixtures were formed on the outermost surface of the modified layer after ion implantation. After modification by the Ta ion, pure iron exhibited a higher corrosion rate due to the formation of severe pitting corrosion. During tests, cells showed an enhanced adhesion and proliferation behaviour on the surfaces of Ta implanted Fe.

In other studies, Huang et al. showed that silver ions were implanted into a pure iron surface by metal vapour vacuum arc technique [51]. The Ag layer was about 60 and the highest content of Ag was 5 at.%. On the outer implantation layer, Ag existed as Ag_2O. With the increase of depth, Ag gradually changed from Ag_2O to Ag elementary substance. In comparison with pure iron, the surface modified by silver exhibited faster and much more uniform corrosion. However, Ag ions slightly decreased the viabilities of experimental cells and increased the risk of thrombosis.

6.4. Coatings

Promising results were given by coating iron devices with bioresorbable polymers because during degradation, they locally acidify the environment around the implant. The study of Yusop et al. demonstrated that iron corrosion could be expedited by the hydrolysis of PLGA (poly(lactic-co-glycolic acid)) that produced soluble monomers consisting of carboxylic acid groups, which could lead to a pH decrease [145]. Qui et al. studied the behaviour of an iron stent covered in polylactic acid (PLA), and a strut with a thickness of 50 μm was completely dissolved after 6 months [18]. Studies by Lin et al. also showed promising results with the use of PLA [113]. Recently, research by Gorejová et al. about coated iron foams with three different concentrations of PEI (polyethyleneimine) showed that the coverage on the iron surface by PEI gives a desirable corrosion rate enhancement mediated through polymer cracking and corrosion medium penetration [146].

In an investigation by Huang et al., they checked the corrosion behaviour of collagen-coated porous and nonporous Fe-Mn [147]. In both porous and non-porous alloys, the collagen coating reduced the corrosion rate as the protein prevented direct contact of oxygen with the iron surface.

Drevet et al. studied the influence of pulsed electrodeposited calcium phosphate coatings on FeMnSi alloys with variable percentage composition [148]. In every example, the corrosion rate of the alloy was reduced. A similar effect was achieved by applying an osteoconductive hydroxyapatite-zirconia coating on FeMnSi by using pulsed laser deposition in research carried out by Cimpoesu et al.—the ceramic coating protected the substrate with a decreased corrosion rate for the alloy [149].

Hydroxyapatite has been used several times for iron coating; however this material appears more often in works on bone implants due to it being known to be biocompatible with bone tissue [34,150].

6.5. Drug Delivery from Iron-Based Stents

A major problem with bare metal stents is restenosis due to neointimal overgrowth. To resolve this problem, drug eluting stents with immunosuppressive or cytotoxic drugs were designed to inhibit neointimal hyperplasia. Usually, drug loading is from a polymeric coating this makes it easier to control the elution kinetics. Controlled and sustained release of the drug in the first 10–30 days after transplantation, during healing wound after surgery is critical to the device's anti-restenotic effectiveness [151]. In the case of iron-based BRS the polymeric coating must also be fully biodegradable.

The most commonly used drugs in cardiac stents, among others, are (1) Curcumin, as an inhibitor platelet aggregation [152]; (2) sirolimus to inhibit inflammation, proliferation and cell migration [153]; (3) estrogen to promote vascular healing [154]; (4) angiopeptin as redactor of neointimal hyperplasia [151]; (5) dexamethasone, to reduce inflammation [155]; (6) tranilast to reduce neointima formation and inflammation [155].

In studies, Yusop et al. used poly-L-lactic acid with different curcumin amounts as coating for iron scaffolds [156]. This research looked at bone implants; hence the role of curcumin is to retard the growth of osteosarcoma cells owing to cancer arrest. It turned out that the iron degradation process may affect the kinetics of curcumin release. From the point of view of mechanics, the CP particle stiffness and interfacial interactions between the CP coating and the Fe surface further enhanced the mechanical strength of CP-Fe. After 28 days, the scaffold showed good strength reduction.

Shi et al. conducted in vivo studies on rats, checking the biocompatibility and corrosion rate of the sirolimus-eluting iron bioresorbable coronary stent system [157]. After 90 days, the iron was at an early stage of corrosion.

Cysewska et al. tested the release of anti-inflammatory salicylate from polypyrrole (PPy)-coated iron [158]. Sodium salicylate was incorporated into PPy by one-step electropolymerisation. As a result, depending on different conditions, they obtained materials with different morphologies and redox properties. The drug-release study showed that during polymer degradation, the release of salicylate proceeds efficiently.

Sharipova et al. took a different approach to the topic because they did not use the polymer as a drug-eluting medium [159]. Instead, the iron powder was mixed with 10% Ag and VH as a model drug (used for the treatment of periprosthetic or implant-related infections). Afterwards, the mixed powder mixtures were cold-sintered in tungsten carbide dies at RT by utilising a single-stage press. In in vitro tests, the tested material killed pathogens associated with chronic bone infections after 5 days. At the same time, studies showed no iron-based material cytotoxicity.

Nitric oxide is among other substances that positively influence the treatment process because it is an important molecule in biological systems and plays a critical role in pathophysiology and disease [160]. This resulted in the development of new therapeutic strategies for nitric oxide. It may mean that nitriding iron stents, in addition to good mechanical and corrosive parameters, can also have a beneficial effect on the adjacent tissue.

7. Conclusions and Future Perspectives

The excellent biocompatibility of iron alloys and their high strength, ductility, and—especially—elastic modulus compared to magnesium and zinc alloys make iron one of the best potential materials for creating biodegradable metallic stents since the beginning of the 21st century. Due to its mechanical properties and workability, it could be used to create lighter and smaller stents.

The problem encountered by researchers working with iron is the slow corrosion time in vivo, which prevents the recognition of iron as fully biodegradable. The challenge will be to create new, iron-based materials that, without losing their advantages, will have higher corrosion rate. For this purpose, research is carried out on alloy additions, the use of various manufacturing techniques to modify the structure and surface morphology of the material or the use of additional materials as a coating to influence the course of corrosion.

The analysis of data on the rate of degradation processes of iron, its alloys, or iron-based systems and a critical comparison of the numerical values of this property allows us to conclude that some of the methods of obtaining potential ferrous biomaterial, such as spark plasma sintering, and vacuum induction melting, as less effective, should be abandoned. And a few others, including vacuum arc melting and electroforming, should be looked at once again in detail to optimize them further. Both powder metallurgy and templat-based porous systems synthesis have the potential to produce almost optimally degrading iron-based systems, and further work to achieve the desired iron-based systems should be planned. Of course, it is also possible to continue the work on introducing an additional phase into the iron and creating new alloys, but then it should be borne in mind that among the degradation products, there will be systems of more than one element, which may lead to a greater probability of negative side effects. In the case of staying with pure iron, it is possible to focus only on optimising the degradation rate related directly to the morphology of the system (its porosity, specific surface), excluding the influence of other elements. However, the correct solution to obtain optimally degraded Fe-based material is still unknown. After reaching described in the review milestones, there is still much work to do. With the development and dissemination of new manufacturing and imaging technologies, it will be possible to produce individual devices that will be perfectly adapted to the needs of a particular patient in terms of their composition and shape. Also, the size and weight of the stents will be possibly decreased without losing their mechanical strength to limit their harmful effects on the blood vessel walls. Another way to go further is to optimise drug release in stents and propose new solutions for personalised therapy.

Author Contributions: Conceptualization, G.G., J.S. and A.R.; writing—original draft preparation, G.G., J.S., A.R.; writing—review and editing, A.R.; supervision, A.R.; project administration, A.R.; funding acquisition, A.R. All authors have read and agreed to the published version of the manuscript.

Funding: The APC was funded within grant Preludium BIS (no. 2019/35/O/ST5/00405) by National Centre of Science.

Institutional Review Board Statement: Not applicable.

Informed Consent Statement: Not applicable.

Data Availability Statement: Not applicable.

Conflicts of Interest: The authors declare no conflict of interest. The funders had no role in the design of the study; in the collection, analyses, or interpretation of data; in the writing of the manuscript, or in the decision to publish the results.

ferences

Lloyd-Jones, D.; Adams, R.; Carnethon, M.; De Simone, G.; Ferguson, T.B.; Flegal, K.; Ford, E.; Furie, K.; Go, A.; Greenlund, K.; et al. Heart Disease and Stroke Statistics—2009 Update: A Report From the American Heart Association Statistics Committee and Stroke Statistics Subcommittee. *Circulation* **2009**, *119*, e21–e181. [CrossRef] [PubMed]

Byrne, R.A.; Stone, G.W.; Ormiston, J.; Kastrati, A. Coronary Balloon Angioplasty, Stents, and Scaffolds. *Lancet* **2017**, *390*, 781–792. [CrossRef]

Garg, S.; Serruys, P.W. Coronary Stents. *J. Am. Coll. Cardiol.* **2010**, *56*, S43–S78. [CrossRef] [PubMed]

Tan, C.; Schatz, R.A. The History of Coronary Stenting. *Interv. Cardiol. Clin.* **2016**, *5*, 271–280. [CrossRef]

Sigwart, U.; Puel, J.; Mirkovitch, V.; Joffre, F.; Kappenberger, L. Intravascular Stents to Prevent Occlusion and Re-Stenosis after Transluminal Angioplasty. *N. Engl. J. Med.* **1987**, *316*, 701–706. [CrossRef]

Mani, G.; Feldman, M.D.; Patel, D.; Agrawal, C.M. Coronary Stents: A Materials Perspective. *Biomaterials* **2007**, *28*, 1689–1710. [CrossRef]

Chesta, F.; Rizvi, Z.H.; Oberoi, M.; Buttar, N. The Role of Stenting in Patients with Variceal Bleeding. *Tech. Innov. Gastrointest. Endosc.* **2020**, *22*, 205–211. [CrossRef]

Law, M.A.; Shamszad, P.; Nugent, A.W.; Justino, H.; Breinholt, J.P.; Mullins, C.E.; Ing, F.F. Pulmonary Artery Stents: Long-Term Follow-Up. *Catheter. Cardiovasc. Interv.* **2010**, *75*, 757–764. [CrossRef]

9. Breinholt, J.P.; Nugent, A.W.; Law, M.A.; Justino, H.; Mullins, C.E.; Ing, F.F. Stent Fractures in Congenital Heart Disease. *Cathet. Cardiovasc. Interv.* **2008**, *72*, 977–982. [CrossRef]
10. Lejay, A.; Koncar, I.; Diener, H.; de Ceniga, V.M.; Chakfé, N. Post-Operative Infection of Prosthetic Materials or Stents Involvi the Supra-Aortic Trunks: A Comprehensive Review. *Eur. J. Vasc. Endovasc. Surg.* **2018**, *56*, 885–900. [CrossRef]
11. Zheng, Y.; Yang, H. Manufacturing of cardiovascular stents. In *Metallic Biomaterials Processing and Medical Device Manufacturi* Elsevier: Amsterdam, The Netherlands, 2020; pp. 317–340. ISBN 978-0-08-102965-7.
12. Reddy, S.R.V.; Welch, T.R.; Nugent, A.W. Biodegradable Stent Use for Congenital Heart Disease. *Prog. Pediatr. Cardiol.* **2021**, 101349. [CrossRef]
13. Tamai, H.; Igaki, K.; Kyo, E.; Kosuga, K.; Kawashima, A.; Matsui, S.; Komori, H.; Tsuji, T.; Motohara, S.; Uehata, H. Initial a 6-Month Results of Biodegradable Poly-*l*-Lactic Acid Coronary Stents in Humans. *Circulation* **2000**, *102*, 399–404. [CrossRef]
14. Campos, C.A.; Zhang, Y.-J.; Bourantas, C.V.; Muramatsu, T.; Garcia-Garcia, H.M.; Lemos, P.A.; Iqbal, J.; Onuma, Y.; Serruys, P. Bioresorbable Vascular Scaffolds in the Clinical Setting. *Interv. Cardiol.* **2013**, *5*, 639–646. [CrossRef]
15. Abizaid, A.; Carrié, D.; Frey, N.; Lutz, M.; Weber-Albers, J.; Dudek, D.; Chevalier, B.; Weng, S.-C.; Costa, R.A.; Anderson, J.; et 6-Month Clinical and Angiographic Outcomes of a Novel Radiopaque Sirolimus-Eluting Bioresorbable Vascular Scaffold. *JAC Cardiovasc. Interv.* **2017**, *10*, 1832–1838. [CrossRef]
16. McMahon, S.; Bertollo, N.; Cearbhaill, E.D.O.; Salber, J.; Pierucci, L.; Duffy, P.; Dürig, T.; Bi, V.; Wang, W. Bio-Resorbable Polym Stents: A Review of Material Progress and Prospects. *Prog. Polym. Sci.* **2018**, *83*, 79–96. [CrossRef]
17. Zhao, F.; Sun, J.; Xue, W.; Wang, F.; King, M.W.; Yu, C.; Jiao, Y.; Sun, K.; Wang, L. Development of a Polycaprolactone/Poly Dioxanone) Bioresorbable Stent with Mechanically Self-Reinforced Structure for Congenital Heart Disease Treatment. *Bioa Mater.* **2021**, *6*, 2969–2982. [CrossRef]
18. Qi, Y.; Qi, H.; He, Y.; Lin, W.; Li, P.; Qin, L.; Hu, Y.; Chen, L.; Liu, Q.; Sun, H.; et al. Strategy of Metal–Polymer Composite Stent Accelerate Biodegradation of Iron-Based Biomaterials. *ACS Appl. Mater. Interfaces* **2018**, *10*, 182–192. [CrossRef]
19. Beshchasna, N.; Saqib, M.; Kraskiewicz, H.; Wasyluk, Ł.; Kuzmin, O.; Duta, O.C.; Ficai, D.; Ghizdavet, Z.; Marin, A.; Ficai, et al. Recent Advances in Manufacturing Innovative Stents. *Pharmaceutics* **2020**, *12*, 349. [CrossRef]
20. Dong, H.; Lin, F.; Boccaccini, A.R.; Virtanen, S. Corrosion Behavior of Biodegradable Metals in Two Different Simulat Physiological Solutions: Comparison of Mg, Zn and Fe. *Corros. Sci.* **2021**, *182*, 109278. [CrossRef]
21. Qin, Y.; Wen, P.; Guo, H.; Xia, D.; Zheng, Y.; Jauer, L.; Poprawe, R.; Voshage, M.; Schleifenbaum, J.H. Additive Manufacturing Biodegradable Metals: Current Research Status and Future Perspectives. *Acta Biomater.* **2019**, *98*, 3–22. [CrossRef]
22. Li, Y.; Jahr, H.; Zhou, J.; Zadpoor, A.A. Additively Manufactured Biodegradable Porous Metals. *Acta Biomater.* **2020**, *115*, 29– [CrossRef] [PubMed]
23. Zartner, P.; Cesnjevar, R.; Singer, H.; Weyand, M. First Successful Implantation of a Biodegradable Metal Stent into the L Pulmonary Artery of a Preterm Baby. *Catheter. Cardiovasc. Interv.* **2005**, *66*, 590–594. [CrossRef] [PubMed]
24. Bowen, P.K.; Shearier, E.R.; Zhao, S.; Guillory, R.J.; Zhao, F.; Goldman, J.; Drelich, J.W. Biodegradable Metals for Cardiovascul Stents: From Clinical Concerns to Recent Zn-Alloys. *Adv. Healthc. Mater.* **2016**, *5*, 1121–1140. [CrossRef] [PubMed]
25. Loffredo, S.; Paternoster, C.; Mantovani, D. Iron-Based Degradable Implants. In *Encyclopedia of Biomedical Engineering*; Elsevi Amsterdam, The Netherlands, 2019; pp. 374–385; ISBN 978-0-12-805144-3.
26. Zhou, Y.; Wu, P.; Yang, Y.; Gao, D.; Feng, P.; Gao, C.; Wu, H.; Liu, Y.; Bian, H.; Shuai, C. The Microstructure, Mechanical Properti and Degradation Behavior of Laser-Melted Mg Sn Alloys. *J. Alloys Compd.* **2016**, *687*, 109–114. [CrossRef]
27. Ng, C.C.; Savalani, M.M.; Lau, M.L.; Man, H.C. Microstructure and Mechanical Properties of Selective Laser Melted Magnesiu *Appl. Surf. Sci.* **2011**, *257*, 7447–7454. [CrossRef]
28. Wen, P.; Voshage, M.; Jauer, L.; Chen, Y.; Qin, Y.; Poprawe, R.; Schleifenbaum, J.H. Laser Additive Manufacturing of Zn Me Parts for Biodegradable Applications: Processing, Formation Quality and Mechanical Properties. *Mater. Des.* **2018**, *155*, 36– [CrossRef]
29. Song, B.; Dong, S.; Deng, S.; Liao, H.; Coddet, C. Microstructure and Tensile Properties of Iron Parts Fabricated by Selective Las Melting. *Opt. Laser Technol.* **2014**, *56*, 451–460. [CrossRef]
30. Song, B.; Dong, S.; Liu, Q.; Liao, H.; Coddet, C. Vacuum Heat Treatment of Iron Parts Produced by Selective Laser Melti Microstructure, Residual Stress and Tensile Behavior. *Mater. Des. 1980–2015* **2014**, *54*, 727–733. [CrossRef]
31. Peuster, M. A Novel Approach to Temporary Stenting: Degradable Cardiovascular Stents Produced from Corrodible Meta Results 6–18 Months after Implantation into New Zealand White Rabbits. *Heart* **2001**, *86*, 563–569. [CrossRef]
32. Revie, R.W.; Uhlig, H.H. *Corrosion and Corrosion Control: An Introduction to Corrosion Science and Engineering*; Wiley: Hoboken, USA, 2008; ISBN 978-0-470-27725-6.
33. Zhu, S.; Huang, N.; Xu, L.; Zhang, Y.; Liu, H.; Sun, H.; Leng, Y. Biocompatibility of Pure Iron: In Vitro Assessment of Degradati Kinetics and Cytotoxicity on Endothelial Cells. *Mater. Sci. Eng. C* **2009**, *29*, 1589–1592. [CrossRef]
34. Ulum, M.F.; Arafat, A.; Noviana, D.; Yusop, A.H.; Nasution, A.K.; Kadir, A.M.R.; Hermawan, H. In Vitro and in Vivo Degradati Evaluation of Novel Iron-Bioceramic Composites for Bone Implant Applications. *Mater. Sci. Eng. C* **2014**, *36*, 336–344. [CrossR
35. Qi, Y.; Li, X.; He, Y.; Zhang, D.; Ding, J. Mechanism of Acceleration of Iron Corrosion by a Polylactide Coating. *ACS Appl. Ma Interfaces* **2019**, *11*, 202–218. [CrossRef]

6. Kraus, T.; Moszner, F.; Fischerauer, S.; Fiedler, M.; Martinelli, E.; Eichler, J.; Witte, F.; Willbold, E.; Schinhammer, M.; Meischel, M.; et al. Biodegradable Fe-Based Alloys for Use in Osteosynthesis: Outcome of an in Vivo Study after 52 weeks. *Acta Biomater.* **2014**, *10*, 3346–3353. [CrossRef]

7. Pierson, D.; Edick, J.; Tauscher, A.; Pokorney, E.; Bowen, P.; Gelbaugh, J.; Stinson, J.; Getty, H.; Lee, C.H.; Drelich, J.; et al. A Simplified in Vivo Approach for Evaluating the Bioabsorbable Behavior of Candidate Stent Materials. *J. Biomed. Mater. Res. B Appl. Biomater.* **2012**, *100B*, 58–67. [CrossRef]

8. Mouzou, E.; Paternoster, C.; Tolouei, R.; Chevallier, P.; Biffi, C.A.; Tuissi, A.; Mantovani, D. CO2-Rich Atmosphere Strongly Affects the Degradation of Fe-21Mn-1C for Biodegradable Metallic Implants. *Mater. Lett.* **2016**, *181*, 362–366. [CrossRef]

9. J01 Committee. *Guide for Laboratory Immersion Corrosion Testing of Metals*; ASTM International: West Conshohocken, PA, USA, 2012.

0. G01 Committee. *Test Method for Conducting Potentiodynamic Polarization Resistance Measurements*; ASTM International: West Conshohocken, PA, USA, 2009.

1. G01 Committee. *Practice for Verification of Algorithm and Equipment for Electrochemical Impedance Measurements*; ASTM International: West Conshohocken, PA, USA, 2015.

2. Papanikolaou, G.; Pantopoulos, K. Iron Metabolism and Toxicity. *Toxicol. Appl. Pharmacol.* **2005**, *202*, 199–211. [CrossRef]

3. Uchmanowicz, I. Oxidative Stress, Frailty and Cardiovascular Diseases: Current Evidence. In *Frailty and Cardiovascular Diseases*; Veronese, N., Ed.; Advances in Experimental Medicine and Biology; Springer International Publishing: Cham, Switzerland, 2020; Volume 1216, pp. 65–77; ISBN 978-3-030-33329-4.

4. Andrews, N.C. Iron Deficiency and Iron Overload. *Annu. Rev. Genom. Hum. Genet.* **2000**, *1*, 75–98. [CrossRef]

5. Andrews, N.C. Disorders of Iron Metabolism. *N. Engl. J. Med.* **1999**, *341*, 1986–1995. [CrossRef]

6. Pivina, L.; Semenova, Y.; Doşa, M.D.; Dauletyarova, M.; Bjørklund, G. Iron Deficiency, Cognitive Functions, and Neurobehavioral Disorders in Children. *J. Mol. Neurosci.* **2019**, *68*, 1–10. [CrossRef]

7. Georgieff, M.K. Iron Deficiency in Pregnancy. *Am. J. Obstet. Gynecol.* **2020**, *223*, 516–524. [CrossRef]

8. Muckenthaler, M.U.; Rivella, S.; Hentze, M.W.; Galy, B. A Red Carpet for Iron Metabolism. *Cell* **2017**, *168*, 344–361. [CrossRef] [PubMed]

9. Boutrand, J.-P. (Ed.) *Biocompatibility and Performance of Medical Devices*; Woodhead Publishing Series in Biomaterials; Woodhead Publishing: Philadelphia, PA, USA, 2012; ISBN 978-0-85709-070-6.

0. Virtanen, S. Corrosion of Biomedical Implant Materials. *Corros. Rev.* **2008**, *26*, 148–171. [CrossRef]

1. Huang, T.; Cheng, Y.; Zheng, Y. In Vitro Studies on Silver Implanted Pure Iron by Metal Vapor Vacuum Arc Technique. *Colloids Surf. B Biointerfaces* **2016**, *142*, 20–29. [CrossRef] [PubMed]

2. Cheng, J.; Zheng, Y.F. In Vitro Study on Newly Designed Biodegradable Fe-X Composites (X = W, CNT) Prepared by Spark Plasma Sintering. *J. Biomed. Mater. Res. B Appl. Biomater.* **2013**, *101B*, 485–497. [CrossRef]

3. Čapek, J.; Msallamová, Š.; Jablonská, E.; Lipov, J.; Vojtěch, D. A Novel High-Strength and Highly Corrosive Biodegradable Fe-Pd Alloy: Structural, Mechanical and in Vitro Corrosion and Cytotoxicity Study. *Mater. Sci. Eng. C* **2017**, *79*, 550–562. [CrossRef]

4. Liu, B.; Zheng, Y.F.; Ruan, L. In Vitro Investigation of Fe30Mn6Si Shape Memory Alloy as Potential Biodegradable Metallic Material. *Mater. Lett.* **2011**, *65*, 540–543. [CrossRef]

5. Paim, T.C.; Wermuth, D.P.; Bertaco, I.; Zanatelli, C.; Naasani, L.I.S.; Slaviero, M.; Driemeier, D.; Schaeffer, L.; Wink, M.R. Evaluation of in Vitro and in Vivo Biocompatibility of Iron Produced by Powder Metallurgy. *Mater. Sci. Eng. C* **2020**, *115*, 111129. [CrossRef]

6. Eliaz, N. Corrosion of Metallic Biomaterials: A Review. *Materials* **2019**, *12*, 407. [CrossRef]

7. Swinney, D.C. Molecular Mechanism of Action (MMoA) in Drug Discovery. In *Annual Reports in Medicinal Chemistry*; Elsevier: Amsterdam, The Netherlands, 2011; Volume 46, pp. 301–317. ISBN 978-0-12-386009-5.

8. Anderson, J.M.; Rodriguez, A.; Chang, D.T. Foreign Body Reaction to Biomaterials. *Semin. Immunol.* **2008**, *20*, 86–100. [CrossRef]

9. Waksman, R.; Pakala, R.; Baffour, R.; Seabron, R.; Hellinga, D.; Tio, F.O. Short-Term Effects of Biocorrodible Iron Stents in Porcine Coronary Arteries. *J. Intervent. Cardiol.* **2008**, *21*, 15–20. [CrossRef]

0. Liu, Y.; Zheng, Y.; Chen, X.; Yang, J.; Pan, H.; Chen, D.; Wang, L.; Zhang, J.; Zhu, D.; Wu, S.; et al. Fundamental Theory of Biodegradable Metals—Definition, Criteria, and Design. *Adv. Funct. Mater.* **2019**, *29*, 1805402. [CrossRef]

1. F04 Committee. *Practice for Assessment of Hemolytic Properties of Materials*; ASTM International: West Conshohocken, PA, USA, 2017.

2. Cheng, J.; Huang, T.; Zheng, Y.F. Microstructure, Mechanical Property, Biodegradation Behavior, and Biocompatibility of Biodegradable Fe-Fe_2O_3 Composites: Biodegradable Fe-Fe_2O_3 Composites. *J. Biomed. Mater. Res. A* **2014**, *102*, 2277–2287. [CrossRef]

3. Cheng, J.; Huang, T.; Zheng, Y.F. Relatively Uniform and Accelerated Degradation of Pure Iron Coated with Micro-Patterned Au Disc Arrays. *Mater. Sci. Eng. C* **2015**, *48*, 679–687. [CrossRef]

4. Sharma, P.; Jain, K.G.; Pandey, P.M.; Mohanty, S. In Vitro Degradation Behaviour, Cytocompatibility and Hemocompatibility of Topologically Ordered Porous Iron Scaffold Prepared Using 3D Printing and Pressureless Microwave Sintering. *Mater. Sci. Eng. C* **2020**, *106*, 110247. [CrossRef]

5. Oriňaková, R.; Gorejová, R.; Králová, Z.O.; Haverová, L.; Oriňak, A.; Maskal'ová, I.; Kupková, M.; Džupon, M.; Baláž, M.; Hrubovčáková, M.; et al. Evaluation of Mechanical Properties and Hemocompatibility of Open Cell Iron Foams with Polyethylene Glycol Coating. *Appl. Surf. Sci.* **2020**, *505*, 144634. [CrossRef]

66. Králová, Z.O.; Gorejová, R.; Oriňaková, R.; Petráková, M.; Oriňak, A.; Kupková, M.; Hrubovčáková, M.; Sopčák, T.; Baláž, M.; Maskal'ová, I.; et al. Biodegradable Zinc-Iron Alloys: Complex Study of Corrosion Behavior, Mechanical Properties and Hemocompatibility. *Prog. Nat. Sci. Mater. Int.* **2021**, *31*, 279–287. [CrossRef]
67. Huang, T.; Zheng, Y.; Han, Y. Accelerating Degradation Rate of Pure Iron by Zinc Ion Implantation. *Regen. Biomater.* **2016**, *3*, 205–215. [CrossRef]
68. Fântânariu, M.; Trincă, L.C.; Solcan, C.; Trofin, A.; Strungaru, Ş.; Şindilar, E.V.; Plăvan, G.; Stanciu, S. A New Fe–Mn–Si Alloplastic Biomaterial as Bone Grafting Material: In Vivo Study. *Appl. Surf. Sci.* **2015**, *352*, 129–139. [CrossRef]
69. Traverson, M.; Eiden, M.; Stanciu, L.; Nauman, E. In Vivo Evaluation of Biodegradability AndBiocompatibility of Fe30Mn Alloy. *Vet. Comp. Orthop. Traumatol.* **2017**, *31*, 10–16. [CrossRef]
70. Lin, W.; Qin, L.; Qi, H.; Zhang, D.; Zhang, G.; Gao, R.; Qiu, H.; Xia, Y.; Cao, P.; Wang, X.; et al. Long-Term in Vivo Corrosion Behavior, Biocompatibility and Bioresorption Mechanism of a Bioresorbable Nitrided Iron Scaffold. *Acta Biomater.* **2017**, *5*4, 454–468. [CrossRef]
71. Drynda, A.; Hassel, T.; Bach, F.W.; Peuster, M. In Vitro and in Vivo Corrosion Properties of New Iron–Manganese Alloys Designed for Cardiovascular Applications. *J. Biomed. Mater. Res. B Appl. Biomater.* **2015**, *103*, 649–660. [CrossRef] [PubMed]
72. Dargusch, M.S.; Venezuela, J.; Dehghan-Manshadi, A.; Johnston, S.; Yang, N.; Mardon, K.; Lau, C.; Allavena, R. In Vivo Evaluation of Bioabsorbable Fe-35Mn-1Ag: First Reports on In Vivo Hydrogen Gas Evolution in Fe-Based Implants. *Adv. Healthc. Mater.* **2021**, *10*, 2000667. [CrossRef] [PubMed]
73. Mandal, S.; Ummadi, R.; Bose, M.; Balla, V.K.; Roy, M. Fe–Mn–Cu Alloy as Biodegradable Material with Enhanced Antimicrobial Properties. *Mater. Lett.* **2019**, *237*, 323–327. [CrossRef]
74. Mitchell, B.S. *An Introduction to Materials Engineering and Science for Chemical and Materials Engineers*; John Wiley: Hoboken, NJ, USA, 2004; ISBN 978-0-471-43623-2.
75. Ratner, B.D. (Ed.) *Biomaterials Science: An Introduction to Materials in Medicine*, 3rd ed.; Elsevier: Amsterdam, The Netherlands; Academic Press: Boston, MA, USA, 2013; ISBN 978-0-12-374626-9.
76. Garbiec, D. Iskrowe Spiekanie Plazmowe (SPS): Teoria i Praktyka. *Inż. Mater.* **2015**, *1*, 10–14. [CrossRef]
77. Huang, T.; Cheng, J.; Zheng, Y.F. In Vitro Degradation and Biocompatibility of Fe–Pd and Fe–Pt Composites Fabricated by Spark Plasma Sintering. *Mater. Sci. Eng. C* **2014**, *35*, 43–53. [CrossRef]
78. Liu, B.; Zheng, Y.F. Effects of Alloying Elements (Mn, Co, Al, W, Sn, B, C and S) on Biodegradability and in Vitro Biocompatibility of Pure Iron. *Acta Biomater.* **2011**, *7*, 1407–1420. [CrossRef]
79. ASM International; Stefanescu, D.M. *ASM Handbook 4: Casting*, 10th ed.; ASM International, Ed.; ASM International: Novelty, OH, USA, 1998; ISBN 978-0-87170-021-6.
80. Schinhammer, M.; Steiger, P.; Moszner, F.; Löffler, J.F.; Uggowitzer, P.J. Degradation Performance of Biodegradable FeMnC(Pd) Alloys. *Mater. Sci. Eng. C* **2013**, *33*, 1882–1893. [CrossRef]
81. Mathabathe, M.N.; Bolokang, A.S.; Govender, G.; Siyasiya, C.W.; Mostert, R.J. Cold-Pressing and Vacuum Arc Melting of γ-TiAl Based Alloys. *Adv. Powder Technol.* **2019**, *30*, 2925–2939. [CrossRef]
82. Wang, Y.B.; Li, H.F.; Zheng, Y.F.; Li, M. Corrosion Performances in Simulated Body Fluids and Cytotoxicity Evaluation of Fe-Based Bulk Metallic Glasses. *Mater. Sci. Eng. C* **2012**, *32*, 599–606. [CrossRef]
83. Drevet, R.; Zhukova, Y.; Malikova, P.; Dubinskiy, S.; Korotitskiy, A.; Pustov, Y.; Prokoshkin, S. Martensitic Transformations and Mechanical and Corrosion Properties of Fe-Mn-Si Alloys for Biodegradable Medical Implants. *Metall. Mater. Trans. A* **2018**, *4*9, 1006–1013. [CrossRef]
84. Moravej, M.; Purnama, A.; Fiset, M.; Couet, J.; Mantovani, D. Electroformed Pure Iron as a New Biomaterial for Degradable Stents: In Vitro Degradation and Preliminary Cell Viability Studies. *Acta Biomater.* **2010**, *6*, 1843–1851. [CrossRef]
85. Hermawan, H.; Dubé, D.; Mantovani, D. Degradable Metallic Biomaterials: Design and Development of Fe-Mn Alloys for Stents. *J. Biomed. Mater. Res. A* **2009**, *93*, 1–11. [CrossRef]
86. Schaffer, J.E.; Nauman, E.A.; Stanciu, L.A. Cold Drawn Bioabsorbable Ferrous and Ferrous Composite Wires: An Evaluation of in Vitro Vascular Cytocompatibility. *Acta Biomater.* **2013**, *9*, 8574–8584. [CrossRef] [PubMed]
87. Alavi, R.; Trenggono, A.; Champagne, S.; Hermawan, H. Investigation on Mechanical Behavior of Biodegradable Iron Foams under Different Compression Test Conditions. *Metals* **2017**, *7*, 202. [CrossRef]
88. Ray, S.; Thormann, U.; Eichelroth, M.; Budak, M.; Biehl, C.; Rupp, M.; Sommer, U.; El Khassawna, T.; Alagboso, F.I.; Kampschulte, M.; et al. Strontium and Bisphosphonate Coated Iron Foam Scaffolds for Osteoporotic Fracture Defect Healing. *Biomaterials* **201**8, *157*, 1–16. [CrossRef] [PubMed]
89. Čapek, J.; Kubasek, J.; Vojtěch, D. Microstructural, Mechanical, Corrosion and Cytotoxicity Characterization of the Hot Forged FeMn30 (Wt.%) Alloy. *Mater. Sci. Eng. C* **2016**, *58*, 900–908. [CrossRef]
90. Li, Y.; Jahr, H.; Lietaert, K.; Pavanram, P.; Yilmaz, A.; Fockaert, L.I.; Leeflang, M.A.; Pouran, B.; Gonzalez-Garcia, Y.; Weinans, H.; et al. Additively Manufactured Biodegradable Porous Iron. *Acta Biomater.* **2018**, *77*, 380–393. [CrossRef]
91. Heiden, M.; Nauman, E.; Stanciu, L. Bioresorbable Fe-Mn and Fe-Mn-HA Materials for Orthopedic Implantation: Enhancing Degradation through Porosity Control. *Adv. Healthc. Mater.* **2017**, *6*, 1700120. [CrossRef]
92. Zhang, Q.; Cao, P. Degradable Porous Fe-35 wt.% Mn Produced via Powder Sintering from NH_4HCO_3 Porogen. *Mater. Chem. Phys.* **2015**, *163*, 394–401. [CrossRef]
93. Panwar, A.; Tan, L. Current Status of Bioinks for Micro-Extrusion-Based 3D Bioprinting. *Molecules* **2016**, *21*, 685. [CrossRef]

Ren, L.; Zhou, X.; Song, Z.; Zhao, C.; Liu, Q.; Xue, J.; Li, X. Process Parameter Optimization of Extrusion-Based 3D Metal Printing Utilizing PW–LDPE–SA Binder System. *Materials* **2017**, *10*, 305. [CrossRef] [PubMed]

Bishop, E.S.; Mostafa, S.; Pakvasa, M.; Luu, H.H.; Lee, M.J.; Wolf, J.M.; Ameer, G.A.; He, T.-C.; Reid, R.R. 3-D Bioprinting Technologies in Tissue Engineering and Regenerative Medicine: Current and Future Trends. *Genes Dis.* **2017**, *4*, 185–195. [CrossRef] [PubMed]

Taylor, S.L.; Jakus, A.E.; Shah, R.N.; Dunand, D.C. Iron and Nickel Cellular Structures by Sintering of 3D-Printed Oxide or Metallic Particle Inks: Iron and Nickel Cellular Structures by Sintering. *Adv. Eng. Mater.* **2017**, *19*, 1600365. [CrossRef]

Oghbaei, M.; Mirzaee, O. Microwave versus Conventional Sintering: A Review of Fundamentals, Advantages and Applications. *J. Alloys Compd.* **2010**, *494*, 175–189. [CrossRef]

Sharma, P.; Pandey, P.M. Rapid Manufacturing of Biodegradable Pure Iron Scaffold Using Amalgamation of Three-Dimensional Printing and Pressureless Microwave Sintering. *Proc. Inst. Mech. Eng. Part C J. Mech. Eng. Sci.* **2019**, *233*, 1876–1895. [CrossRef]

Asgari, M.; Hang, R.; Wang, C.; Yu, Z.; Li, Z.; Xiao, Y. Biodegradable Metallic Wires in Dental and Orthopedic Applications: A Review. *Metals* **2018**, *8*, 212. [CrossRef]

0. Crangle, J. Ferromagnetism in Pd-Rich Palladium-Iron Alloys. *Philos. Mag.* **1960**, *5*, 335–342. [CrossRef]

1. Geurtsen, W. Biocompatibility of Dental Casting Alloys. *Crit. Rev. Oral Biol. Med.* **2002**, *13*, 71–84. [CrossRef]

2. Gao, C.; Yao, M.; Li, S.; Feng, P.; Peng, S.; Shuai, C. Highly Biodegradable and Bioactive Fe-Pd-Bredigite Biocomposites Prepared by Selective Laser Melting. *J. Adv. Res.* **2019**, *20*, 91–104. [CrossRef]

3. Hulander, M.; Hong, J.; Andersson, M.; Gervén, F.; Ohrlander, M.; Tengvall, P.; Elwing, H. Blood Interactions with Noble Metals: Coagulation and Immune Complement Activation. *ACS Appl. Mater. Interfaces* **2009**, *1*, 1053–1062. [CrossRef]

4. Okamoto, H.; Massalski, T.B. Methods for Phase Diagram Determination. In *Methods for Phase Diagram Determination*; Elsevier: Amsterdam, The Netherlands, 2007; pp. 51–107; ISBN 978-0-08-044629-5.

5. Bosetti, M.; Massè, A.; Tobin, E.; Cannas, M. Silver Coated Materials for External Fixation Devices: In Vitro Biocompatibility and Genotoxicity. *Biomaterials* **2002**, *23*, 887–892. [CrossRef]

6. Li, J.; Liu, X.; Qiao, Y.; Zhu, H.; Ding, C. Antimicrobial Activity and Cytocompatibility of Ag Plasma-Modified Hierarchical TiO_2 Film on Titanium Surface. *Colloids Surf. B Biointerfaces* **2014**, *113*, 134–145. [CrossRef] [PubMed]

7. Huang, T.; Cheng, J.; Bian, D.; Zheng, Y. Fe-Au and Fe-Ag Composites as Candidates for Biodegradable Stent Materials: FE-AU AND FE-AG COMPOSITES FOR STENT MATERIALS. *J. Biomed. Mater. Res. B Appl. Biomater.* **2016**, *104*, 225–240. [CrossRef] [PubMed]

8. Loffredo, S.; Paternoster, C.; Giguère, N.; Barucca, G.; Vedani, M.; Mantovani, D. The Addition of Silver Affects the Deformation Mechanism of a Twinning-Induced Plasticity Steel: Potential for Thinner Degradable Stents. *Acta Biomater.* **2019**, *98*, 103–113. [CrossRef]

9. Čapek, J.; Stehlíková, K.; Michalcová, A.; Msallamová, Š.; Vojtěch, D. Microstructure, Mechanical and Corrosion Properties of Biodegradable Powder Metallurgical Fe-2 Wt% X (X = Pd, Ag and C) Alloys. *Mater. Chem. Phys.* **2016**, *181*, 501–511. [CrossRef]

0. Wang, H.; Zheng, Y.; Liu, J.; Jiang, C.; Li, Y. In Vitro Corrosion Properties and Cytocompatibility of Fe-Ga Alloys as Potential Biodegradable Metallic Materials. *Mater. Sci. Eng. C* **2017**, *71*, 60–66. [CrossRef]

1. Feng, Q.; Zhang, D.; Xin, C.; Liu, X.; Lin, W.; Zhang, W.; Chen, S.; Sun, K. Characterization and in Vivo Evaluation of a Bio-Corrodible Nitrided Iron Stent. *J. Mater. Sci. Mater. Med.* **2013**, *24*, 713–724. [CrossRef]

2. Lin, W.; Zhang, G.; Cao, P.; Zhang, D.; Zheng, Y.; Wu, R.; Qin, L.; Wang, G.; Wen, T. Cytotoxicity and Its Test Methodology for a Bioabsorbable Nitrided Iron Stent: Cytocompatibility of Bioabsorbable Iron-Based Materials. *J. Biomed. Mater. Res. B Appl. Biomater.* **2015**, *103*, 764–776. [CrossRef]

3. Lin, W.-J.; Zhang, D.-Y.; Zhang, G.; Sun, H.-T.; Qi, H.-P.; Chen, L.-P.; Liu, Z.-Q.; Gao, R.-L.; Zheng, W. Design and Characterization of a Novel Biocorrodible Iron-Based Drug-Eluting Coronary Scaffold. *Mater. Des.* **2016**, *91*, 72–79. [CrossRef]

4. Hermawan, H.; Moravej, M.; Dubé, D.; Fiset, M.; Mantovani, D. Degradation Behaviour of Metallic Biomaterials for Degradable Stents. *Adv. Mater. Res.* **2006**, *15–17*, 113–118. [CrossRef]

5. O'Neal, S.L.; Zheng, W. Manganese Toxicity upon Overexposure: A Decade in Review. *Curr. Environ. Health Rep.* **2015**, *2*, 315–328. [CrossRef]

6. Rutchik, J.S.; Zheng, W.; Jiang, Y.; Mo, X. How Does an Occupational Neurologist Assess Welders and Steelworkers for a Manganese-Induced Movement Disorder? An International Team's Experiences in Guanxi, China, Part I. *J. Occup. Environ. Med.* **2012**, *54*, 1432–1434. [CrossRef]

7. Rutchik, J.S.; Zheng, W.; Jiang, Y.; Mo, X. How Does an Occupational Neurologist Assess Welders and Steelworkers for a Manganese-Induced Movement Disorder? An International Team's Experiences in Guanxi, China Part II. *J. Occup. Environ. Med.* **2012**, *54*, 1562–1564. [CrossRef]

8. Chen, P.; Bornhorst, J.; Aschner, M. Manganese Metabolism in Humans. *Postprints Univ. Potsdam Math. Nat. Reihe* **2019**, *711*, 25. [CrossRef]

9. Carluccio, D.; Xu, C.; Venezuela, J.; Cao, Y.; Kent, D.; Bermingham, M.; Demir, A.G.; Previtali, B.; Ye, Q.; Dargusch, M. Additively Manufactured Iron-Manganese for Biodegradable Porous Load-Bearing Bone Scaffold Applications. *Acta Biomater.* **2020**, *103*, 346–360. [CrossRef]

0. Hermawan, H. *Biodegradable Metals*; Springer: Berlin/Heidelberg, Germany, 2012; ISBN 978-3-642-31169-7.

121. Field, D.M.; Baker, D.S.; Van Aken, D.C. On the Prediction of α-Martensite Temperatures in Medium Manganese Steels. *Meta Mater. Trans. A* **2017**, *48*, 2150–2163. [CrossRef]
122. Chou, D.-T.; Wells, D.; Hong, D. Novel Processing of Iron–Manganese Alloy-Based Biomaterials by Inkjet 3-D Printing. *Ac Biomater.* **2013**, *9*, 8593–8603. [CrossRef]
123. Safaie, N.; Khakbiz, M.; Sheibani, S. Synthesizing of Nanostructured Fe-Mn Alloys by Mechanical Alloying Process. *Proce Mater. Sci.* **2015**, *11*, 381–385. [CrossRef]
124. Huang, S.M.; Nauman, E.A.; Stanciu, L.A. Investigation of Porosity on Mechanical Properties, Degradation and in-Vit Cytotoxicity Limit of Fe30Mn Using Space Holder Technique. *Mater. Sci. Eng. C* **2019**, *99*, 1048–1057. [CrossRef]
125. Sotoudehbagha, P.; Sheibani, S.; Khakbiz, M.; Ebrahimi-Barough, S.; Hermawan, H. Novel Antibacterial Biodegradable Fe-Mn- Alloys Produced by Mechanical Alloying. *Mater. Sci. Eng. C* **2018**, *88*, 88–94. [CrossRef]
126. Schinhammer, M.; Hänzi, A.C.; Löffler, J.F.; Uggowitzer, P.J. Design Strategy for Biodegradable Fe-Based Alloys for Medic Applications. *Acta Biomater.* **2010**, *6*, 1705–1713. [CrossRef] [PubMed]
127. Moszner, F.; Sologubenko, A.S.; Schinhammer, M.; Lerchbacher, C.; Hänzi, A.C.; Leitner, H.; Uggowitzer, P.J.; Löffler, J Precipitation Hardening of Biodegradable Fe–Mn–Pd Alloys. *Acta Mater.* **2011**, *59*, 981–991. [CrossRef]
128. Mandal, S.; Kishore, V.; Bose, M.; Nandi, S.K.; Roy, M. In Vitro and in Vivo Degradability, Biocompatibility and Antimicrob Characteristics of Cu Added Iron-Manganese Alloy. *J. Mater. Sci. Technol.* **2021**, *84*, 159–172. [CrossRef]
129. Liu, R.-Y.; He, R.-G.; Chen, Y.-X.; Guo, S.-F. Effect of Ag on the Microstructure, Mechanical and Bio-Corrosion Properties Fe–30Mn Alloy. *Acta Metall. Sin. Engl. Lett.* **2019**, *32*, 1337–1345. [CrossRef]
130. Liu, R.-Y.; He, R.-G.; Xu, L.-Q.; Guo, S.-F. Design of Fe–Mn–Ag Alloys as Potential Candidates for Biodegradable Metals. *Ac Metall. Sin. Engl. Lett.* **2018**, *31*, 584–590. [CrossRef]
131. Spandana, D.; Desai, H.; Chakravarty, D.; Vijay, R.; Hembram, K. Fabrication of a Biodegradable Fe-Mn-Si Alloy by Field Assist Sintering. *Adv. Powder Technol.* **2020**, *31*, 4577–4584. [CrossRef]
132. Trincă, L.C.; Burtan, L.; Mareci, D.; Fernández-Pérez, B.M.; Stoleriu, I.; Stanciu, T.; Stanciu, S.; Solcan, C.; Izquierdo, J.; Souto, R. Evaluation of in Vitro Corrosion Resistance and in Vivo Osseointegration Properties of a FeMnSiCa Alloy as Potential Degradab Implant Biomaterial. *Mater. Sci. Eng. C* **2021**, *118*, 111436. [CrossRef]
133. Hong, D.; Chou, D.-T.; Velikokhatnyi, O.I.; Roy, A.; Lee, B.; Swink, I.; Issaev, I.; Kuhn, H.A.; Kumta, P.N. Binder-Jetting 3 Printing and Alloy Development of New Biodegradable Fe-Mn-Ca/Mg Alloys. *Acta Biomater.* **2016**, *45*, 375–386. [CrossRef]
134. Schinhammer, M.; Pecnik, C.M.; Rechberger, F.; Hänzi, A.C.; Löffler, J.F.; Uggowitzer, P.J. Recrystallization Behavior, Microstru ture Evolution and Mechanical Properties of Biodegradable Fe–Mn–C(–Pd) TWIP Alloys. *Acta Mater.* **2012**, *60*, 2746–27 [CrossRef]
135. Gebert, A.; Kochta, F.; Voß, A.; Oswald, S.; Fernandez-Barcia, M.; Kühn, U.; Hufenbach, J. Corrosion Studies on Fe-30Mn- Alloy in Chloride-Containing Solutions with View to Biomedical Application. *Mater. Corros.* **2018**, *69*, 167–177. [CrossRef]
136. Harjanto, S.; Pratesa, Y.; Suharno, B.; Syarif, J. Corrosion Behavior of Fe-Mn-C Alloy as Degradable Materials Candidate Fabricat via Powder Metallurgy Process. *Adv. Mater. Res.* **2012**, *576*, 386–389. [CrossRef]
137. Fiocchi, J.; Biffi, C.A.; Gambaro, S.; Paternoster, C.; Mantovani, D.; Tuissi, A. Effect of Laser Welding on the Mechanical a Degradation Behaviour of Fe-20Mn-0.6C Bioabsorbable Alloy. *J. Mater. Res. Technol.* **2020**, *9*, 13474–13482. [CrossRef]
138. Venezuela, J.; Dargusch, M.S. Addressing the Slow Corrosion Rate of Biodegradable Fe-Mn: Current Approaches and Futu Trends. *Curr. Opin. Solid State Mater. Sci.* **2020**, *24*, 100822. [CrossRef]
139. Zhu, S.; Huang, N.; Xu, L.; Zhang, Y.; Liu, H.; Lei, Y.; Sun, H.; Yao, Y. Biocompatibility of Fe–O Films Synthesized by Plasr Immersion Ion Implantation and Deposition. *Surf. Coat. Technol.* **2009**, *203*, 1523–1529. [CrossRef]
140. Zhou, J.; Yang, Y.; Frank, A.M.; Detsch, R.; Boccaccini, A.R.; Virtanen, S. Accelerated Degradation Behavior and Cytocompatibil of Pure Iron Treated with Sandblasting. *ACS Appl. Mater. Interfaces* **2016**, *8*, 26482–26492. [CrossRef]
141. Huang, T.; Zheng, Y. Uniform and Accelerated Degradation of Pure Iron Patterned by Pt Disc Arrays. *Sci. Rep.* **2016**, *6*, 236 [CrossRef]
142. Chen, H.; Zhang, E.; Yang, K. Microstructure, Corrosion Properties and Bio-Compatibility of Calcium Zinc Phosphate Coating Pure Iron for Biomedical Application. *Mater. Sci. Eng. C* **2014**, *34*, 201–206. [CrossRef]
143. Wang, H.; Zheng, Y.; Li, Y.; Jiang, C. Improvement of in Vitro Corrosion and Cytocompatibility of Biodegradable Fe Surfa Modified by Zn Ion Implantation. *Appl. Surf. Sci.* **2017**, *403*, 168–176. [CrossRef]
144. Wang, H.; Zheng, Y.; Jiang, C.; Li, Y.; Fu, Y. In Vitro Corrosion Behavior and Cytocompatibility of Pure Fe Implanted with Ta. *Sı Coat. Technol.* **2017**, *320*, 201–205. [CrossRef]
145. Yusop, A.H.M.; Daud, N.M.; Nur, H.; Kadir, M.R.A.; Hermawan, H. Controlling the Degradation Kinetics of Porous Iron Poly(Lactic-Co-Glycolic Acid) Infiltration for Use as Temporary Medical Implants. *Sci. Rep.* **2015**, *5*, 11194. [CrossRef]
146. Gorejová, R.; Oriňaková, R.; Králová, O.Z.; Baláž, M.; Kupková, M.; Hrubovčáková, M.; Haverová, L.; Džupon, M.; Oriňak, Kaľavský, F.; et al. In Vitro Corrosion Behavior of Biodegradable Iron Foams with Polymeric Coating. *Materials* **2020**, *13*, 1 [CrossRef] [PubMed]
147. Huang, S.; Ulloa, A.; Nauman, E.; Stanciu, L. Collagen Coating Effects on Fe–Mn Bioresorbable Alloys. *J. Orthop. Res.* **2020**, 523–535. [CrossRef] [PubMed]
148. Drevet, R.; Zhukova, Y.; Kadirov, P.; Dubinskiy, S.; Kazakbiev, A.; Pustov, Y.; Prokoshkin, S. Tunable Corrosion Behavior Calcium Phosphate Coated Fe-Mn-Si Alloys for Bone Implant Applications. *Metall. Mater. Trans. A* **2018**, *49*, 6553–6560. [CrossR

49. Cimpoeşu, N.; Trincă, L.C.; Dascălu, G.; Stanciu, S.; Gurlui, S.O.; Mareci, D. Electrochemical Characterization of a New Biodegradable FeMnSi Alloy Coated with Hydroxyapatite-Zirconia by PLD Technique. *J. Chem.* **2016**, *2016*, 1–9. [CrossRef]
50. Daud, M.N.; Sing, N.B.; Yusop, A.H.; Majid, A.F.A.; Hermawan, H. Degradation and in Vitro Cell–Material Interaction Studies on Hydroxyapatite-Coated Biodegradable Porous Iron for Hard Tissue Scaffolds. *J. Orthop. Transl.* **2014**, *2*, 177–184. [CrossRef]
51. Majewska, P.; Oledzka, E.; Sobczak, M. Overview of the Latest Developments in the Field of Drug-Eluting Stent Technology. *Biomater. Sci.* **2020**, *8*, 544–551. [CrossRef]
52. Pan, C.J.; Tang, J.J.; Weng, Y.J.; Wang, J.; Huang, N. Preparation, Characterization and Anticoagulation of Curcumin-Eluting Controlled Biodegradable Coating Stents. *J. Control. Release* **2006**, *116*, 42–49. [CrossRef]
53. Forrestal, B.J.; Case, B.C.; Yerasi, C.; Garcia-Garcia, H.M.; Waksman, R. The Orsiro Ultrathin, Bioresorbable-Polymer Sirolimus-Eluting Stent: A Review of Current Evidence. *Cardiovasc. Revasc. Med.* **2020**, *21*, 540–548. [CrossRef]
54. Joung, Y.K.; Kim, H.I.; Kim, S.S.; Chung, K.H.; Jang, Y.S.; Park, K.D. Estrogen Release from Metallic Stent Surface for the Prevention of Restenosis. *J. Control. Release* **2003**, *92*, 83–91. [CrossRef]
55. Yang, X.; Yang, Y.; Guo, J.; Meng, Y.; Li, M.; Yang, P.; Liu, X.; Aung, L.H.H.; Yu, T.; Li, Y. Targeting the Epigenome in In-Stent Restenosis: From Mechanisms to Therapy. *Mol. Ther. Nucleic Acids* **2021**, *23*, 1136–1160. [CrossRef]
56. Yusop, A.H.; Sarian, M.N.; Januddi, F.S.; Ahmed, Q.U.; Kadir, M.R.; Hartanto, D.; Hermawan, H.; Nur, H. Structure, Degradation, Drug Release and Mechanical Properties Relationships of Iron-Based Drug Eluting Scaffolds: The Effects of PLGA. *Mater. Des.* **2018**, *160*, 203–217. [CrossRef]
57. Shi, J.; Miao, X.; Fu, H.; Jiang, A.; Liu, Y.; Shi, X.; Zhang, D.; Wang, Z. In Vivo Biological Safety Evaluation of an Iron-Based Bioresorbable Drug-Eluting Stent. *BioMetals* **2020**, *33*, 217–228. [CrossRef]
58. Cysewska, K.; Karczewski, J.; Jasiński, P. Influence of the Electrosynthesis Conditions on the Spontaneous Release of Anti-Inflammatory Salicylate during Degradation of Polypyrrole Coated Iron for Biodegradable Cardiovascular Stent. *Electrochim. Acta* **2019**, *320*, 134612. [CrossRef]
59. Sharipova, A.; Unger, R.E.; Sosnik, A.; Gutmanas, E. Dense Drug-Eluting Biodegradable Fe-Ag Nanocomposites. *Mater. Des.* **2021**, *204*, 109660. [CrossRef]
60. Cooper, C.E. Nitric oxide and iron proteins. *Biochim. Biophys. Acta BBA Bioenerg.* **1999**, *1411*, 290–309. [CrossRef]

·ticle

·n–0.8Mg–0.2Sr (wt.%) Absorbable Screws—An In-Vivo iocompatibility and Degradation Pilot Study on a ·abbit Model

·arel Klíma [1], Dan Ulmann [1], Martin Bartoš [1], Michal Španko [1,2], Jaroslava Dušková [3], Radka Vrbová [1], ·n Pinc [4], Jiří Kubásek [5], Tereza Ulmannová [1], René Foltán [1], Eitan Brizman [1], Milan Drahoš [1], Michal Beňo [1] ·d Jaroslav Čapek [4,*]

1 Department of Stomatology—Maxillofacial Surgery, General Teaching Hospital and First Faculty of Medicine, Charles University, 121 08 Prague, Czech Republic; karel.klima@vfn.cz (K.K.); dan.ulmann@vfn.cz (D.U.); martin.bartos@vfn.cz (M.B.); michal.spanko@vfn.cz (M.Š.); radka.vrbova@vfn.cz (R.V.); tereza.ulmannova@vfn.cz (T.U.); Rene.Foltan@vfn.cz (R.F.); eitan.brizman@vfn.cz (E.B.); drahomil@gmail.com (M.D.); michal.beno@vfn.cz (M.B.)

2 Department of Anatomy, First Faculty of Medicine, Charles University, 121 08 Prague, Czech Republic

3 Department of Pathology, First Faculty of Medicine, Charles University, 121 08 Prague, Czech Republic; jaroslava.duskova@lf1.cuni.cz

4 Department of Functional Materials, FZU The Institute of Physics of the Czech Academy of Sciences, Na Slovance 1999/2, 182 21 Prague, Czech Republic; pinc@fzu.cz

5 Department of Metals and Corrosion Engineering, University of Chemistry and Technology, Technická 6, 166 28 Prague, Czech Republic; Jiri.Kubasek@vscht.cz

* Correspondence: capekj@fzu.cz; Tel.: +42-02-6605-2604

·ation: Klíma, K.; Ulmann, D.; ·toš, M.; Španko, M.; Dušková, J.; ·ová, R.; Pinc, J.; Kubásek, J.; ·nannová, T.; Foltán, R.; et al. ·0.8Mg–0.2Sr (wt.%) Absorbable ·ews—An In-Vivo Biocompatibility · Degradation Pilot Study on a ·bit Model. *Materials* **2021**, *14*, 3271. ·os://doi.org/10.3390/ma14123271

·demic Editors: ·ndra Hermawan and ·hdi Razavi

·eived: 25 May 2021
·epted: 9 June 2021
·lished: 13 June 2021

·lisher's Note: MDPI stays neutral ·h regard to jurisdictional claims in ·lished maps and institutional affil·ons.

Abstract: In this pilot study, we investigated the biocompatibility and degradation rate of an extruded Zn–0.8Mg–0.2Sr (wt.%) alloy on a rabbit model. An alloy screw was implanted into one of the tibiae of New Zealand White rabbits. After 120 days, the animals were euthanized. Evaluation included clinical assessment, microCT, histological examination of implants, analyses of the adjacent bone, and assessment of zinc, magnesium, and strontium in vital organs (liver, kidneys, brain). The bone sections with the implanted screw were examined via scanning electron microscopy and energy dispersive spectroscopy (SEM-EDS). This method showed that the implant was covered by a thin layer of phosphate-based solid corrosion products with a thickness ranging between 4 and 5 µm. Only negligible changes of the implant volume and area were observed. The degradation was not connected with gas evolution. The screws were fibrointegrated, partially osseointegrated histologically. We observed no inflammatory reaction or bone resorption. Periosteal apposition and formation of new bone with a regular structure were frequently observed near the implant surface. The histological evaluation of the liver, kidneys, and brain showed no toxic changes. The levels of Zn, Mg, and Sr after 120 days in the liver, kidneys, and brain did not exceed the reference values for these elements. The alloy was safe, biocompatible, and well-tolerated.

Keywords: bioabsorbable metals; in-vivo biocompatibility; magnesium; zinc; strontium; toxicity; biocompatibility; systemic reactions; alloy accumulation; internal organs

1. Introduction

A vast improvement in living standards over the past few decades, has brought with it increasing demands to medicine. The classical concept for treatment of fractures utilizing steel or titanium plates and screws to stabilize a fractured bone brought with it the need for a secondary procedure to remove these metallic materials from the body. To avoid this, experimentation began with the use of resorbable polymer materials. They are based on poly-lactic or poly-glycolic acids and their co-polymers. Units of polymers are broken down into water and carbon dioxide in the body. Some patients exhibit an inflammatory reaction during the process of resorption. In normal conditions, bending these polymer plates not possible. Only hot water or ultrasound can be used for the bending and thus bett alignment of polymeric materials. The load-bearing ability of polymers is also low, whi represents another disadvantage. Absorbable metals could solve these problems wi (i) better bending for individualization of bone shape, (ii) better mechanical streng (iii) possibly better degradation parameters after fulfilling their function by differe principles of absorption [1–3].

Biodegradable metals (BMs) are defined by gradual corrosion in-vivo with appropria reaction of a host body. BMs should dissolve completely when fulfilling their missic Therefore, a major component of BMs should be elements that could be metabolized a human body without toxic, carcinogenic, teratogenic, and allergic reactions [4]. T international organization for standards (ASTM) has defined standards for bioabsorpti of absorbable metals (F3160 and F3268) [5,6].

The history of resorbable metals began in 1878 when Edward C. Huse used magr sium wires for ligature of bleeding vessels [7,8]. He described that the resorption ra depended upon the diameter of the Mg wire. Austrian physician Erwin Payr conti ued by using magnesium in various surgical procedures. He developed a magnesiu connector for end-to-end vessel anastomosis. Payr also described an intramedullary stab ising rod for the treatment of non-union fractures [7]. Magnesium-based materials wi absorbable properties have been extensively studied since the turn of 20th century ar have been used for the treatment of (i) vessel stenting [9], (ii) intestinal anastomoses [7], a (iii) bone implants [1,7,10], both in experimental animals [2,3,11,12] and also humans [1,10]. Even though Mg-based materials are now readily used in surgical pr tice [10,13,14], they still possess some drawbacks—preventing their wide use in surgery. particular, their corrosion rate is too high for many applications. Moreover, the corrosi process is accompanied by hydrogen production, which may have a detrimental effect the healing process [15,16]. Due to that side effect, iron- and zinc-based materials have be suggested as alternative potential metallic absorbable implants for both cardiovascular a bone implants [15–17]. In contrast to magnesium, iron and its alloys degrade very slow Except for the austenitic alloys, the Fe-based alloys are ferromagnetic, which preclud patients from undergoing magnetic resonance imaging. This limits the application Fe-based alloys in implantology as well [18,19]. In 2011, Vojtěch et al. introduced zinc a candidate for fabrication of biodegradable implants [13,14,20]. Since that time, ma Zn-based materials containing a large variety of different alloy elements have been vestigated. Based on the obtained results, it was suggested that Zn-based materials c adequately fulfil all requirements for use in implantology [16,18,19,21].

In this paper, we show the results of in-vivo testing of Zn–0.8Mg–0.2Sr (wt. implants—in the form of maxillofacial screws in rabbit tibias—left exposed to their en ronment for 120 days. The biocompatibility, degradation, and implant–bone interacti were investigated. To allow the readers an easy comparison with those properties of oth Zn- and Mg-based absorbable materials, a short review of the results obtained by vario scientific teams was performed. The results of this review are summarized in Table 1 a in the following paragraphs.

The In-Vivo Biological Behaviour of Zn- and Mg-Based Materials

Table 1. A summary of the results of the in-vivo tests on Zn/Mg alloys performed by various scientific groups.

No. of Animals, Experimental Time	Setting	Zn/Mg Alloy	Methods of Analysis	Results	Ref.
30 rats, 4, 8, 12 weeks	Femoral condyle defect, with 99.99% Ti (N = 15), Zn–0.8Sr (wt.%), N = 15	Zn–0.8Sr (wt.%) Ti 99.99%	microCT, Histomorphometry, SEM	Good biosafety. Osteogenesis time related: greater after 12 weeks than after 4 weeks. Biodegradation products greater around Zn–0.8Sr alloy. Heart, liver, spleen, lung, and kidneys—no abnormalities in the Zn–0.8Sr alloy group in comparison with the pure Ti group. The concentration of Zn^{2+} and Sr^{2+} in the blood and organs of the Zn–0.8Sr group was not higher than the pure Ti implant group	[22]
54 rats, 8 weeks	Experimental alloys implanted into femoral bone of rats	Zn 99.99% Zn–0.1Sr Zn–0.1Mn Zn–0.4Cu Zn–0.4Fe Zn–0.2Li Zn–0.8Mg Zn–0.8Ca Zn–2.0Ag	microCT, Histology, SEM, blood analysis of Zn concentration	No gas, no obvious degradation after 8 weeks. Circumferential osteogenesis. Volume of pure Zn decreased to 95.12 ± 1.39% after 8 weeks and degradation rate was 0.14 ± 0.05 mm/year. Zn–0.4Cu alloys had higher rate of degradation: 0.26 ± 0.03 mm/year. Faster degradation: Zn–2.0Ag, Zn–0.4Li, Zn–0.4Fe, and Zn–0.8Mg. Same degradation rate as pure Zn: Zn–0.1Mn, Zn–0.8Ca, and Zn–0.8Sr	[23]
27 beagle dogs, 8, 12, 24 weeks	Experimental alloy used for treatment of mandibular fractures of beagle dogs. Control group: Ti 99.99%, PLLA (poly-L-lactic acid)	Zn–2.0Mg–2.0Fe PPLA Ti 99.99%	3-point bending test, X-ray, microCT, histology, analysis of Zn concentration in blood, before surgery, 4, 12, and 24 weeks after surgery	Zinc alloy: good mechanical properties. Increasing in vivo degradation rate of the zinc alloy implants: 0.033 ± 0.015 mm/year at 4 weeks, 0.079 ± 0.009 mm/year at 12 weeks, 0.095 ± 0.009 mm/year at 24 weeks. Good biocompatibility: no difference between the pre-op and post-op blood and biochemical results. Serum zinc value in the zinc alloy group was slightly higher after implantation than before, no statistically significant difference	[24]

Table 1. *Cont.*

No. of Animals, Experimental Time	Setting	Zn/Mg Alloy	Methods of Analysis	Results	Ref.
12 rabbits, 24 operated sites, 16 weeks	Fracture of ulnar bone and its osteosynthesis	99.9% Mg (wt.%)	Mechanical testing, X-ray, microCT, histology	Well tolerated. Corrosion product formation and gas formation. Histologically normal bone properties. All fractures healed. Bone growth over and around degrading Mg devices. Good mechanical properties of healed bone. Facilitated fracture healing while stimulating local bone growth	[3]
32 rats, 24 weeks	Femoral pin implanted bilaterally	ZX50 *, WZ21 **	Histology, microCT	Volume loss: ZX50: 1.2%/day, WZ21: 0.5%/day. ZX50: massive gas production. WZ21: new bone production, good osteoconductivity and osteoinductivity of Mg. Excellent bone recovery	[2]
36 rats, 4, 8, 12 weeks	Ulnar bone defect replaced by a metallic experimental alloy tube	Mg–3Zn–1Ca–0.5REE ***/ hydrothermally (HT) coated and uncoated alloy	Histology, microCT, X-ray, serum Mg^{2+} and Ca^{2+} concentrations	Mg^{2+} normal levels in all groups. Ca^{2+} levels higher than normal, but not significantly. No organ histological pathology. HT coated alloy with better biocompatibility and higher resistance to corrosion	[11]
8 rabbits, 9 months to 3.5 years	Cylindrical alloy implant in a medullary cavity	LAE442 ****	Histology, microCT, mechanical testing, organ corrosion products testing	Good biocompatibility, no gas formation, no toxicity in vital organs. Slow resorption: 2–2.5% in 9 months, 5% in 3.5 years. Accumulation of REE *** in implant site and vital organs	[12]
18 rabbits, 16 weeks	Alloy plate and 2 screws implanted into femoral bone	99.9% Mg Mg–1Sr (wt.%)	Histology, blood count, Sr concentration in implanted site and vital organs	Corrosion rate slower in group with Sr. Good biocompatibility in the group with Sr. Highest increase of Sr^{2+} concentration in liver (388 ± 25 μg/kg) at 4w post-implantation and returned to 32 ± 2 μg/kg after 16 weeks implantation. Increase in the Sr^{2+} ion concentration in the blood from 2–8 weeks post-implantation	[25]

Table 1. *Cont.*

No. of Animals, Experimental Time	Setting	Zn/Mg Alloy	Methods of Analysis	Results	Ref.
48 humans, 12 months	Treatment of osteonecrosis of femoral head by vascularized bone grafting. Two groups: (i) Mg screw (ii) without fixation	99.9 Mg (wt.%) §	X-ray, CT, functional recovery, Harris hip score (HHS), serum levels of Mg, Ca, and P	Mg screw group 25% volume reduction in 12 months. Normal levels of Mg, Ca, P in both groups. HHS was significantly improved in the Mg screw group	[26]
23 goats, 4, 8, 12, 48 weeks	Femoral neck fractures. Control group: no treatment/empty defect	Mg 99.99 wt.% §	Histology, microCT, X-ray, serum Mg and Ca ion, liver and kidney functions pre-op and at 2, 4, 8, 12, and 48 weeks after surgery.	Normal levels of ions all weeks/all groups. Normal organ function & histology. All fractures healed. Degradation of metal: 10% at 4 weeks, 38.8% at 12 weeks and 45.3% at 48 weeks. No cytotoxicity. Good bone production around screws. No gas production observed	[27]
6 rabbits, 6, 12, 24 weeks	Reconstruction of anterior cruciate knee ligament by Mg alloy and titanium screws	MgYREEZr and Ti–6Al–4V (wt.%)	Histology, microCT, blood testing for alloy elements	Good biocompatibility, no inflammatory changes, no necrosis. Similar results to titanium alloy. Gas production decreased within 24 weeks. Very low levels of alloy elements in blood	[28]
18 rats, 4, 26, and 52 weeks	Pins into femoral bone of a rat, one group treated by ZX10, other by ZX20	ZX10 and ZX20 §§ ZX20	Histology, microCT	Higher degradation rate of ZX20. Histologically ZX10 and ZX20 are well tolerated. Good implant-bone contact from 4 weeks post-op Gas production: insignificant between pure Mg (99.999%) and ZX10, significantly higher gas production of ZX20, compared to ZX10 and pure Mg	[29]

* ZX50 = Mg–5.0Zn–0.25Ca–0.15Mn (wt.%)—early degradation. ** WZ21 = Mg–1.0ZN–0.25Ca–0.15Mn (wt.%)—longer degradation. *** Rare earth elements (REE). **** LAE442: LAE442 (Mg–4%Li–3.6%Al–2.4%REE. ****, in wt.%) § Mg 99.99%; Al 0.002; Si < 0.001; Ca < 0.001; Ti < 0.0001; Mn 0.002; Fe 0.001; Ni < 0.0001; Cu 0.0002; Zn 0.0028; Pb 0.0008 (wt.%). MgYREEZr: Mg–4.3Y–0.4Zr–4.4REE (wt.%). §§ ZX10: Mg-1.0Zn-0.3Ca (in wt.%)/ZX20: Mg-1.5Zn-0.25Ca (in wt.%).

From the aforementioned studies, which were analysed in detail and whose results are summarized in Table 1, magnesium can be characterized as having very good biocompatibility with the surrounding bone—expressed by good new-bone production, good osteoconductivity and osteoinductivity, and good mechanical properties [2,3,27,28,30]. In the case of pure magnesium, hydrogen pockets were often formed [2,28,29]. Those gas pockets did not interfere with normal bone healing or lead to any significant inflammation or necrosis of the surrounding soft tissues [3,28,29]. In one study that used pure Mg, contamination around the bone by corrosion products was observed [3], and in a study where the author used an alloy of Mg–4Li–3.6Al–2.4REE (wt.%), REE contamination around the implanted alloy and in vital organs was reported [12]. The use of pure Mg metal did not si nificantly increase the level of Mg in the blood or surrounding organs [11,26–28]. The ra of corrosion and resorption of the experimental metal was faster for pure magnesium [compared to magnesium alloys with the addition of Sr [25].

Zinc alloys also show good mechanical properties [22–24] with good biocompatibili and no increase in Zn and Sr levels in vital organs compared to the use of pure Ti [2 The corrosion rate of a zinc alloys is slower compared to magnesium alloys. Pure M resorbs at a rate of 25% volume reduction in 12 months [26], whereas pure Zn resor at a rate of 10% per year, equivalent to only 0.14 ± 0.05 mm/year [21,23]. Zinc allo exhibited similarly slow resorption rates: Zn–2.0Mg–2.0Fe alloy had a degradation ra of 0.095 ± 0.009 mm/year [24]. One study observed an insignificant increase in seru Zn levels compared to the post-op and pre-op levels [24]. In recent years, Zn alloys ha been preferred for their slower degradation rate, as well as not having any production hydrogen gas pockets and their good mechanical and biological properties. We decid to use a Zn–Mg–Sr biodegradable alloy for our experimental rabbit study. This alloy w used because of the results that we obtained in our previous studies that described t mechanical, corrosion, and in-vitro biological behaviour of a Zn–0.8Mg–0.2Sr extrud alloy [31,32]. Those studies showed that this alloy possesses promising behaviour f application as a bone implant [31,32].

2. Materials and Methods

2.1. Implant Material

The extruded Zn–0.8Mg–0.2Sr (wt.%) alloy was prepared by melting a mixture appropriate amounts of pure elements followed by gravity casting. Subsequently, t ingot was annealed for 24 h at a temperature of 350 °C to homogenize its compositi and microstructure. The annealed material was extruded at the temperature 200 °C usi an extrusion ratio of 25:1. The preparation procedure is described in more detail in o previous study [32]. The exact chemical composition of the alloy was measured via atom absorption spectroscopy (AAS) using an Agilent 280FS AA spectrometer (Agilent, San Clara, CA, USA) with flame atomization. For this analysis, samples from several locatio of the extruded rod were dissolved in HNO_3 and the obtained solutions were diluted wi deionized water to concentrations suitable for analysis. The mean composition of the all was as follows: 0.83 wt.% of Mg, 0.17 wt.% of Sr, and 99 wt.% of Zn.

The extruded rods were machined using a CNC Fanuc Robodrill α-T21iFa machi (Fanuc, Tsukuba, Japan) into screws whose shape and dimensions were inspired by called "micro-maxillofacial" screws. The shape of the screws was designed accordi to the mechanical properties of the selected alloy to withstand expected loading duri implantation and subsequent exposure in tissue. A scheme of the screw is shown Figure 1.

2.2. Animals

We used 3 male New Zealand rabbits (Velaz, Prague, Czech Republic) with bo weights 750–850 g. We only included males in our study to minimize the effects of h mone levels on the variability of the healing as well as bone regeneration [33,34]. Th study was performed in accordance with the European Communities Council Directive 24 November 1986 (86/609/EEC) regarding the use of animals in research and was a proved by the Ethical Committee of the First Faculty of Medicine, Charles Universi Prague, Czech Republic. All effort was made to minimize the number of animals used the study.

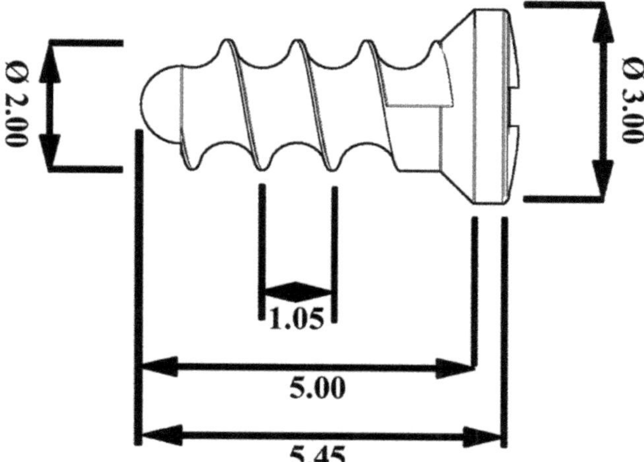

Figure 1. A schematic figure of the screws used in this study. The dimensions are listed in mm.

2.3. Experimental Group

Three laboratory rabbits were chosen at random to have the experimental alloy screw implanted into their tibia bone.

2.4. Surgery

The implant procedure was performed on the rabbits under general anaesthesia. All surgery was performed in a certified veterinary operating theatre. Anaesthesia was induced with 5% isoflurane (Isoflurane Piramal, Piramal Healthcare UK Ltd., Morpeth, UK) at a flow rate of 300 mL/min. The animals were maintained with ketamine 20 mg/kg and xylazine 3 mg/kg. Hair was shaven from the tibia bone region under aseptic conditions and then a 2 cm skin incision at the proximal part of diaphysis of the tibia bone was made. The muscles and periosteum were reflected from the bone. A hole was drilled into the proximal diaphysis of the tibia bone to a depth of 6 mm with a diameter of 1.6 mm. The drilling was accompanied by sterile cooling utilising a physiological solution. The three rabbits were implanted with an experimental alloy screw measuring 2 mm in diameter and 5 mm in length. All layers were sutured by resorbable sutures. The rabbits were euthanized 120 days after the surgery.

2.5. Euthanasia

Euthanasia was performed by inhalation of the anaesthetic isoflurane (Isoflurane Piramal, Piramal Healthcare UK Ltd., Morpeth, UK) that is otherwise used for general anaesthesia. Subsequently, 360 mg sodium thiopental (Sandoz GmbH, Vienna, Austria) was injected directly into the heart, leading to immediate cardiac failure. After reflection of the soft tissues, we harvested the bone specimens containing the experimental screws. The bone, together with specimens of the liver, kidneys, and brain, were fixed in 4% formalin solution and processed for histological examination. The bone was also examined via microCT. Liver, kidney, and brain specimens were frozen to −40 °C for analysis of strontium, zinc, and magnesium levels.

2.6. Histological Methods for Analysis of the Bone Specimens

The tibiae were fixed in a solution of 36–38% formaldehyde and 80% ethanol (ratio 1:2). The samples were dehydrated for 1 week in a series of ethanol solutions ranging from 70 to 100%, 1 concentration/1–2 days, culminating with 1 day in a solution of 100% ethanol and acetone (ratio 1:1). The bones were twice immersed in destabilized methyl methacrylate (MMA; 1 MMA/2–3 days), and finally in an embedding medium (100 g MMA + 12 mL dibutyl phthalate + 1.8 g benzoyl peroxide). Penetration of the medium was aided by vacuum pump (ILMVAC GmbH, Ilmenau, Germany). Polymerization was induced usin a water bath (thermostat F25-HE, Julabo, Seelbach, Germany) with the temperature risin by one degree every 2–3 days from 24 to 36 °C. The bone blocks were cut using a la saw ISOMETTM with a diamond disc (Buehler, Lake Bluff, IL, USA). The surfaces wer ground down using wet silicon–carbide papers P1200, P2500, and P4000 and polishe with 1 mm and 0.3 mm Al_2O_3 suspensions used in combination with TexMet and Micr Cloth polishing cloths (Buehler, Lake Bluff, IL, USA) and the equipment MetaServ 25 (Buehler, Lake Bluff, IL, USA). The sections of 60 μm thickness were stained for 5 mi with 1% solution of toluidine blue in 30% ethanol and heated to 60 °C. After rinsing wit running distilled water and differentiation in 96% ethanol, the slides were stained fc 12 min with 0.2% solution of toluidine blue in a phosphate buffer (pH = 9.1) and heated t 60 °C. The slides were then washed with distilled water and dried. The healing process c the tested materials in the artificially created holes and grooves was examined via optica microscopy using a Nikon Eclipse 80i microscope (Nikon Instruments Inc, Melville, N USA), Jenoptik camera (Jenoptik, Jena, Germany), and an image analysis system by NI Elements, Nikon AR (Nikon Instruments Inc, Melville, NY, USA). Any bone reaction, it morphology, and the presence of fibrous tissue and cells were studied and evaluated o the histopathology of specimens using a semiquantitative scoring system with parameter described by Reifenrath et al., 2011 [35]. The evaluated features were: gas bubble forma tion, overall bone structure, bone cavities, periosteum remodelling, endosteal remodellin periosteum apposition, peri-implant bone formation, peri-implant fibrosis, lymphoplasm cellular reaction, presence of macrophages, and giant cells [35]. Two to five histologica slides from each implant were examined. Any morphological signs of possible damag and bone tissue response were monitored at the level of: (i) periosteum, (ii) endosteun (iii) bone–implant contact (BIC) [36,37], (iv) connective tissue formation, (v) inflammator response, and (vi) ossification in the newly formed connective tissue [35].

2.7. Histopathological Methods of Parenchymal Organ Processing

The organs of the sacrificed rabbits were fixed in a buffered formalin solution. Macro scopically, they did not exhibit any pathology changes. Two excisions from the kidney an liver and a part of the brain tissue were embedded in paraffin. Standard haematoxylin eosin staining was used, and in some samples Pearls' reaction for iron was also performe In all animals, kidney, liver, and brain specimens were investigated for any possible tox influence of the implanted material.

2.8. Methods of X-ray Examination

All rabbits underwent X-ray examination under general anaesthesia 120 days afte implantation. All surgery was performed in a certified veterinary operating theatre. Afte induction of anaesthesia with 5% isoflurane (flow 300 mL/min), we made two projection of each rabbit's tibia, one projection perpendicular to the other. All X-rays were mad using an In-Vivo Xtreme BI 4MP (Bruker BioSpin, Rheinstetten, Germany). In order not t disturb the bone around the experimental screw, the X-ray examination was performe immediately prior to sacrificing the animals.

2.9. MicroCT Examination

The bone specimens (n = 3) containing the implants were scanned using a deskto SkyScan 1272 Micro-CT (Bruker, Kontich, Belgium). The specimens were immersed i 70% ethanol solution and scanned in plastic tubes with the following parameters: pixe size 15 μm, source voltage 100 kV, source current 100 μA, 0.11 mm Cu filter, frame a eraging 3, rotation step 0.1°, rotation 180°. The scanning time was approximately 2 for each specimen. The flat-field correction was updated before each acquisition. Imag data were reconstructed and processed using NRecon, DataViewer, CTVox, and CTA software (Bruker BioSpin, Rheinstetten, Germany). Prior to 3D analysis of the implant the parametric data for the volume and surface values were image-processed (with respect to improvement of the signal-to-noise ratio) and subsequently binarized. Data processing was optimized using TelGen software, which is described in [38]. Bone–implant contact was quantified via manual measurements of the 2D cross-section images in each specimen (BIC = implant perimeter in contact with bone-implant perimeter). The bone contact was only evaluated in the cervical region of the screw because the implant apex position was quite variable amongst the specimens and the screw head was excluded from this evaluation.

2.10. Analysis of the Solid Corrosion Products

To investigate the extent and uniformity of the corrosive process and the chemical composition of the solid corrosion products, sections of the bones with implanted screws were observed using a scanning electron microscope FEI Quanta 3D FEG (ThermoFisher Scientific, Waltham, MA, USA) equipped with an energy dispersive spectrometer EDAX Apollo 40 (Ametek, Berwyn, PA, USA) (SEM-EDS).

2.11. Analysis of Systemic Toxicity in the Vital Organs

Liver, kidney, and brain specimens were analysed for their content of strontium, magnesium, and zinc.

The tissue samples from the experimental animals were mineralized utilizing nitric acid and hydrogen peroxide. The mineralization process would cause decomposition of the biological and inorganic matrix, enabling the transfer of the analytes into a solution which would allow for their analysis. A Milestone MLS 1200 MEGA microwave (Milestone Inc., Shelton, CT, USA) mineralization device with a 6-position high-pressure decomposition rotor and an evaporating rotor was used to mineralize the samples. Control reference materials and blank samples were mineralized together (in parallel) with the actual samples.

2.12. Strontium and Zinc Analysis

The content of Sr and Zn elements in the samples was determined via inductively coupled plasma mass spectrometry (ICP-MS) on an ELAN DRC-e instrument (Perkin Elmer SCIEX, Waltham, MA, USA) in which the concentrations of the given analytes in the sample solution were selectively determined. Strontium ions ^{88}Sr and zinc ions ^{66}Zn were used for quantification. The sample mineral was adjusted, if necessary, by dilution and addition of a ^{74}Ge internal standard for both monitored elements before final analysis via the ICP-MS method. Quantitative evaluation of the analytes was performed with external calibration. Calibration for Sr was in the range 0–10 µg/L and for Zn the range 0–500 µg/L. A Seronorm WB L-2 whole blood sample was used as a control. The limit of detection for Sr was 0.003 µg/g tissue and for Zn 0.1 µg/g tissue.

2.13. Magnesium Analysis

The magnesium content was determined in the sample via flame atomic absorption spectrometry (F-AAS) on an AAnalyst 400 (Perkin Elmer, Waltham, MA, USA). Before the F-AAS analysis, the sample was treated by diluting and adding ionic buffers Cs and La so that their final concentration in the solution was 2000 mg/L. Quantitative evaluation of the analytes was performed with a method of external calibration. Calibration for Mg was in the range 0–2 mg/L. A Seronorm WB L-2 whole blood sample was used as a control. The limit of detection for Mg was 0.2 µg/g tissue.

3. Results

3.1. Mechanical Properties of the Alloy

The mechanical behaviour of the investigated alloy has been studied and was discussed in our previous work [32]. It is worth mentioning that the tensile yield strength, ultimate tensile strength, ductility, and Young's modulus of the alloy were 244, 324 MP 20%, and 104 GPa, respectively.

3.2. X-ray Examination

The specimens were visualised using a standardised 2D X-ray imaging techniq in two planes 120 days after implantation. The images obtained via X-ray examinati are shown in Figure 2. The screws appeared to be well integrated, showing successf osseointegration. Microscopic interposition of fibrous tissue (between the experiment screws and the adjacent bone) could of course not be excluded. No osteolytic changes we seen around the experimental screws indicating that no inflammatory response took pla

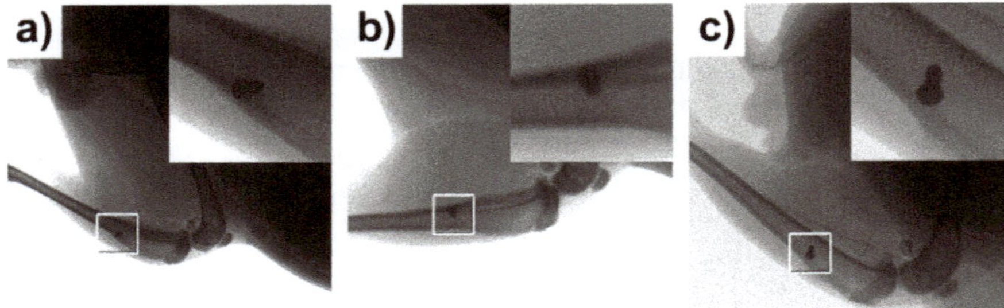

Figure 2. X-ray of the tibial bone of the three experimental rabbits 120 days after experimental screw implantation. (**a**) rabbit no. 1, (**b**) rabbit no. 2 and (**c**) rabbit no. 3. There are no osteolytic changes seen between the bones and experimental screws.

3.3. MicroCT Examination

The specimens were visualized using a standardized 2D cross-sectional image 3 perpendicular planes (Figure 3a). The microCT image data enabled quantitative evalu tion of bone–implant contact (manual 2D measurement), the quantitative 3D analysis implant structure, and qualitative evaluation of bone surrounding the implants. Howev metal-induced artefacts in the experimental group did not allow quantitative 3D analys of bone–implant contact (Figure 3a,c).

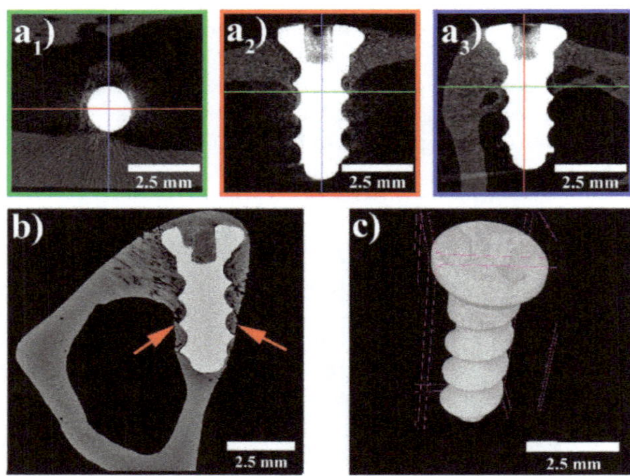

Figure 3. MicroCT images; (**a1–a3**) standardised cross-section images in three perpendicular plan (**b**) implant completely surrounded by bone—arrows; (**c**) 3D visualization of the implanted screw

All the evaluated implants appeared to be in contact with the surrounding bone without any signs of a foreign body reaction or fibrointegration. The thickness of the bone surrounding the implants was found to be greater than the average cortical thickness. Histomorphometric analysis of bone–implant contact was established as the most accurate method for evaluation of the percentage of the implant in contact with bone [36,37]. The mean BIC value for the implants at 120 days was 22%. There were no signs of implant degradation. Only minor changes of volume and surface were found in the screws.

3.4. SEM-EDS Observations of the Implant–Bone Interface

The results of SEM-EDS observations of a section of a bone with an implanted screw are shown in Figure 4. In Figure 4, it is visible that new bone has overgrown the head of the screw (Figure 4a), which suggests good biocompatibility of the implanted material. At higher magnifications, it was observed that the implant was surrounded by tissue rich in carbon. As will be shown later, this tissue was identified as fibrous tissue. In Figure 4d, it is clearly visible that the implanted screw was covered by a layer of solid corrosion products, which consisted of Zn, O, Ca, and P. This suggests that the corrosion products were mainly based on phosphates. The thickness of the layer of corrosion products ranged between 4–5 μm, corresponding to an average degradation rate of approximately 13.5 μm/year.

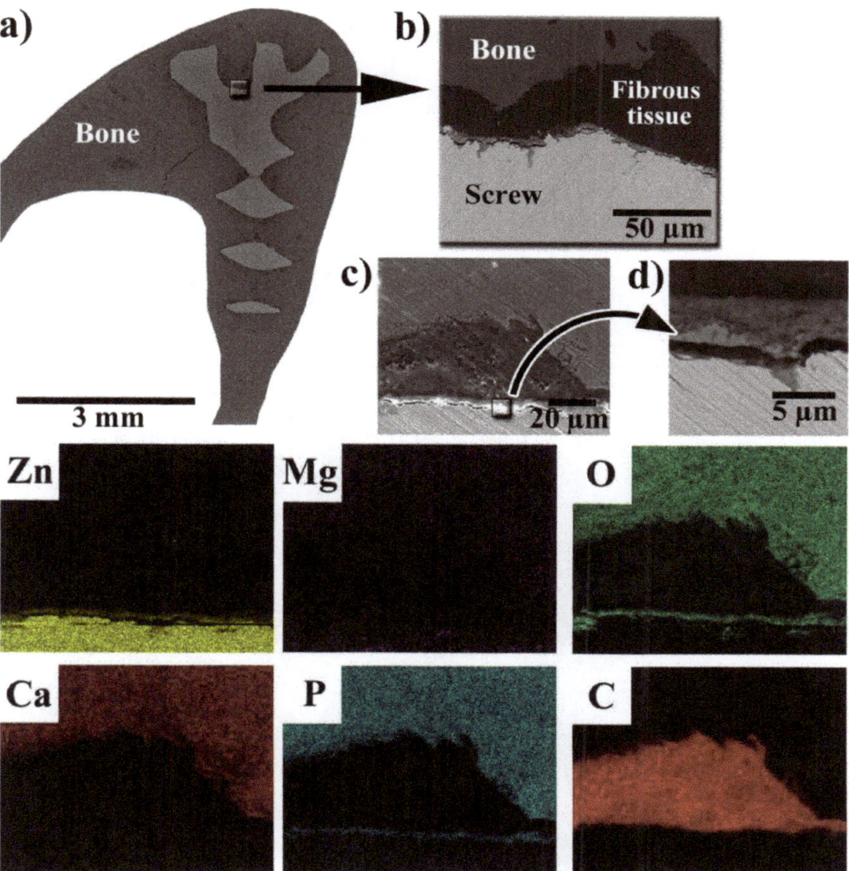

Figure 4. SEM-EDS observation of the bone section with the implanted screw. (**a**) SEM picture of the overview; (**b**) SEM picture of the screw–bone interface; (**c**) SEM picture of the area mapped via EDS (the elemental maps are shown below); and (**d**) a detailed view (SEM) of the layer of corrosion products.

3.5. Histopathological Examination of Bone Specimens Containing Implants

The presence of gas bubbles was not detected in any of the examined samples. As i shown in Figure 5, the irregular structure of the bone was observed rather rarely in th vicinity of the implants (Figure 5a) and the majority of the bone surrounding the implar was of a regular structure (Figure 5b).

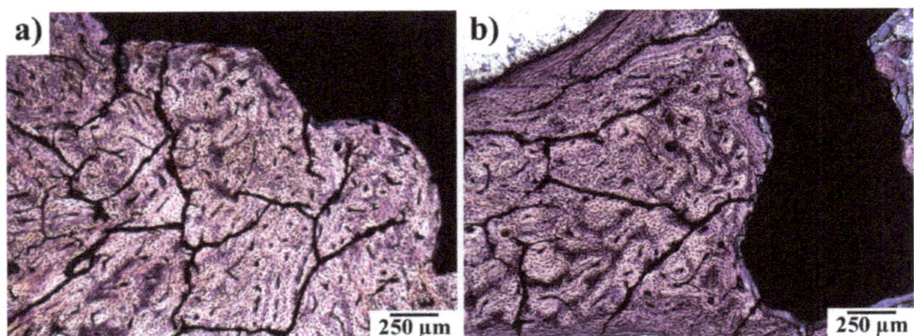

Figure 5. Histological examination of implant site with experimental screw. (**a**) Irregular osteon arrangement; (**b**) regular osteon arrangement. Note: the black field represents a histological cut of an experimental screw. Toluidine blue stained slides.

In Figure 6, periosteal apposition and endosteal remodelling in the vicinity of a implanted screw are shown. The periosteal apposition was connected with the remodellin of new bone tissue. Bundle bone was formed more often than the lamellar one. Th intensity of endosteal remodelling was less frequent than periosteal apposition, although similar extent of those events was observed in individual experimental animals.

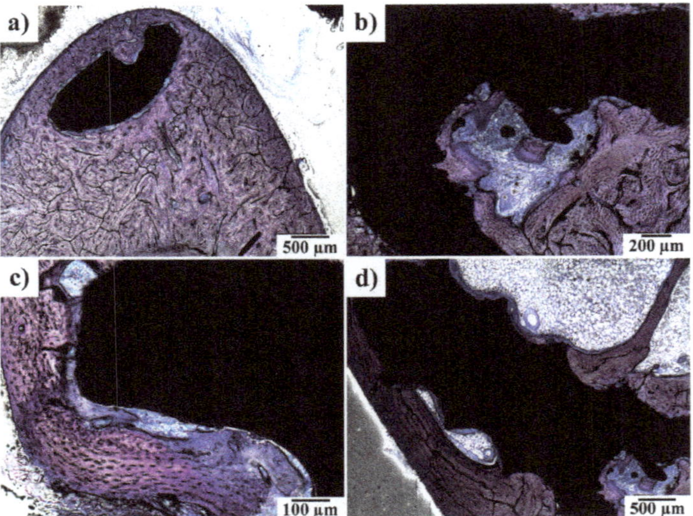

Figure 6. (**a**) Head of the implanted screw (black) embedded in the newly formed bone–perioste apposition here is followed by regular bone formation; (**b**) periosteal apposition and bone remodellin over an experimental screw; (**c**) bone remodelling with bundle bone formation in the presence of a experimental screw; and (**d**) endosteal remodelling–intrabony segment of the experimental screw. A 120 days after implantation. Toluidine blue staining.

In Figure 7, one can clearly see that peri-implant fibrosis took place. The fibrous tissue covered more than 51% of the implant surface but with variable thickness. The thickness of the fibrous tissue ranged between 0.01 mm and 0.3 mm (see Figure 7 and Table 2.). At 120 days after implantation, two out of the three rabbits had no detectable inflammatory reaction. The third rabbit reacted, presenting with chronic lymphoplasmocellular infiltration in the peri-implant fibrous tissue. There were scattered macrophages as well; no giant cells were found. As is shown in Figure 7c, the inflammatory response in the connective tissue surrounding the implant was of moderate to high intensity. In contrast, the presence of macrophages was mostly subthreshold (<3 in section) or sparse (3–20 macrophages in section), as is visible in Figure 7d. Neither phagocytosed material in the cytoplasm nor substantial irregularities of the experimental screw surface were observed. Nevertheless, their role in triggering the fibroproductive inflammatory process is undeniable. There were no signs of active osteolysis beyond the described peri-implant fibrosis. Giant multinucleated cells were only found exceptionally—in one solitary case.

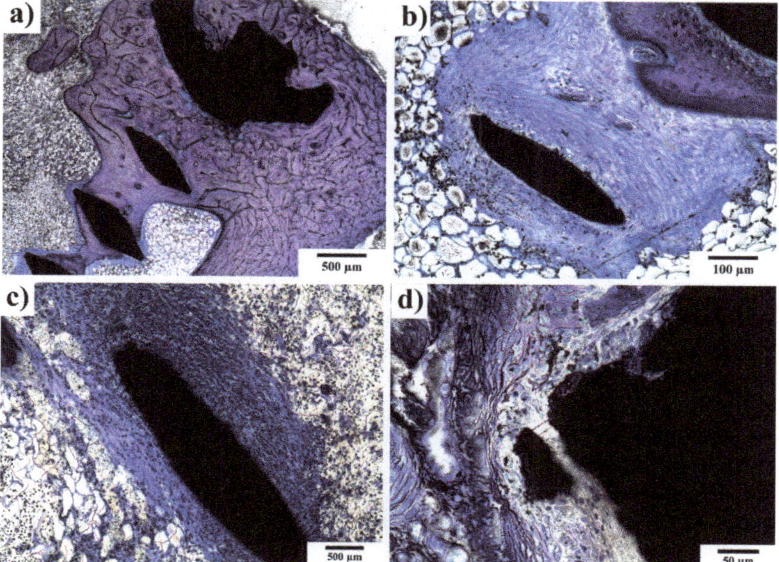

Figure 7. Peri-implant fibrosis around experimental screw 120 days after implantation. (**a**) The whole surface of the experimental screw was covered with fibrosis; (**b**) under high magnification, remnants of fibroproductive inflammation, represented by scattered lymphocytes and macrophages in the fibrous tissue, can be traced; (**c**) peri-implant fibrosis with dense lymphocytic infiltrate surrounding the experimental screw; and (**d**) macrophages in close vicinity to the implanted screw. At 120 days after implantation. Toluidine blue staining.

Table 2. Evaluation of the healing process according to parameters on histology of specimens (Reifenrath et al.) [35].

Parameter	Score	Interpretation	120 Days of Implantation						
Gas bubbles	0	No	0	0	0	0	0	0	NI
—	1	Yes	—	—	—	—	—	—	—
Overall impression of bone structure (BS)	0	Smooth	0	0	—	—	0	—	—
	1	Irregular	—	—	1	1	—	1	NI
Bone cavities (BC)	0	≤3 osteon-like cavities	0	0	0	0	0	0	NI
	1	4–6 osteon-like cavities or ≤10 smaller	—	—	—	—	—	—	—
	2	7–10 osteon-like cavities or 11–20 smaller	—	—	—	—	—	—	—
Periosteal remodelling (PR)	0	No	—	—	NA	1	—	—	NI
	1	≥1/4 periosteal bone 1 osteon thick	2	2	—	—	2	2	—
	2	≥1/4 periosteal bone 2 osteon thick	—	—	—	—	—	—	—
	3	≥1/4 periosteal bone 3 osteon thick	—	—	—	—	—	—	—
Endosteal remodelling (ER)	0	No	—	0	—	NC	NC	NC	NI
	1	≥1/4 endosteal bone 1 osteon thick	—	—	—	—	—	—	—
	2	≥1/4 endosteal bone 2 osteons thick	—	—	2	—	—	—	—
	3	≥1/4 endosteal bone 3 osteons thick	3	—	—	—	—	—	—
Periosteal apposition (PA)	0	No	—	—	NA	—	—	—	NI
	1	Yes	1	1		1	1	1	—
Peri-implant bone formation (PIF)	0	No	—	—	—	—	—	—	NI
	1	Yes	1	1	NI	1	1	1	
Peri-implant fibrosis (PF)	0	No	—	—	—	—	—	—	NI
	1	≤25% implant surface	—	—	—	—	—	—	—
	2	25–50% implant surface	—	—	—	—	2	—	—
	3	≥51% implant surface	3	3	NI	3	—	3	—
	[mm]	max. thickness	0.10	0.10	0.05	0.04	0.05	0.05	
Lymphoplasmacellular reaction (LYM)	0	<30 cells per section	0	0	0	0	0	0	N
	1	30–50 cells per section	—	—	—	—	—	—	—
	2	51–100 cells per section	—	—	—	—	—	—	—
	3	>100 cells per section	—	—	—	—	—	—	—
Macrophages (MPH)	0	<3 cells per section	—	0	0	0	0	0	N
	1	3–20 cells per section	1	—	—	—	—	—	—
	2	>20 cells per section	—	—	—	—	—	—	—
Giant cells (GC)	0	No	0	0	0	0	0	0	N
	1	1–10 cells per section	—	—	—	—	—	—	—
	2	>10 cells per section	—	—	—	—	—	—	—
Interface—features of material corrosion	0	No	0	0	0	0	0	0	0
	1	Yes	—	—	—	—	—	—	—

Note: NA = not applicable, NI = no implant visible, NC = not applicable a screw in compacta.

As is visible in Figure 8, the implanted material showed (at the level of light microscopy) only minimal disturbance to the contour sharpness—as a considered sign of resorption. No metallosis (expressed by a granulomatous reaction) was observed.

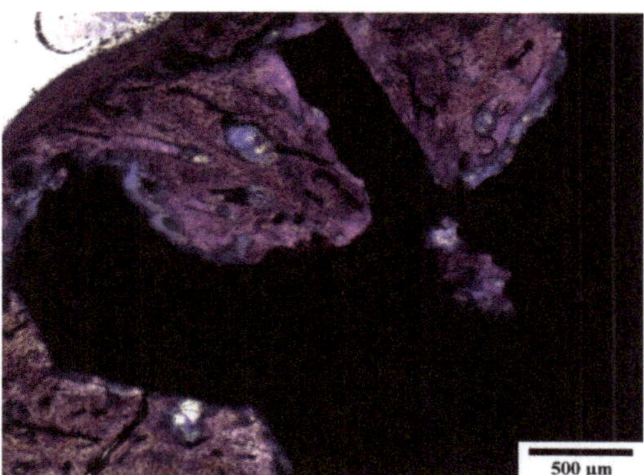

Figure 8. Very minimal disturbance of the contour sharpness showing a low level of resorption.

Despite the presence of the fibrous tissue, peri-implant bone formation was always expressed but with markedly variable intensity (Figure 9). The formation of new bone indicates that the experimental alloy has biocompatible properties to bone. Bone regenerates at a slow rate. Thus, 120 days from the insertion of the screw represents an adequate increase of bone around the implants.

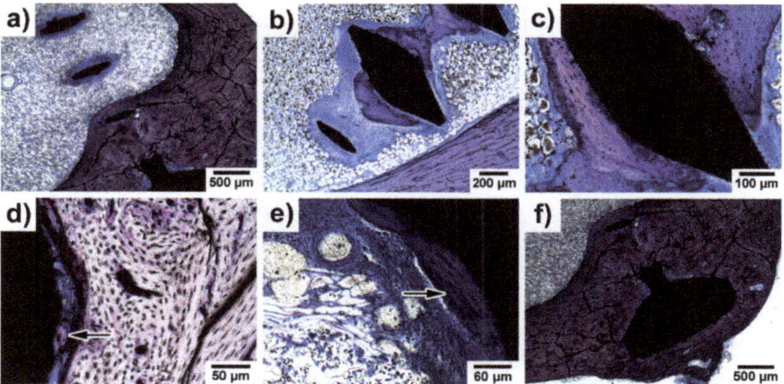

Figure 9. (**a**) Peri-implant fibrosis; (**b**,**c**) bone formation with tight apposition of the newly created bone to the surface of the experimental screw; (**d**) initial bone formation—detailed view; (**e**) fibroproductive inflammation and initial bone formation; and (**f**) thickened compact bone with the experimental screw embedded. Peri-implant bone formation around the experimental screw is marked by arrows in (**d**,**e**). At 120 days of implantation. Toluidine blue staining.

A summary of the morphological findings on the tibia of the rabbits is shown in Table 2. The material was well-tolerated. No gas bubbles were observed. Regular structure of the bone surrounding the implanted material in the majority of rabbits was seen, with only focal irregularities observed in 3 out of the 11 slides examined. Bone cavities were not found. Periosteal remodelling and apposition were present with mild or medium intensity. Medium-to-high intensity of endosteal remodelling was also a constant finding in samples with implant/endosteal contact. Peri-implant fibrosis was presented with high intensity (score 3) in all slides with implant/bone interface.

3.6. Systemic Toxicity in the Liver, Kidneys and Brain Specimen

The kidneys, liver, and brain of the experimental rabbits were histologically examine 120 days after experimental alloy implantation. Histological evaluation with haematoxyli & eosin staining of those organs is shown in Figure 10. No morphological signs of toxi damage to the kidneys, liver, and brain were observed.

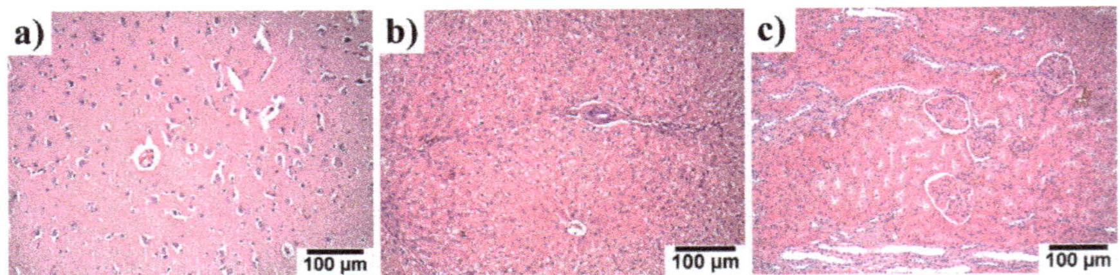

Figure 10. Histological examination of (**a**) brain, (**b**) liver, and (**c**) kidneys of rabbits 120 days after implantation with experimental screws. After 120 days there was no visible morphological/histopathological evidence of any toxic effect in the examined organs from the absorbable metals used. All specimens were processed in the routine way, i.e., paraffin-embedded, 5-μm-thick sections stained with haematoxylin & eosin.

The content of the elements contained in the implanted material (Zn, Mg, and Sr) i the organs was analysed and the results can be seen in Table. 3. In Table 3, the content of the elements found in the organs of control groups used in other studies are also liste The concentrations of Zn and Mg found in the organs of our experimental group wa comparable with those of the control groups. For strontium, we found a common S content in organs only for rabbit females, whose organ composition could slightly diffe from that of rabbit males [39]. In spite of that, the Sr content in livers was comparabl to that found in the control group in [25]. The Sr content in the kidneys was higher i our experimental group, but this could be caused by the differences in the gender of th experimental animals [39]. Due to the fact no toxic effects were observed (see Figure 1(and considering the results obtained by Jia et al. [22], who found no increase of Sr in organ of rats after the implantation of a Zn–Sr alloy, we can suppose that only negligible or rathe no accumulation of strontium in the organs took place in our case.

Table 3. Organ analysis of Magnesium, Zinc and Strontium content, mean ± SMODCH.

	Experimental Group—This Study			Control Groups (For Details See References)		
Tissue	**Zinc [mg/kg] of Tissue**	**Magnesium [mg/kg] of Tissue**	**Strontium [μg/kg] of Tissue**	**Zinc [mg/kg] of Tissue**	**Magnesium [mg/kg] of Tissue**	**Strontium [μg/kg] of Tissue**
Brain	10.4 ± 0.99	144 ± 5	120 ± 35	~11.3 [40] ~11.0 [41] ~16.7 [42]	~152 [40] ~257 [42]	-
Kidneys	24.27 ± 1.18	177 ± 14	50 ± 5	~23.4 [40] ~10.2 [43] ~44.3 [42]	~182 [40] ~345 [42]	~19 [25]
Liver	31.57 ± 1.97	174 ± 6	28 ± 19	~32.1 [40] ~24.6 [41] ~14.7 [43] ~41.9 [42]	~151 [40] ~349 [42]	~32 [25]

4. Discussion

Age-related fractures are projected to increase in the U.S. to over 3 million fractur per year by 2025 [44]. In human traumatology, especially paediatric traumatology, th use of absorbable metals represents an interesting alternative to conventional titanium or steel plates and screws. Conventional fracture management often requires removal of the implanted material at a later date. Minimizing the number of operations would be beneficial for the patients and could also bring significant financial savings [1,26]. As shown in our previous study [32], the mechanical and tensile strength of the investigated material fulfilled all the basic criteria for the fabrication of bone implants [45]. Magnesium alloys also have good mechanical properties [3,12], are well tolerated by bone and soft tissues [2,3], but produce gas bubbles [2,3,28], diminishing the possibilities of their practical use in human and veterinary medicine. Therefore, we have chosen a Zn–0.8Mg–0.2Sr alloy, which is known to corrode without hydrogen evolution and fulfil the basic mechanical criteria for implantology [31,32]. Moreover, this alloy was found to possess acceptable cytocompatibility in-vitro and enhanced antibacterial activity [32]. After 120 days of implantation, no volume changes of the screw or formation of voluminous corrosion products were observed via microCT and light microscopy, suggesting a low degradation rate. As was proved by the SEM-EDS analyses, the corrosion was connected by a formation of solid corrosion products based on phosphates. The corrosion products formed a relatively uniform and dense layer, which acts as a barrier and slows the degradation process [31]. Formation of Zn-based phosphates can be beneficial because, as was proved by Su et al., they can show antibacterial behaviour and enhance the biocompatibility of the material [46]. Bonding of the Zn^{2+} ions into solid phosphates also decreases the risks connected with the transport of Zn^{2+} ions to surrounding tissue and to important organs. As a consequence, no osteolysis and no enhanced concentrations of Zn, Mg, or Sr in the examined organs was observed (see Table 3). The degradation rate was comparable with those observed in other in-vivo studies dealing with Zn-based biodegradable alloys for bone applications [21,23,24]. The degradation rate was rather slow considering the basic requirements for bioabsorbable metals. The absorbable implant should degrade 1–2 years after the implantation [47], which would not be fulfilled in our case—considering a constant degradation rate of about 13.5 μm/year. Histologically, no phagocytosed material was observed in the cytoplasm, or were there any substantial irregularities found in the surface of the experimental screws.

Böstman et al., in their human study, observed that absorbable materials from polyglycolide rods would often lead to sterile inflammatory reactions, visible via X-ray as osteolysis—the destruction of bone around an implant [48]. Osteolysis is generally recognised as an indirect indicator of irritant or toxic changes to bone from a foreign material [49]. In our experiment on extruded Zn–0.8Mg–0.2Sr alloy implanted in rabbits, there were no osteolytic changes observed via X-ray. Via microCT, all implants were seen to be in contact with the surrounding bone, with no signs of any adverse reactions. The thickness of the bone surrounding the implants was found to be greater than average cortical thickness, proving biocompatibility. Histologically, we used a proven scoring method [35] to evaluate the regeneration of hard and soft tissues around the experimental screw. Histologically, peri-implant bone formation was observed after 120 days together with peri-implant fibrosis. Zinc supplementation stimulates the osteoblast bone formation and inhibits osteoclast differentiation and results in increased bone strength [50,51]. A large amount of newly formed bone tissue in the zinc alloy group was described in the systematic review and meta-analysis based on animal studies. In addition, the newly formed trabecular bone was also thicker than that in the pure titanium group [52]. Based on that, we can deduce that local zinc enrichment may promote osteogenesis. This could explain the thickened compact bone in the vicinity of the experimental screws (Figure 9f).

Osseointegration means that bone cells (osteoblasts) grow directly onto the surface of an osseointegrated implant [53]. Titanium, for example, has this ability [36,53]. Bone–implant contact was observed in the experimental screws, with peri-implant fibrosis histologically described in more than 51% of surfaces in all experimental specimens. Fibrosis is the reaction of the body to a foreign object [54]. It can be described as a "foreign body reaction ". This is the end-stage response to inflammation, composed of macrophages and foreign body giant cells reacting to wound healing following implantation of a medical device, prosthesis, or biomaterial [44]. Engraftment of an implant in soft tissue is charact ized by fibrointegration. The presence of fibrous tissue surrounding Zn-based implants often observed [22,23,55]. We can conclude that biodegradable alloys could osseointegra which means that they have a good biocompatibility. We observed good osteogenes around the implanted screws. Partial fibrointegration was observed in our study, too. Th presence of fibrous tissue around an implant was found in fast-degraded ones [56]. By oth authors, the bone–implant contact ratio varied depending on the degradation behavio of implants [23]. Slow corrosion rate resulted in improved BIC, while severe localiz corrosion provoked a thick fibrotic layer surrounding the implants [23]. We can conclu that absorbability requires reaction from surrounding tissues. This reaction should not aggressive and therefore not result in osteolytic changes of the bone or a massive significa inflammatory response. It should be a mild one, which was observed in our study.

We did not detect any white blood cells, indicative of an inflammatory reaction, the majority of specimens. Only one histological slide (from 15 slides) showed a possib chronic inflammatory reaction. We can therefore conclude that our pilot study show histologically, after 120 days, that there was very good tolerance to the implanted materi This finding, however, may have been slightly underestimated as we did not use a spec staining (TRAP—tartrate-resistant acid phosphatase staining) [57]. TRAP staining wou possibly have contributed to a better analysis of the inflammatory response to an implant experimental alloy.

Similarly, as in other studies investigating Zn-based biodegradable alloys, we observ no damage to the examined organs of the experimental animals. This can be attributed the low degradation rate and to the fact that the material transformed into solid corrosi products. As a result, only a minimal amount of free metallic ions, which could be tra ported to the organs and cause their damage, was formed. Such a small amount of fr metallic ions was most likely metabolized without any adverse effect on the host organis or accumulation of metallic ions in organs, as was shown in Table 3). The recommend dietary allowances (RDAs) of zinc are 11 mg/day for males and 8 mg/day for females [5 Tolerable upper intake levels (ULs) for zinc are 40 mg/day [58]. For experimental rabbits diet containing at least 50 mg/kg is usually fed [42]. Even if the material was fully solub and did not form any solid corrosion products, the ion release at the measured corrosi rate (~3.5 μm/year) would be approximately 5 mg/year (considering the surface ar of the screw being 52.5 mm^2, which was estimated from the CAD model of the scre Such an amount is approximately thousand times lower than the RDA for human ma and also lower than the Zn content in the recommended rabbit diet. The recommend dietary allowances (RDAs) for magnesium are 420 mg/day for males and 320 mg/day f females [59]. For rabbits, the usual diet contains about 2.5 g/kg of magnesium. In hum adults, the total daily intake of strontium is estimated to be about 4 mg/day [60]. Drinki water contributes about 0.7–2 mg/day and food (mainly leafy vegetables, grains, and dai products) another 1.2–2.3 mg/day [60]. A maximum acceptable concentration (MAC) 7.0 mg/L is proposed for total strontium in drinking water [39]. The content of magnesiu and strontium in the organs related to other studies [25,40] and was within normal leve All three used elements are known for promoting bone regeneration [3,22,61–63]; therefo their slow release during the implant degradation could be beneficial for the healing of t injured bone. Based on the obtained data, we can conclude that the implanted materi should not cause any systemic toxicity and organ damage in humans because the amou of released ions were negligible compared to the common daily intake of Zn, Mg, and S

5. Conclusions

In this study, Zn–0.8Mg–0.2Sr alloy was successfully implanted into rabbits and mo tored for 16 weeks (120 days), which presented a longer interval than in other publish studies related to the application of zinc-based materials as bone implants. The screws we regularly dissolved at a corrosion rate of 0.014 mm/year without the production of a gas bubbles, which make this alloy superior to magnesium-based biodegradable materia

However, the degradation should be enhanced to fulfil the basic criteria for absorbable implants. The material did not induce any inflammatory reaction locally (around the experimental screw) or distally, e.g., in the vital organs such as the liver, kidneys, and brain. Neither bone resorption nor systemic toxicity were observed. Periosteal apposition and endosteal remodelling were observed in the implant vicinity. More than 50% of implant surfaces were covered by fibrous tissue, while osteolysis of the surrounding bone was not observed. Histologically, it was proven that the experimental alloy represents a promising prospect for future applications of Zn-based materials in clinical use.

Author Contributions: Conceptualization, K.K.; data curation, D.U., M.B. (Martin Bartoš), M.Š., J.D., M.D., R.V., T.U., M.B. (Michal Beňo) and J.Č.; funding acquisition, K.K., J.K. and J.Č.; investigation, K.K., D.U., M.B. (Martin Bartoš), M.Š., J.D., R.V., T.U., M.B. (Michal Beňo) and J.Č.; methodology, K.K., R.F., M.D. and J.Č.; project administration, J.Č.; resources, K.K., J.K. and J.Č.; supervision, K.K. and J.Č.; validation, E.B.; visualization, J.P.; writing—original draft, K.K. and J.Č.; writing—review and editing, K.K., J.P., J.K., R.F., E.B. and J.Č. All authors have read and agreed to the published version of the manuscript.

Funding: This research was funded by the Czech Science Foundation, grant number 18-06110S, Operational Programme Research, Development and Education financed by European Structural and Investment Funds and the Czech Ministry of Education, Youth and Sports (Project No. SOLID21-CZ.02.1.01/0.0/0.0/16_019/0000760), and by the CzechNanoLab Research Infrastructure supported by MEYS CR (LM2018110).

Institutional Review Board Statement: The study was conducted according to the guidelines of the Declaration of Helsinki and approved by the Ethics Committee of Ministry of Education, Youth and Sports of the Czech Republic (protocol code: MSMT-7025/2018-6, date of approval: 9 May 2018).

Informed Consent Statement: Not applicable.

Data Availability Statement: The data presented in this study are available on request from the corresponding author.

Acknowledgments: The authors would like to thank Stanislav Habr for the fabrication of the screws, Jan Maňák for the SEM analyses and acquisition of EDS maps, and Viktor Sýkora and his team for taking care of the experimental animals.

Conflicts of Interest: The authors declare no conflict of interest.

References

Kose, O.; Turan, A.; Ünal, M.; Acar, B.; Guler, F. Fixation of medial malleolar fractures with magnesium bioabsorbable headless compression screws: Short-term clinical and radiological outcomes in eleven patients. *Arch. Orthop. Trauma Surg.* **2018**, *138*, 1069–1075. [CrossRef]

Kraus, T.; Fischerauer, S.F.; Hänzi, A.C.; Uggowitzer, P.J.; Löffler, J.F.; Weinberg, A.M. Magnesium alloys for temporary implants in osteosynthesis: In vivo studies of their degradation and interaction with bone. *Acta Biomater.* **2012**, *8*, 1230–1238. [CrossRef]

Chaya, A.; Yoshizawa, S.; Verdelis, K.; Myers, N.; Costello, B.J.; Chou, D.-T.; Pal, S.; Maiti, S.; Kumta, P.N.; Sfeir, C. In vivo study of magnesium plate and screw degradation and bone fracture healing. *Acta Biomater.* **2015**, *18*, 262–269. [CrossRef]

Zheng, Y.; Gu, X.; Witte, F. Biodegradable metals. *Mater. Sci. Eng. R Rep.* **2014**, *77*, 1–34. [CrossRef]

ASTM F3160-21, Standard Guide for Metallurgical Characterization of Absorbable Metallic Materials for Medical Implants. Available online: https://www.astm.org/Standards/F3160.htm (accessed on 2 June 2021).

ASTM F3268-18a, Standard Guide for In Vitro Degradation Testing of Absorbable Metals. Available online: https://www.astm.org/Standards/F3268.htm (accessed on 2 June 2021).

Witte, F. The history of biodegradable magnesium implants: A review. *Acta Biomater.* **2010**, *6*, 1680–1692. [CrossRef]

Huse, E.C. A new ligature? *Chic. Med. J. Exam.* **1878**, 171–172.

Yue, Y.; Lei, L.; Ye, H.; Yang, S.; Wang, L.; Yang, N.; Huang, J.; Ren, L. Effectiveness of Biodegradable Magnesium Alloy Stents in Coronary Artery and Femoral Artery. *J. Interv. Cardiol.* **2015**, *28*, 358–364. [CrossRef] [PubMed]

Zhao, D.; Witte, F.; Lu, F.; Wang, J.; Li, J.; Qin, L. Current status on clinical applications of magnesium-based orthopaedic implants: A review from clinical translational perspective. *Biomaterials* **2017**, *112*, 287–302. [CrossRef] [PubMed]

Xi, Z.; Wu, Y.; Xiang, S.; Sun, C.; Wang, Y.; Yu, H.; Fu, Y.; Wang, X.; Yan, J.; Zhao, D.; et al. Corrosion Resistance and Biocompatibility Assessment of a Biodegradable Hydrothermal-Coated Mg–Zn–Ca Alloy: An in vitro and in vivo Study. *ACS Omega* **2020**, *5*, 4548–4557. [CrossRef]

12. Angrisani, N.; Reifenrath, J.; Zimmermann, F.; Eifler, R.; Meyer-Lindenberg, A.; Herrera, K.V.; Vogt, C. Biocompatibility and degradation of LAE442-based magnesium alloys after implantation of up to 3.5years in a rabbit model. *Acta Biomater.* **2016**, *4*… 355–365. [CrossRef] [PubMed]
13. Mei, D.; Lamaka, S.V.; Lu, X.; Zheludkevich, M.L. Selecting medium for corrosion testing of bioabsorbable magnesium and other metals—A critical review. *Corros. Sci.* **2020**, *171*, 108722. [CrossRef]
14. Windhagen, H.; Radtke, K.; Weizbauer, A.; Diekmann, J.; Noll, Y.; Kreimeyer, U.; Schavan, R.; Stukenborg-Colsman, C.; Waizy, H. Biodegradable magnesium-based screw clinically equivalent to titanium screw in hallux valgus surgery: Short term results of the first prospective, randomized, controlled clinical pilot study. *Biomed. Eng. Online* **2013**, *12*, 62. [CrossRef]
15. Su, Y.; Cockerill, I.; Wang, Y.; Qin, Y.-X.; Chang, L.; Zheng, Y.; Zhu, D. Zinc-Based Biomaterials for Regeneration and Therapy. *Trends Biotechnol.* **2019**, *37*, 428–441. [CrossRef]
16. Venezuela, J.J.D.; Johnston, S.; Dargusch, M.S. The Prospects for Biodegradable Zinc in Wound Closure Applications. *Adv. Healthc. Mater.* **2019**, *8*, e1900408. [CrossRef] [PubMed]
17. Schinhammer, M.; Hänzi, A.C.; Löffler, J.F.; Uggowitzer, P.J. Design strategy for biodegradable Fe-based alloys for medical applications. *Acta Biomater.* **2010**, *6*, 1705–1713. [CrossRef] [PubMed]
18. Hernández-Escobar, D.; Champagne, S.; Yilmazer, H.; Dikici, B.; Boehlert, C.J.; Hermawan, H. Current status and perspectives of zinc-based absorbable alloys for biomedical applications. *Acta Biomater.* **2019**, *97*, 1–22. [CrossRef]
19. Li, G.; Yang, H.; Zheng, Y.; Chen, X.-H.; Yang, J.-A.; Zhu, D.; Ruan, L.; Takashima, K. Challenges in the use of zinc and its alloys as biodegradable metals: Perspective from biomechanical compatibility. *Acta Biomater.* **2019**, *97*, 23–45. [CrossRef]
20. Vojtěch, D.; Kubásek, J.; Šerák, J.; Novák, P. Mechanical and corrosion properties of newly developed biodegradable Zn-based alloys for bone fixation. *Acta Biomater.* **2011**, *7*, 3515–3522. [CrossRef] [PubMed]
21. 21. Kubásek, J.; Dvorský, D.; Šedý, J.; Msallamová, Š.; Levorová, J.; Foltán, R.; Vojtěch, D. The Fundamental Comparison of Zn–2Mg and Mg–4Y–3RE Alloys as a Perspective Biodegradable Materials. *Materials* **2019**, *12*, 3745. [CrossRef]
22. Jia, B.; Yang, H.; Zhang, Z.; Qu, X.; Jia, X.; Wu, Q.; Han, Y.; Zheng, Y.; Dai, K. Biodegradable Zn–Sr alloy for bone regeneration in rat femoral condyle defect model: In vitro and in vivo studies. *Bioact. Mater.* **2021**, *6*, 1588–1604. [CrossRef]
23. Yang, H.; Jia, B.; Zhang, Z.; Qu, X.; Li, G.; Lin, W.; Zhu, D.; Dai, K.; Zheng, Y. Alloying design of biodegradable zinc as promising bone implants for load-bearing applications. *Nat. Commun.* **2020**, *11*, 1–16. [CrossRef]
24. Wang, X.; Shao, X.; Dai, T.; Xu, F.; Zhou, J.G.; Qu, G.; Tian, L.; Liu, B.; Liu, Y. In vivo study of the efficacy, biosafety, and degradation of a zinc alloy osteosynthesis system. *Acta Biomater.* **2019**, *92*, 351–361. [CrossRef]
25. Tie, D.; Guan, R.; Liu, H.; Cipriano, A.; Liu, Y.; Wang, Q.; Huang, Y.; Hort, N. An in vivo study on the metabolism and osteogenic activity of bioabsorbable Mg–1Sr alloy. *Acta Biomater.* **2016**, *29*, 455–467. [CrossRef]
26. Zhao, D.; Huang, S.; Lu, F.; Wang, B.; Yang, L.; Qin, L.; Yang, K.; Li, Y.; Li, W.; Wang, W.; et al. Vascularized bone grafting fixed by biodegradable magnesium screw for treating osteonecrosis of the femoral head. *Biomaterials* **2016**, *81*, 84–92. [CrossRef] [PubMed]
27. Huang, S.; Wang, B.; Zhang, X.; Lu, F.; Wang, Z.; Tian, S.; Li, D.; Yang, J.; Cao, F.; Cheng, L.; et al. High-purity weight-bearing magnesium screw: Translational application in the healing of femoral neck fracture. *Biomaterials* **2020**, *238*, 119829. [CrossRef]
28. Diekmann, J.; Bauer, S.; Weizbauer, A.; Willbold, E.; Windhagen, H.; Helmecke, P.; Lucas, A.; Reifenrath, J.; Nolte, I.; Ezechieli, M. Examination of a biodegradable magnesium screw for the reconstruction of the anterior cruciate ligament: A pilot in vivo study in rabbits. *Mater. Sci. Eng. C* **2016**, *59*, 1100–1109. [CrossRef] [PubMed]
29. Cihova, M.; Martinelli, E.; Schmutz, P.; Myrissa, A.; Schäublin, R.; Weinberg, A.; Uggowitzer, P.; Löffler, J. The role of zinc in the biocorrosion behavior of resorbable Mg-Zn-Ca alloys. *Acta Biomater.* **2019**, *100*, 398–414. [CrossRef] [PubMed]
30. Yamaguchi, M. Role of nutritional zinc in the prevention of osteoporosis. *Mol. Cell. Biochem.* **2010**, *338*, 241–254. [CrossRef] [PubMed]
31. Hybasek, V.; Kubasek, J.; Capek, J.; Alferi, D.; Pinc, J.; Jiru, J.; Fojt, J. Influence of model environment complexity on corrosion mechanism of biodegradable zinc alloys. *Corros. Sci.* **2021**, *187*, 109520. [CrossRef]
32. Čapek, J.; Kubásek, J.; Pinc, J.; Fojt, J.; Krajewski, S.; Rupp, F.; Li, P. Microstructural, mechanical, in vitro corrosion and biological characterization of an extruded Zn-0.8Mg-0.2Sr (wt%) as an absorbable material. *Mater. Sci. Eng. C* **2021**, *122*, 111924. [CrossRef]
33. Luize, D.S.; Bosco, A.F.; Bonfante, S.; De Almeida, J.M. Influence of ovariectomy on healing of autogenous bone block grafts in the mandible: A histomorphometric study in an aged rat model. *Int. J. Oral Maxillofac. Implant.* **2008**, *23*, 207–214.
34. Šedý, J.; Urdzíková, L.; Jendelová, P.; Syková, E. Methods for behavioral testing of spinal cord injured rats. *Neurosci. Biobehav. Rev.* **2008**, *32*, 550–580. [CrossRef] [PubMed]
35. Reifenrath, J.; Bormann, D.; Meyer-Lindenberg, A. Magnesium Alloys as Promising Degradable Implant Materials in Orthopaedic Research. In *Magnesium Alloys—Corrosion and Surface Treatments*; IntechOpen: London, UK, 2011; p. 10577214143.
36. Johansson, C.; Hansson, H.; Albrektsson, T. Qualitative interfacial study between bone and tantalum, niobium or commercially pure titanium. *Biomaterials* **1990**, *11*, 277–280. [CrossRef]
37. Bernhardt, R.; Kuhlisch, E.; Schulz, M.C.; Eckelt, U.; Stadlinger, B. Comparison of bone-implant contact and bone-implant volume between 2D-histological sections and 3D-SRμCT slices. *Eur. Cells Mater.* **2012**, *23*, 237–248. [CrossRef]
38. Jiřík, M.; Bartoš, M.; Tomášek, P.; Malečková, A.; Kural, T.; Horáková, J.; Lukáš, D.; Suchý, T.; Kochová, P.; Kalbacova, M.H.; et al. Generating standardized image data for testing and calibrating quantification of volumes, surfaces, lengths, and object counts in fibrous and porous materials using X-ray microtomography. *Microsc. Res. Tech.* **2018**, *81*, 551–568. [CrossRef]

Passlack, N.; Mainzer, B.; Lahrssen-Wiederholt, M.; Schafft, H.; Palavinskas, R.; Breithaupt, A.; Zentek, J. Concentrations of strontium, barium, cadmium, copper, zinc, manganese, chromium, antimony, selenium, and lead in the liver and kidneys of dogs according to age, gender, and the occurrence of chronic kidney disease. *J. Vet. Sci.* **2015**, *16*, 57–66. [CrossRef]

Zhang, J.; Li, H.; Wang, W.; Huang, H.; Pei, J.; Qu, H.; Yuan, G.; Li, Y. The degradation and transport mechanism of a Mg-Nd-Zn-Zr stent in rabbit common carotid artery: A 20-month study. *Acta Biomater.* **2018**, *69*, 372–384. [CrossRef] [PubMed]

Bentley, P.J.; Grubb, B.R. Effects of a zinc-deficient diet on tissue zinc concentrations in rabbits. *J. Anim. Sci.* **1991**, *69*, 4876–4882. [CrossRef] [PubMed]

Bulat, Z.; Đukić-Ćosić, D.; Antonijević, B.; Bulat, P.; Vujanović, D.; Buha, A.; Matovic, V. Effect of Magnesium Supplementation on the Distribution Patterns of Zinc, Copper, and Magnesium in Rabbits Exposed to Prolonged Cadmium Intoxication. *Sci. World J.* **2012**, *2012*, 1–9. [CrossRef]

Lin, S.; Ran, X.; Yan, X.; Yan, W.; Wang, Q.; Yin, T.; Zhou, J.G.; Hu, T.; Wang, G. Corrosion behavior and biocompatibility evaluation of a novel zinc-based alloy stent in rabbit carotid artery model. *J. Biomed. Mater. Res. Part B Appl. Biomater.* **2019**, *107*, 1814–1823. [CrossRef]

Burge, R.; Dawson-Hughes, B.; Solomon, D.H.; Wong, J.B.; King, A.; Tosteson, A. Incidence and Economic Burden of Osteoporosis-Related Fractures in the United States, 2005–2025. *J. Bone Miner. Res.* **2006**, *22*, 465–475. [CrossRef] [PubMed]

Čapek, J.; Kubásek, J.; Pinc, J.; Drahokoupil, J.; Čavojský, M.; Vojtěch, D. Extrusion of the biodegradable ZnMg0.8Ca0.2 alloy—The influence of extrusion parameters on microstructure and mechanical characteristics. *J. Mech. Behav. Biomed. Mater.* **2020**, *108*, 103796. [CrossRef] [PubMed]

Su, Y.; Wang, K.; Gao, J.; Yang, Y.; Qin, Y.-X.; Zheng, Y.; Zhu, D. Enhanced cytocompatibility and antibacterial property of zinc phosphate coating on biodegradable zinc materials. *Acta Biomater.* **2019**, *98*, 174–185. [CrossRef] [PubMed]

Venezuela, J.; Dargusch, M. The influence of alloying and fabrication techniques on the mechanical properties, biodegradability and biocompatibility of zinc: A comprehensive review. *Acta Biomater.* **2019**, *87*, 1–40. [CrossRef]

Bostman, O. Osteolytic changes accompanying degradation of absorbable fracture fixation implants. *J. Bone Jt. Surgery. Br. Vol.* **1991**, *73-B*, 679–682. [CrossRef] [PubMed]

Patil, N.; Goodman, S. Wear particles and osteolysis. In *Orthopaedic Bone Cements*; Elsevier: Amsterdam, The Netherlands, 2008; pp. 140–163.

Seo, H.-J.; Cho, Y.-E.; Kim, T.; Shin, H.-I.; Kwun, I.-S. Zinc may increase bone formation through stimulating cell proliferation, alkaline phosphatase activity and collagen synthesis in osteoblastic MC3T3-E1 cells. *Nutr. Res. Pract.* **2010**, *4*, 356–361. [CrossRef]

Moonga, B.S.; Dempster, D.W. Zinc is a potent inhibitor of osteoclastic bone resorption in vitro. *J. Bone Miner. Res.* **2009**, *10*, 453–457. [CrossRef]

Zhang, J.; Jiang, Y.; Shang, Z.; Zhao, B.; Jiao, M.; Liu, W.; Cheng, M.; Zhai, B.; Guo, Y.; Liu, B.; et al. Biodegradable metals for bone defect repair: A systematic review and meta-analysis based on animal studies. *Bioact. Mater.* **2021**, *6*, 4027–4052. [CrossRef]

Brånemark, P.I.; Hansson, B.O.; Adell, R.; Breine, U.; Lindström, J.; Hallén, O.; Ohman, A. Osseointegrated implants in the treatment of the edentulous jaw. Experience from a 10-year period. *Scand. J. Plast. Reconstr. Surg. Suppl.* **1977**, *16*, 1–132. [PubMed]

Anderson, J.M.; Rodriguez, A.; Chang, D.T. Foreign body reaction to biomaterials. *Semin. Immunol.* **2008**, *20*, 86–100. [CrossRef]

Su, Y.; Yang, H.; Gao, J.; Qin, Y.; Zheng, Y.; Zhu, D. Interfacial Zinc Phosphate is the Key to Controlling Biocompatibility of Metallic Zinc Implants. *Adv. Sci.* **2019**, *6*, 1900112. [CrossRef] [PubMed]

Gu, X.; Xie, X.; Li, N.; Zheng, Y.; Qin, L. In vitro and in vivo studies on a Mg–Sr binary alloy system developed as a new kind of biodegradable metal. *Acta Biomater.* **2012**, *8*, 2360–2374. [CrossRef]

Toyosaki-Maeda, T.; Takano, H.; Tomita, T.; Tsuruta, Y.; Maeda-Tanimura, M.; Shimaoka, Y.; Takahashi, T.; Itoh, T.; Suzuki, R.; Ochi, T. Differentiation of monocytes into multinucleated giant bone-resorbing cells: Two-step differentiation induced by nurse-like cells and cytokines. *Arthritis Res.* **2001**, *3*, 306–310. [CrossRef] [PubMed]

Zinc, Fact Sheet for Health Professionals. Available online: Https://Ods.Od.Nih.Gov/Factsheets/Zinc-HealthProfessional (accessed on 2 June 2021).

Magnesium, Fact Sheet for Health Professionals. Available online: https://ods.od.nih.gov/factsheets/Magnesium-HealthProfessional/ (accessed on 2 June 2021).

Strontium in Drinking Water—Guideline Technical Document for Public Consultation. Available online: Ttps://Www.Canada.ca/En/Health-Canada/Programs/Consultation-Strontium-Drinking-Water/Document.Html (accessed on 2 June 2021).

Vormann, J. Magnesium: Nutrition and Homoeostasis. *AIMS Public Health* **2016**, *3*, 329–340. [CrossRef] [PubMed]

Qu, X.; Yang, H.; Yu, Z.; Jia, B.; Qiao, H.; Zheng, Y.; Dai, K. Serum zinc levels and multiple health outcomes: Implications for zinc-based biomaterials. *Bioact. Mater.* **2020**, *5*, 410–422. [CrossRef]

Fonseca, J.E.; Brandi, M.L. Mechanism of action of strontium ranelate: What are the facts? *Clin. Cases Miner. Bone Metab.* **2010**, *7*, 17–18. [PubMed]

 materials

Article

High Magnesium and Sirolimus on Rabbit Vascular Cells—An In Vitro Proof of Concept

Giorgia Fedele [1], Sara Castiglioni [1], Jeanette A. Maier [1,2,*] and Laura Locatelli [1]

[1] Department of Biomedical and Clinical Sciences L. Sacco, Università di Milano, Via GB Grassi 74, 20157 Milano, Italy; giorgia.fedele@unimi.it (G.F.); sara.castiglioni@unimi.it (S.C.); laura.locatelli@unimi.it (L.L.)
[2] Interdisciplinary Centre for Nanostructured Materials and Interfaces (CIMaINa), Università di Milano, 20133 Milan, Italy
* Correspondence: jeanette.maier@unimi.it

Abstract: Drug-eluting bioresorbable scaffolds represent the last frontier in the field of angioplasty and stenting to treat coronary artery disease, one of the leading causes of morbidity and mortality worldwide. In particular, sirolimus-eluting magnesium-based scaffolds were recently introduced in clinical practice. Magnesium alloys are biocompatible and dissolve in body fluids, thus determining high concentrations of magnesium in the local microenvironment. Since magnesium regulates cell growth, we asked whether high levels of magnesium might interfere with the antiproliferative action of sirolimus. We performed in vitro experiments on rabbit coronary artery endothelial and smooth muscle cells (rCAEC and rSMC, respectively). The cells were treated with sirolimus in the presence of different concentrations of extracellular magnesium. Sirolimus inhibits rCAEC proliferation only in physiological concentrations of magnesium, while high concentrations prevent this effect. On the contrary, high extracellular magnesium does not rescue rSMC growth arrest by sirolimus and accentuates the inhibitory effect of the drug on cell migration. Importantly, sirolimus and magnesium do not impair rSMC response to nitric oxide. If translated into a clinical setting, these results suggest that, in the presence of sirolimus, local increases of magnesium concentration maintain normal endothelial proliferative capacity and function without affecting rSMC growth inhibition and response to vasodilators.

Keywords: magnesium; sirolimus; rabbit coronary artery endothelial cells; smooth muscle cells

 check for updates

Citation: Fedele, G.; Castiglioni, S.; Maier, J.A.; Locatelli, L. High Magnesium and Sirolimus on Rabbit Vascular Cells—An In Vitro Proof of Concept. *Materials* **2021**, *14*, 1970. https://doi.org/10.3390/ma14081970

Academic Editors: Hendra Hermawan and Mehdi Razavi

Received: 18 March 2021
Accepted: 10 April 2021
Published: 14 April 2021

Publisher's Note: MDPI stays neutral with regard to jurisdictional claims in published maps and institutional affiliations.

1. Introduction

Coronary artery disease exacts a very high toll in terms of morbidity and mortality all over the world [1]. While prevention and lifestyle changes remain the best weapons to control the disease, it might be necessary to resort to invasive procedures that increase blood supply to the heart. Since the implantation of the first bare metal stents in 1986, innovative solutions have been adopted to limit in-stent restenosis, stent thrombosis and, as recently reported, neoatherosclerosis [2]. Antiproliferative drug eluting stents were generated and percutaneous coronary intervention involving these devices is now widely utilized to treat artery stenosis and prevent re-stenosis [3,4]. The last frontier for coronary stenting was the development of bioresorbable scaffolds (BRS), which provide transient vessel support and are gradually dissolved to avoid the permanent caging of the vessel, thus granting the restoration of vasomotion [5]. In this field, magnesium (Mg) alloys raised major interest because of their mechanical and biological features [6]. Indeed, they possess an adequate radial force and are biocompatible, since magnesium is an essential element involved in myriads of physiological processes [7]. In addition, Mg scaffolds are resorbed by the body with no adverse effects. Moreover, Mg is beneficial for the arteries and, in particular, it positively impacts endothelial function by preventing oxidative stress and inhibiting pro-inflammatory events [8–10]. Sirolimus eluting biodegradable Mg scaffold have recently been introduced into clinical practice, with favorable outcomes [11].

Sirolimus (also known as rapamycin) is a macrolide with potent immunosuppressiv and anti-proliferative properties because it inhibits the mammalian target of rapamyci (mTOR), a highly conserved serine/threonine kinase that regulates crucial cell function such as metabolism, growth, survival, autophagy and transcription [12]. Sirolimus i used for coronary stent coating, since it brakes the proliferation of medial smooth muscl cells, reducing in-stent restenosis [13]. However, vascular dysfunction has been reporte in rats after continuous sirolimus infusion at a rate of 5 mg/kg/day, reaching a bloo concentration as high as 14.2 $\pm$ 5.7 µg/L [14]. Similarly, in isolated human thoracic arterie sirolimus impaired endothelial function [15]. In cultured human umbilical vein endotheli cells, a widely used in vitro model of macrovascular endothelial cells, conflicting result are available. A first study demonstrated an impairment of cell viability and functio with concentrations of sirolimus higher than 1 nM [16]. In the same cells, another grou reported the beneficial effect of sirolimus (100 ng/mL) in inhibiting TNFα-induced VCA upregulation, thus reducing the recruitment of leukocytes into inflamed sites [17].

Studies from preclinical models implanted with Mg-based sirolimus eluting sca folds yielded promising results. Mg-based BRS are safe, induce healing and show lo thrombogenicity in rabbits and pigs [2]. These scaffolds also reduce neo-atherosclerosis i rabbits [2].

The purpose of the present in vitro study was to assess the effects of different conce trations of sirolimus and Mg in rabbit coronary artery endothelial and smooth muscle cel (rCAEC and rSMC, respectively). Since Mg stimulates cell growth and prevents apoptosi we wondered whether an increase in extracellular Mg concentration might reduce the ant proliferative effects of sirolimus on rSMC and counterbalance possible detrimental effect of sirolimus in rCAEC. A two-fold increase in Mg concentration was measured in the wa of pig coronary arteries, after the implantation of a Mg-alloy scaffold [18]. Therefore, w cultured the cells in medium containing from 1, which is the physiological concentration, t 3 mM Mg. To our knowledge, no data are available about the concentration of sirolimus i the vessel wall after scaffold implantation. Studies in the literature indicate that sirolimu is about 1.5 µg/mm^2 of scaffold surface and it degrades in about 90 days [19]. We decide to utilize 2 to 50 ng/mL of sirolimus.

2. Materials and Methods

2.1. Cell Culture

Primary rCAEC and rSMC were purchased from Creative Bioarray (Shirley, NY, US and expanded for 3 population doublings, according to the manufacturer's instruction To increase Mg concentration from 1 to 1.5, 2 and 3 mM Mg, $MgSO_4$ was added to th media. The cells were also treated with different concentrations of sirolimus: 2 ng/m 10 ng/mL or 50 ng/mL. The experiments were conducted for 24, 48, 72 and/or 144 Different batches of cells were utilized to overcome possible differences due to individu variability. The cells were trypsinized and stained with trypan blue solution (0.4%). Viab cells were counted using a Luna Automated Cell Counter (Logos Biosystems, Anyan Korea). All the experiments were performed at least three times.

2.2. In Vitro Wound Assay

Primary rCAEC and rSMC were seeded on 24-well plates (Greiner bio-one, Fricke hausen, Germany) and cultured in the presence of 1 or 3 mM of extracellular Mg with without sirolimus (2 or 50 ng/mL). After 48 h of treatment, the monolayers were scrape washed with PBS twice to remove debris and maintained in media containing Mg and/ sirolimus for an additional 24 h. Then, the cells were fixed and colored with Crystal Viol for 20 min, washed with PBS twice and visualized by optical microscope (4X magnificatio The percentage of migration vs. CTR (1 mM Mg with no sirolimus) was analyzed usin ImageJ [20].

2.3. Fluorescence Microscopy

The cells were grown on microscope glasses, fixed in PBS containing 3% paraformaldehyde and 2% sucrose (pH 7.6), and stained with fluorescent phalloidin to visualize the cytoskeleton or with 4′,6-diamidino-2-phenylindole (DAPI) to detect the nuclei. Finally, cells were mounted with moviol, and images were acquired using a 63X objective in oil by an SP8 confocal microscope (Leica Microsystems, Buffalo Grove, IL, USA). Magnification was 40X.

2.4. Enzyme-Linked Immunosorbent Assay–ELISA

The cells were seeded on 24-well plates and treated for 72 and 144 h in the presence of 1 or 3 mM of extracellular Mg with or without sirolimus (2 or 50 ng/mL). For the quantitative determination of rabbit tissue factor (tF), P-selectin, Vascular Cell Adhesion Protein 1 (VCAM) (Cloud-clone corp., Katy, TX, USA) and endothelial Nitric Oxide Synthase (eNOS) (Creative Diagnostics, New York, NY, USA) ELISA kits were used according to the datasheet instructions. All the ELISA were performed on 30 μg of cell extracts in triplicates for 3 times.

2.5. Cyclic GMP Measurement

rSMC were seeded on 24-well plates and cultured for 72 h in the presence of 1 or 3 mM Mg, combined or not with sirolimus (2 and 50 ng/mL). The cells were then treated for 30 min with Sodium Nitroprusside (SNP 10 μM) (Sigma Aldrich, St. Louis, MO, USA), a donor of nitric oxide (NO). The Cyclic GMP XP Assay Kit (Cell Signaling, Danvers, MA, USA) was utilized according to the manufacturer's instructions. The percentage of activity was calculated as follows: % activity = 100 × [(A − Abasal)/(Amax − Abasal)], where A is the sample absorbance, Amax is the absorbance at maximum stimulation (high cyclic GMP concentration), and Abasal is the absorbance at basal level (no cGMP). The experiment was performed 3 times in triplicate.

2.6. Statistical Analysis

Data are reported as means ± SD. The data were normally distributed and they were analyzed using one-way repeated measures ANOVA. The p-values deriving from multiple pairwise comparisons were corrected by the Bonferroni method. Statistical significance was defined for p-value ≤ 0.05. * $p \leq 0.05$; ** $p \leq 0.01$; *** $p \leq 0.001$; **** $p \leq 0.0001$.

3. Results

3.1. The Effect of Sirolimus and Mg on rCAEC Growth, Migration and Morphology

rCAEC were seeded at low density and cultured in the presence of 2, 10 and 50 ng/mL of sirolimus in media containing different concentrations of Mg (1 to 3 mM). After 72 or 144 h the cells were trypsinized and counted. Figure 1A shows that sirolimus inhibits rCAEC proliferation only in 1 mM Mg, while concentrations higher than 1 mM Mg prevent this inhibitory effect.

We then performed all the following experiments utilizing media containing 1 or 3 mM, with or without 2 and 50 ng/mL of sirolimus. We also found that culture for 72 h in 3 mM Mg with or without sirolimus enhances rCAEC migration (Figure 1B). Moreover, no significant differences were detected in cell morphology, nuclei and cytoskeleton in rCAEC treated with 2 and 50 ng/mL of sirolimus in medium containing 1 or 3 mM Mg for 72 h, after staining with fluorescent phalloidin and DAPI to visualize the cytoskeleton and the nuclei (Figure 1C).

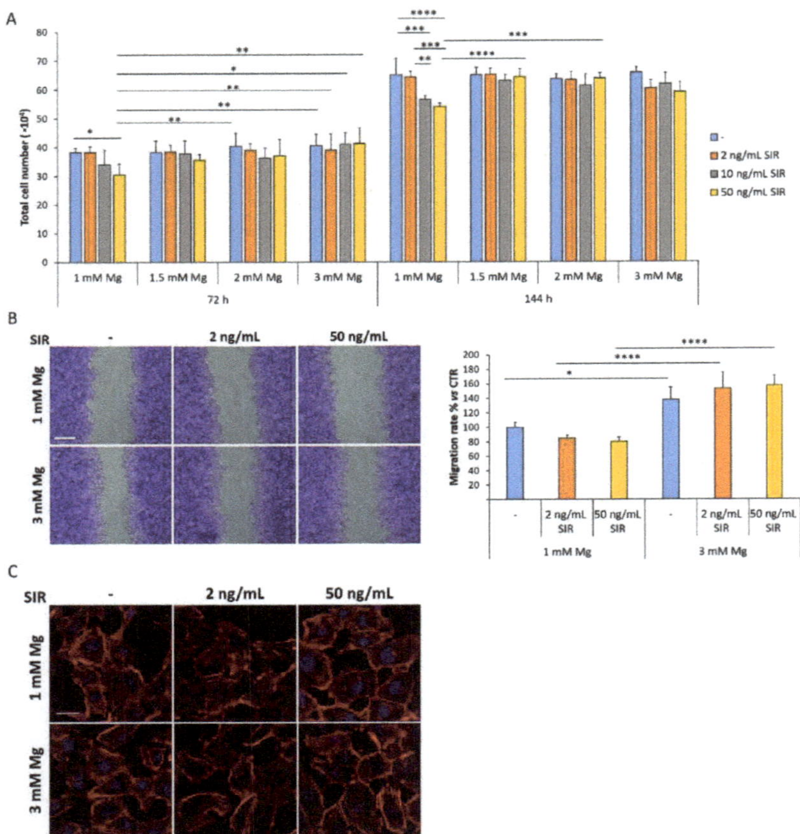

Figure 1. The effects of sirolimus and Mg on rCAEC. (**A**) rCAEC were counted after 72 and 144 h of culture in medium containing 1, 1.5, 2 or 3 mM Mg with or without different concentrations of sirolimus (SIR, 2, 10, 50 ng/mL). -SIR refers to untreated controls. (**B**) Confluent rCAEC were cultured in the presence of 1 or 3 mM Mg combined with 2 or 50 ng/mL of sirolimus for 48 h. Then wound assay was performed. 24 h later the width of the wound was visualized and measured. Representative images of the scratches are shown (left panel). Scale bar: 125 µm. The migration rate was analyzed using ImageJ and data are shown as the percentage of migration compared to the control (CTR, 1 mM Mg without sirolimus, right panel). (**C**) Confluent rCAEC were cultured in culture medium containing 1 or 3 mM Mg with or without sirolimus (2 and 50 ng/mL), stained with phalloidin and DAPI and visualized by fluorescence microscopy. Representative images are shown. Scale bar: 20 µm. Data are expressed as the mean ± standard deviation of three independent assays (* $p \leq 0.05$; ** $p \leq 0.01$; *** $p \leq 0.001$; **** $p \leq 0.0001$).

3.2. The Effect of Sirolimus and Mg on the Amounts of VCAM and P-Selectin in rCAEC

When activated or dysfunctional, endothelial cells upregulate adhesion molecules recruit leukocytes [16]. The cells were treated with 2 or 50 ng/mL of sirolimus in med containing 1 or 3 mM Mg for 72 and 144 h. By ELISA, we measured the amounts VCAM and P-selectin, both involved in atherogenesis. While P-selectin is not modulat by extracellular Mg and sirolimus (Figure 2A), VCAM is downregulated in 3 mM Mg al in the presence of sirolimus at 72 h and upregulated in rCAEC cultured in 3 mM Mg wi 50 ng/mL sirolimus for 144 h (Figure 2B).

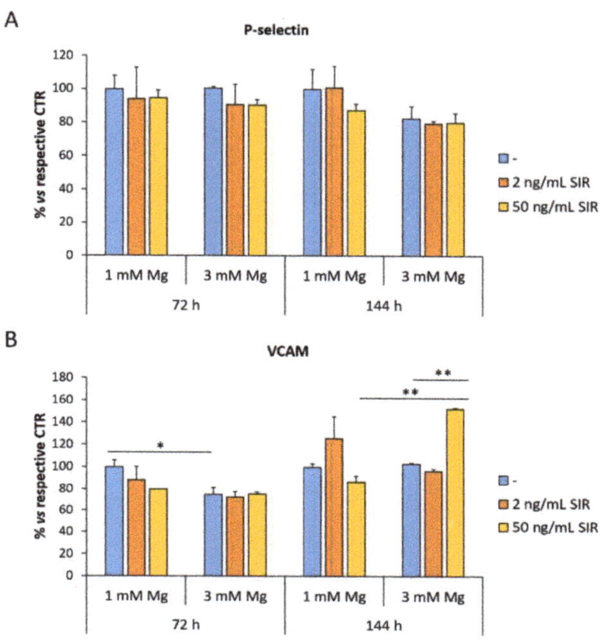

Figure 2. The effects of sirolimus and Mg on the total amounts of P-selectin and VCAM. ELISA was performed on cell extracts after 72 and 144 h of culture in medium containing 1 or 3 mM Mg in the presence of 2 or 50 ng/mL of sirolimus. (**A**) P-selectin (**B**) VCAM. Data are expressed as the mean ± standard deviation of three independent assays (* $p \leq 0.05$; ** $p \leq 0.01$).

3.3. The Effect of Sirolimus and Mg on the Amounts of tF and eNOS in rCAEC

Dysfunctional endothelial cells upregulate tF, which activates the extrinsic coagulation pathway and triggers thrombosis [21]. The total amounts of tF were measured by ELISA on rCAEC cultured as described above. Figure 3A shows no modulation of tF after 72 h of culture in the presence of sirolimus or 3 mM Mg alone, while the total amount of tF is significantly reduced in cells cultured in 3 mM Mg with sirolimus for 144 h. Moreover, since aberrant levels of eNOS have been described in atherosclerosis [22], we analyzed the levels of the enzyme in rCAEC under the same experimental conditions and did not detect significant differences (Figure 3B).

3.4. The Effect of Sirolimus and Mg on rSMC Growth and Migration

rSMC were cultured with or without 2, 10, 50 ng/mL sirolimus in the presence of different concentrations of extracellular Mg, from 1 to 3 mM, for 72 and 144 h. After 72 h control cells were confluent, while a marked growth-inhibitory effect by sirolimus was observed at all time points tested (Figure 4A). Indeed, 2 ng/mL of sirolimus suffice to maximally inhibit cell growth. Importantly, high concentrations of extracellular Mg do not rescue the growth-inhibitory effect of sirolimus. We also found that culture for 72 h with sirolimus inhibits rSMC migration (Figure 4B). 3 mM Mg alone reduces migration compared to 1 mM Mg, in accordance with previous reports [8], and does not interfere with the inhibitory effect of sirolimus.

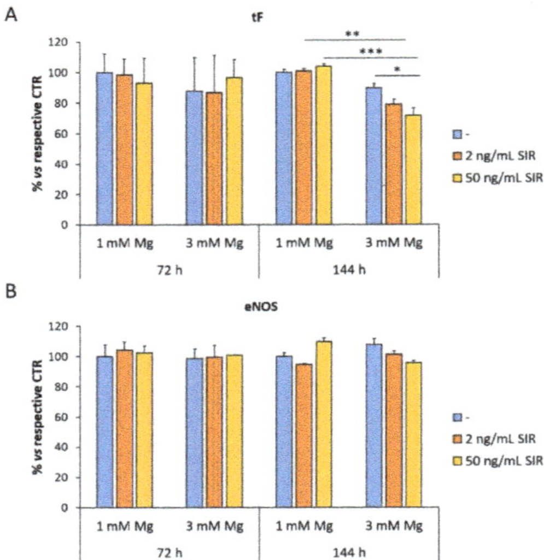

Figure 3. The effects of sirolimus and Mg on the total amounts of tF and eNOS. ELISA for tF (**A**) and eNOS (**B**) was performed as described above. Data are expressed as the mean ± standard deviation of three independent assays (* $p \leq 0.05$; ** $p \leq 0.01$; *** $p \leq 0.001$).

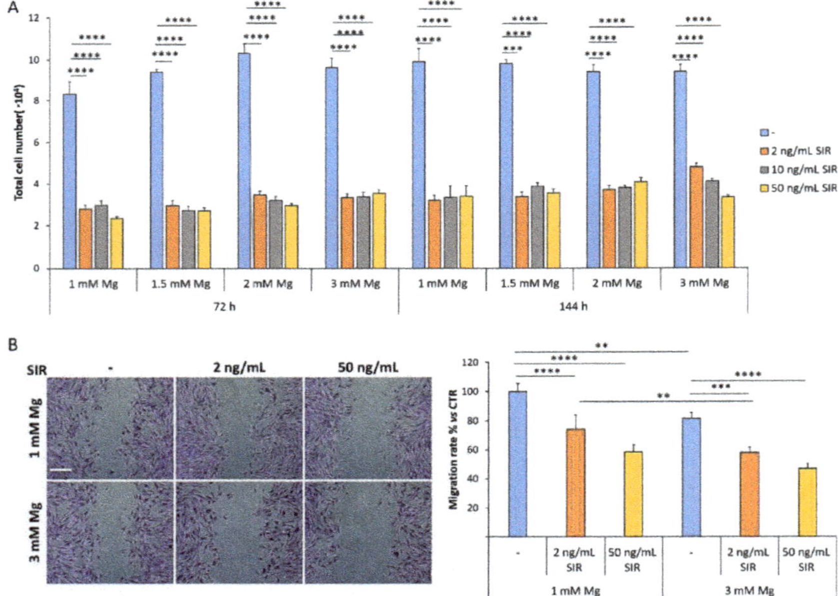

Figure 4. The effects of sirolimus and Mg on rSMC. (**A**) The cells were counted after 72 and 144 h of culture in medium containing 1, 1.5, 2 or 3 mM Mg with or without different concentrations of sirolimus (2, 10, 50 ng/mL). (**B**) Confluent rSMC were cultured in the presence of 1 or 3 mM Mg combined with 2 or 50 ng/mL of sirolimus for 72 h and the wound assay was performed. 24 h later, the width of the wound was visualized and measured. Representative images of the scratches are shown (left panel). Scale bar: 125 µm. The migration rate was analyzed using ImageJ and data are shown as the percentage of migration compared to the control (CTR, 1 mM Mg without sirolimus, right panel). Data are expressed as the mean ± standard deviation of three independent assays (* $p \leq 0.05$; ** $p \leq 0.01$; *** $p \leq 0.001$; **** $p \leq 0.0001$).

3.5. The Effect of Sirolimus and Mg on the Response of rSMC Growth to NO Donor

We then tested whether sirolimus alters rSMC response to NO, which regulates the degree of contraction of vascular smooth muscle cells by stimulating the production of cyclic GMP (cGMP).

rSMC were cultured in medium containing 1 or 3 mM Mg with or without sirolimus (2 and 50 ng/mL) for 72 h. Then, sodium nitroprusside (SNP 10 μM), a vasodilator which acts as a NO donor, was added for 30 min. Therefore, we measured cGMP as a marker of the sensitivity of rSMC to NO [23]. ELISA shows no alterations in the response to SNP of rSMC cultured in 1 and 3 mM Mg, with or without sirolimus (Figure 5).

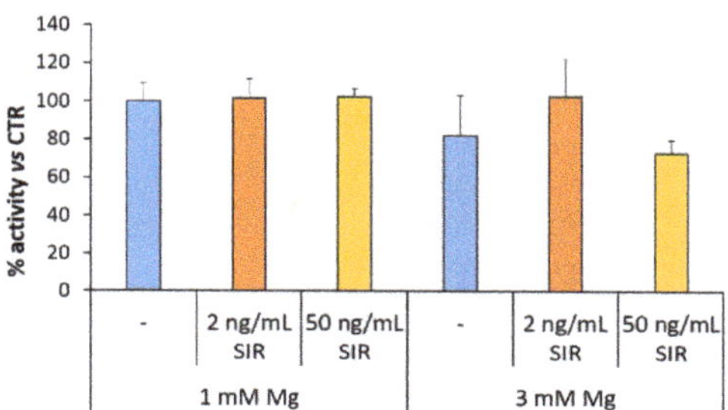

Figure 5. The effects of sirolimus and Mg on the response of rSMC to SNP. The cells were cultured in medium containing 1 or 3 mM Mg with or without sirolimus (2 and 50 ng/mL). After 72 h, the cells were treated with SNP for 30 min. cGMP was measured as described in the methods. Data are expressed as the mean ± standard deviation of three independent assays.

4. Discussion

Sirolimus-eluting resorbable Mg alloy scaffolds have recently entered clinical practice to treat coronary lesions. The choice of Mg alloy to generate the scaffold is based on the mechanical and biological properties of the metal. From a mechanical point-of-view, Mg provides a radial strength which is adequate to support the vascular wall and offers a high collapse pressure with low elastic recoil [24]. Most of the Mg of the alloy is completely resorbed in about 12 months without side effects [25]. Actually, an increase in Mg concentration in the vessel wall might be beneficial, because of its anti-arrhythmic, anti-thrombotic and anti-inflammatory effects [7–9]. Moreover, high extracellular Mg induces quiescence in human coronary SMC, while it stimulates the proliferation of human coronary endothelial cells [18].

Sirolimus, which possesses potent anti-proliferative properties, is frequently used to coat scaffolds. The potential reciprocal interaction between Mg and sirolimus has not been investigated yet. In rabbits, the implantation of Mg-based bioresorbable scaffolds accelerated re-endothelialization and decreased macrophage infiltration in the neointima [2]. Our in vitro results in rCAEC and rSMC might offer some hints to interpret these results. A first issue to consider is that in the in vitro studies performed on endothelial and smooth muscle cells available in the literature, high concentrations of extracellular Mg were used [9,18,26], i.e., 5 mM or more, which is unlikely to be reached even in strict proximity to the scaffold. In addition, in pigs, a Mg-based scaffold increases Mg concentration in the vascular wall up to ~2 mM [18]. This is the reason why we cultured rCAEC and rSMC in a range between 1 and 3 mM Mg. A second point to raise is that our experiments were performed for long times, up to 144 h, while most studies investigated cellular events in the short term.

An interesting finding is that concentrations higher than 1 mM of Mg prevent the inhibitory effect of sirolimus on endothelial proliferation. Moreover, 3 mM Mg with or without sirolimus stimulate migration. These results suggest that re-endothelialization should be granted by a modest increase in extracellular Mg, which counterbalances the inhibitory effects of sirolimus on the endothelium. Turning our attention to endothelial function, we found that high extracellular Mg downregulates VCAM after 72 h of culture, a beneficial event that might reduce the adhesion of leukocytes to the endothelium. Howev after 144 h, sirolimus (50 ng/mL) increases the total amounts of VCAM. At the moment, w have no explanation for these results, which raise the possibility of a late sirolimus-induce enhancement of leukocyte adhesion. It would be interesting to investigate VCAM in anim models at various times after the implant of vascular scaffolds. We also measured th amounts of tF, the critical initiator of blood coagulation in acute coronary syndrome, ar found no upregulation by sirolimus neither alone nor in combination with Mg. Surprising we found a reduction of tF after 144 h in rCAEC cultured in 3 mM of Mg and treated wi sirolimus. In general, these results support the in vivo evidence that Mg-based sirolim eluting scaffolds are less thrombogenic than other absorbable devices in rabbits [27] and humans [11].

In rSMC, high extracellular Mg does not interfere with the anti-proliferative ar anti-migratory effects of sirolimus. Actually, high Mg potentiates sirolimus in inhibitir rSMC migration. Morever, sirolimus and Mg do not alter rSMC sensitivity to NO dono thereby suggesting that vasomotion in vivo might be maintained. It should be recalled th Mg might favor vasodilation also by antagonizing Ca [28]. Again, these results should b confirmed in preclinical animal models.

In brief, these in vitro experiments show that 3 mM of extracellular Mg suffice prevent events associated with neointima formation and in-stent restenosis, contra endothelial acquisition of pro-thrombotic phenotype and stimulate endothelial grow and migration.

Author Contributions: Conceptualization, L.L., J.A.M. and S.C.; methodology, G.F.; formal analys all authors.; investigation, G.F. and L.L.; resources, J.A.M.; writing—original draft preparation, J.A.M writing—review and editing, L.L., S.C. and J.A.M.; funding acquisition, J.A.M. All authors have re and agreed to the published version of the manuscript.

Funding: This research received no external funding.

Institutional Review Board Statement: Not applicable.

Informed Consent Statement: Not applicable.

Data Availability Statement: Data available in a publicly accessible repository. The data present in this study are openly available in Dataverse at https://dataverse.unimi.it/dataverse/materi (accessed on 14 April 2021).

Acknowledgments: The authors acknowledge support from the University of Milan through t APC initiative.

Conflicts of Interest: The authors declare no conflict of interest.

References

1. Joseph, P.; Leong, D.; McKee, M.; Anand, S.S.; Schwalm, J.D.; Teo, K.; Mente, A.; Yusuf, S. Reducing the global burden cardiovascular disease, part 1: The epidemiology and risk factors. *Circ. Res.* **2017**, *121*, 677–694. [CrossRef]
2. Nicol, P.; Bulin, A.; Castellanos, M.I.; Stöger, M.; Obermeier, S.; Lewerich, J.; Lenz, T.; Hoppmann, P.; Baumgartner, C.; Fischer et al. Preclinical investigation of neoatherosclerosis in magnesium-based bioresorbable scaffolds versus thick-strut drug-eluti stents. *EuroIntervention J. Eur. Collab. with Work. Gr. Interv. Cardiol. Eur. Soc. Cardiol.* **2020**, *16*, e922–e929. [CrossRef]
3. Simard, T.; Hibbert, B.; Ramirez, F.D.; Froeschl, M.; Chen, Y.X.; O'Brien, E.R. The Evolution of Coronary Stents: A Brief Revi *Can. J. Cardiol.* **2014**, *30*, 35–45. [CrossRef]
4. Canfield, J.; Totary-Jain, H. 40 years of percutaneous coronary intervention: History and future directions. *J. Pers. Med.* **2018**, *8*, [CrossRef] [PubMed]
5. Testa, L.; Latib, A.; Montone, R.A.; Colombo, A.; Bedogni, F. Coronary Bioresorbable Vascular Scaffold Use in the Treatment Coronary Artery Disease. *Circ. Cardiovasc. Interv.* **2016**, *9*. [CrossRef]
6. Qiu, T.; Zhao, L. Research into biodegradable polymeric stents: A review of experimental and modelling work. *Vessel Plus* **2018** 12. [CrossRef]
7. de Baaij, J.H.F.; Hoenderop, J.G.J.; Bindels, R.J.M. Magnesium in man: Implications for health and disease. *Physiol. Rev.* **2015**, 1–46. [CrossRef]
8. Castiglioni, S.; Maier, J.A.M. Magnesium alloys for vascular stents: The biological bases: A focus on the effects of magnesium vascular cells. *BioNanoMaterials* **2015**, *16*, 23–29. [CrossRef]

Maier, J.A.M.; Bernardini, D.; Rayssiguier, Y.; Mazur, A. High concentrations of magnesium modulate vascular endothelial cell behaviour in vitro. *Biochim. Biophys. Acta Mol. Basis Dis.* **2004**, *1689*, 6–12. [CrossRef]

Maier, J.A.; Castiglioni, S.; Locatelli, L.; Zocchi, M.; Mazur, A. Magnesium and inflammation: Advances and perspectives. *Semin. Cell Dev. Biol.* **2020**. [CrossRef] [PubMed]

Haude, M.; Ince, H.; Kische, S.; Toelg, R.; Van Mieghem, N.M.; Verheye, S.; von Birgelen, C.; Christiansen, E.H.; Barbato, E.; Garcia-Garcia, H.M.; et al. Sustained Safety and Performance of the Second-Generation Sirolimus-Eluting Absorbable Metal Scaffold: Pooled Outcomes of the BIOSOLVE-II and -III Trials at 3 Years. *Cardiovasc. Revasc. Med.* **2020**, *21*, 1150–1154. [CrossRef]

Laplante, M.; Sabatini, D.M. mTOR signaling in growth control and disease. *Cell* **2012**, *149*, 274–293. [CrossRef] [PubMed]

Daniel, J.-M.; Dutzmann, J.; Brunsch, H.; Bauersachs, J.; Braun-Dullaeus, R.; Sedding, D.G. Systemic application of sirolimus prevents neointima formation not via a direct anti-proliferative effect but via its anti-inflammatory properties. *Int. J. Cardiol.* **2017**, *238*, 79–91. [CrossRef]

Jabs, A.; Göbel, S.; Wenzel, P.; Kleschyov, A.L.; Hortmann, M.; Oelze, M.; Daiber, A.; Münzel, T. Sirolimus-Induced Vascular Dysfunction. Increased Mitochondrial and Nicotinamide Adenosine Dinucleotide Phosphate Oxidase-Dependent Superoxide Production and Decreased Vascular Nitric Oxide Formation. *J. Am. Coll. Cardiol.* **2008**, *51*, 2130–2138. [CrossRef]

Reineke, D.C.; Müller-Schweinitzer, E.; Winkler, B.; Kunz, D.; Konerding, M.A.; Grussenmeyer, T.; Carrel, T.P.; Eckstein, F.S.; Grapow, M.T.R. Rapamycin impairs endothelial cell function in human internal thoracic arteries. *Eur. J. Med. Res.* **2015**, *20*, 59. [CrossRef] [PubMed]

Barilli, A.; Visigalli, R.; Sala, R.; Gazzola, G.C.; Parolari, A.; Tremoli, E.; Bonomini, S.; Simon, A.; Closs, E.I.; Dall'Asta, V.; et al. In human endothelial cells rapamycin causes mTORC2 inhibition and impairs cell viability and function. *Cardiovasc. Res.* **2008**, *78*, 563–571. [CrossRef]

Wang, C.; Qin, L.; Manes, T.D.; Kirkiles-Smith, N.C.; Tellides, G.; Pober, J.S. Rapamycin antagonizes TNF induction of VCAM-1 on endothelial cells by inhibiting mTORC2. *J. Exp. Med.* **2014**, *211*, 395–404. [CrossRef]

Sternberg, K.; Gratz, M.; Koeck, K.; Mostertz, J.; Begunk, R.; Loebler, M.; Semmling, B.; Seidlitz, A.; Hildebrandt, P.; Homuth, G.; et al. Magnesium used in bioabsorbable stents controls smooth muscle cell proliferation and stimulates endothelial cells in vitro. *J. Biomed. Mater. Res. Part. B Appl. Biomater.* **2012**, *100 B*, 41–50. [CrossRef]

Cerrato, E.; Barbero, U.; Gil Romero, J.A.; Quadri, G.; Mejia-Renteria, H.; Tomassini, F.; Ferrari, F.; Varbella, F.; Gonzalo, N.; Escaned, J. Magmaris^{TM} resorbable magnesium scaffold: State-of-art review. *Future Cardiol.* **2019**, *15*, 267–279. [CrossRef] [PubMed]

Baldoli, E.; Castiglioni, S.; Maier, J.A.M. Regulation and Function of TRPM7 in Human Endothelial Cells: TRPM7 as a Potential Novel Regulator of Endothelial Function. *PLoS ONE* **2013**, *8*, 1–7. [CrossRef]

Yau, J.W.; Teoh, H.; Verma, S. Endothelial cell control of thrombosis. *BMC Cardiovasc. Disord.* **2015**, *15*, 130. [CrossRef]

Hong, F.-F.; Liang, X.-Y.; Liu, W.; Lv, S.; He, S.-J.; Kuang, H.-B.; Yang, S.-L. Roles of eNOS in atherosclerosis treatment. *Inflamm. Res. Off. J. Eur. Histamine Res. Soc.* **2019**, *68*, 429–441. [CrossRef] [PubMed]

Kraehling, J.R.; Sessa, W.C. Contemporary approaches to modulating the nitric oxide-cGMP pathway in cardiovascular disease. *Circ. Res.* **2017**, *120*, 1174–1182. [CrossRef] [PubMed]

Schmidt, W.; Behrens, P.; Brandt-Wunderlich, C.; Siewert, S.; Grabow, N.; Schmitz, K.-P. In vitro performance investigation of bioresorbable scaffolds—Standard tests for vascular stents and beyond. *Cardiovasc. Revasc. Med.* **2016**, *17*, 375–383. [CrossRef]

Rapetto, C.; Leoncini, M. Magmaris: A new generation metallic sirolimus-eluting fully bioresorbable scaffold: Present status and future perspectives. *J. Thorac. Dis.* **2017**, *9*, S903–S913. [CrossRef]

Farruggia, G.; Castiglioni, S.; Sargenti, A.; Marraccini, C.; Cazzaniga, A.; Merolle, L.; Iotti, S.; Cappadone, C.; Maier, J.A.M. Effects of supplementation with different Mg salts in cells: Is there a clue? *Magnes. Res.* **2014**, *27*, 25–34. [CrossRef] [PubMed]

Waksman, R.; Lipinski, M.J.; Acampado, E.; Cheng, Q.; Adams, L.; Torii, S.; Gai, J.; Torguson, R.; Hellinga, D.M.; Westman, P.C.; et al. Comparison of Acute Thrombogenicity for Metallic and Polymeric Bioabsorbable Scaffolds: Magmaris Versus Absorb in a Porcine Arteriovenous Shunt Model. *Circ. Cardiovasc. Interv.* **2017**, *10*. [CrossRef]

Altura, B.M.; Altura, B.T.; Carella, A.; Gebrewold, A.; Murakawa, T.; Nishio, A. Mg^{2+}-Ca^{2+} interaction in contractility of vascular smooth muscle: Mg2+ versus organic calcium channel blockers on myogenic tone and agonist-induced responsiveness of blood vessels. *Can. J. Physiol. Pharmacol.* **1987**, *65*, 729–745. [CrossRef]

MDPI

·ticle

·he Video Microscopy-Linked Electrochemical Cell: An ·nnovative Method to Improve Electrochemical Investigations ·f Biodegradable Metals

·cho Zimmermann [1], Norbert Hort [2], Yuqiuhan Zhang [1], Wolf-Dieter Müller [1,*] and Andreas Schwitalla [1]

1 Dental Materials and Biomaterial Research, Department of Prosthodontics, School of Dentistry, Charité-University Medicine Berlin, Aßmannshauser Str. 4–6, 14197 Berlin, Germany; tycho.zimmermann@charite.de (T.Z.); yuqiuhan.zhang@charite.de (Y.Z.); andreas.schwitalla@charite.de (A.S.)
2 Helmholtz Zentrum Geesthacht, 21502 Geesthacht, Germany; norbert.hort@hzg.de
* Correspondence: wolf-dieter.mueller@charite.de; Fax: +49-30-450562902

Abstract: An innovative, miniature video-optical-electrochemical cell was developed and tested that allows for the conducting of electrochemical corrosion measurements and simultaneous microscopic observations over a small, well-defined surface area of corroding or degrading samples. The setup consisted of a miniature electrochemical cell that was clamped onto the metal sample and fixed under a video microscope before being filled with electrolyte. The miniature cell was comprised of afferent/efferent electrolyte ducts as well as a connection to the Mini Cell System (MCS) for electrochemical measurements. Consequently, all measured and induced currents and voltages referred to the same small area corroding completely within the field of view of the microscope, thus allowing for real-time observation and linking of surface phenomena such as hydrogen evolution and oxide deposition to electrochemical data. The experimental setup was tested on commercial purity (cp) and extra-high purity (XHP) magnesium (Mg) samples using open circuit potential and cyclic voltammetry methods under static and flowing conditions. The corrosion potential was shifted more anodically for cp Mg in comparison to XHP Mg under dynamic conditions. The corrosion current assessed from the cyclic voltametric curves were higher for the cp Mg in comparison to XHP Mg. However, there were no differences between static and flow conditions in the case of XHP Mg in contrast to cp Mg, where the current density was two times higher at dynamic conditions. The measurements and observations with this new method pave the way for a more detailed understanding of magnesium corrosion mechanisms, thus improving predictive power of electrochemical corrosion measurements on newly developed magnesium or other biodegradable alloys applied for medical devices. Different electrochemical tests can be run under various conditions, while being easy to set up and reproduce as well as being minimally destructive to the sample.

Keywords: magnesium; biomaterials; electrochemistry; corrosion; degradation; hydrogen evolution; microscopy

·ation: Zimmermann, T.; Hort, N.; ·ng, Y.; Müller, W.-D.; Schwitalla, ·The Video Microscopy-Linked ·ctrochemical Cell: An Innovative ·hod to Improve Electrochemical ·estigations of Biodegradable ·als. *Materials* **2021**, *14*, 1601. ·s://doi.org/10.3390/ma14071601

·demic Editor: Digby Macdonald

·eived: 5 February 2021
·epted: 16 March 2021
·lished: 25 March 2021

1. Introduction

Magnesium displays promising potential as implantable biodegradable biomaterial for cardiovascular stents, bone fixation devices, and tissue engineering scaffolds and has been an object of intensive research [1]. Apart from being biodegradable, it is an important trace mineral in human organisms and, thus, fully biocompatible in its unalloyed form [2–4]. Furthermore, magnesium displays a favorable bone response [5,6]. Among implant metals, the elastic modulus of magnesium is the most similar to that of bone (41–45 GPa for magnesium vs. 3–20 GPa for bone [7]), thus minimizing stress shielding [3,8].

The predominant drawbacks of magnesium when used as a biomaterial relate to its corrosion characteristics. In its unalloyed form, magnesium displays relatively rapid, localized, and difficult-to-predict degradation when implanted. The unique environment of the human body, comprising physiological temperature, perfusion, abundance of inorgan ions, amino acids, proteins, and buffering agents, affects the degradation rate and mech nism in as yet unpredictable ways [9]. This may lead to premature failure of the implante magnesium structures bringing about grave ramifications to patients' health [10].

Furthermore, magnesium produces large volumes of hydrogen gas during corrosic If rapid corrosion occurs, the hydrogen promotes the accumulation of gas bubbles in tl tissue surrounding the implant site [11]. Because of the degradation reaction, local p values are increased in the tissue surrounding the implant. This can negatively affe physiological reactions [7].

Current scientific efforts aim for the goal of slowing down magnesium corrosic and achieving a more uniform and predictable corrosion mode, while at the same tin improving mechanical properties and host response [12]. Common methods incluc alloying [13,14], grain refining, surface modifications, and coating with layers of absorbab materials such as calcium phosphate or biocompatible polymers [15–20]. However, ea of these methods comes with its own set of challenges, first among which is a decline biocompatibility [21]. Whereas countless promising advancements have been made in tl lab, only few magnesium-based implants have made it to the stage of animal and huma trials. A further problem standing in the way of clinical translation is the difficulty accurately predicting in vivo magnesium degradation through in vitro testing [22,23], part due to the absence of a commonly accepted methodology [23]. This leads to a wic range of magnesium alloys having to be considered for costly and ethically challengir animal trials [9]. So far, only a few magnesium-based products are commercially availab in the form of magnesium stents made of proprietary magnesium alloy (Biotronik SE & C Kg, Berlin, Germany) and orthopedic screws made of MgYREZr (Syntellix AG, Hanov Germany) [1].

The dissolution of magnesium in aqueous media is described by the following chen cal reactions [24–26]:

Anodic reaction:

$$Mg \rightarrow Mg^{2+} + 2e^{-}$$

Cathodic reaction:

$$2H_2O + 2e^{-} \rightarrow 2OH^{-} + H_2\uparrow$$

Overall reaction:

$$Mg + 2H_2O \rightarrow Mg^{2+} + 2OH^{-} + H_2$$

Product formation:

$$Mg^{2+} + 2OH^{-} \rightarrow Mg(OH)_2$$

Hydrogen gas can be observed developing on the surface of corroding magnesiu samples in the form of gas bubbles.

$Mg(OH)_2$ is not soluble and forms a partially protective layer on the magnesium magnesium alloy surface [25]. The stability of this layer is dependent on local pH value. neutral and alkaline media, this layer is porous and only offers limited protection [24,2 In the presence of chlorides in a concentration of more than 30 mmol/L, $Mg(OH)_2$ conve to soluble $MgCl_2$ [25,27,28]. This is the case in body fluids, where a chloride concentrati of about 150 mmol/L exists. Other inorganic ions found in body fluids such as phospha and carbonate further complicate the corrosion process [7,9].

The corrosion mechanisms and corrosion speed are strongly influenced by the comp nents, surface and bulk treatment of the respective alloy, and by trace impurities in pu magnesium samples [12,24,26,28,29].

In physiological media, magnesium usually corrodes in the form of pitting corrosi or extremely localized corrosion, as opposed to general corrosion [9,26,30]. These corrosi modes are the most problematic for structural integrity and predictability.

In vitro corrosion measurement methods play an important role in choosing among t wide variety of magnesium alloys and surface modifications those with the most potent for further development and in vivo testing. Furthermore, in vitro methods allow for the elucidation of the particular corrosion mechanisms displayed by different modifications.

The most common in vitro corrosion testing methods for magnesium include hydrogen collection measurements, weight/volume loss measurements, and electrochemical testing. Each of these test methods comes with its unique set of challenges [31–33].

In practice, the hydrogen collection is far from efficient and the ratio of mass loss as estimated by hydrogen collection to actual mass loss is much lower than can theoretically be expected. In a review [31], the range of reported ratios was found to lie between 0.22:1 and 1.31:1. This was attributed to a number of reasons, mainly concerning experimental setup.

In mass loss experiments, the correct removal of corrosion product is crucial to obtain a valid result. In the absence of an established standard of conducting mass loss measurement on magnesium, and in recognition of the fact that results are highly dependent on experimental setup, especially sample surface-to-electrolyte volume ratio, a wide range of corrosion rates has been reported [31].

Electrochemical methods (ECM) permit meaningful data to be gleaned by perturbing the relative equilibrium at the surface of the degrading sample by electrically charging the sample surface in the cathodic or anodic direction and measuring the resulting current. Commonly employed measurement protocols include potentiodynamic polarization (PDP), in which a range of DC (dircet current) potentials' cathodic and anodic to the OCP (Open circuit potential) is applied. This permits the isolation and interpretation of the kinetics of cathodic and anodic processes. The voltammetric potentials can be applied as a linear sweep (LSV) or in repeating cycles (CV) and result in PDP plots of I (current) vs. E (potential). These plots provide information on cathodic, anodic, and corrosion current densities as well as corrosion potentials [31,34–36].

Another commonly employed method is electrochemical impedance spectroscopy (EIS) [37–40]. The choice of the correct equivalent circuit for the particular corroding system presents one of the uncertainties of this method.

The interpretation of ECM data is based on a simplified corrosion model and the assumption of uniform corrosion, which is rarely encountered in magnesium and its alloys. Changes to the corroding surface induced by the polarization are the source of measurement artifacts. PDP evokes unusual phenomena such as negative difference effect (NDE). The obtained corrosion rate values are strongly dependent on the chosen measurement protocol and experimental setup, as well as on the mathematical interpretation of the results.

In sum, none of the abovementioned methods alone provides a complete picture of the particular corrosion processes of the corroding system in question, nor are obtained values for corrosion speed necessarily comparable [9,33]. For this reason, a complementary use of these and additional methods is often recommended [2,31,33]. Further challenges arise through the choice of electrolyte and its chemical and physical parameters and the difficulties of replicating the conditions in the human body. Electrolyte components and parameters play a considerable role in determining corrosion speed, especially when electrolytes are chosen that approximate in vivo conditions [9,41]. Among other things, this is attributed to the fact that secondary reactions such as the production of insoluble compounds (phosphates, carbonates) and their interactions with chloride ions and enzymes are not properly taken into account. Additionally, physical conditions such as electrolyte temperature, pH value, flow rate, electrode surface-to-electrolyte ratio, immersion time, and overall experimental setup lead to large variations in reported corrosion rates [22,23,27,30,42]. Different electrochemical techniques were developed with the intention of overcoming these limitations. For example, the development of the Minicell [39] and Scanning Electrochemical Microscopy (SECM) in lieu of the classical cell permits a reduction of the area of analysis and, hence, a higher local resolution [36]. This miniaturization, in turn, raises the question to what extent the results can be extrapolated to the larger surfaces of real components such as stents and orthopedic screws.

These inaccuracies in measurement and interpretation lead to the consequence that results from different in vitro corrosion experiments do not necessarily agree nor can they accurately predict the in vivo corrosion rates of magnesium alloys [22,31,43]. This ha been demonstrated by comparing the predicted degradation rates derived from in vitr experiment to those actually achieved in animal studies [23]. In the most studies, in vitr measurement methods overestimate the actual in vivo degradation rate

In sum, the experimental parameters that most adequately reproduce the in vitro co ditions are not known. However, some agreement exists that the experimental setup need to simulate the in vivo application as fully as possible [32]. Additionally, the informativ value and validity of electrochemical measurements for the assessment of the degradatio of magnesium (-alloys) and other biometals is strongly dependent on the possibility c correlation of measurements to the surface events actually occurring. So far, this is mainl attempted by examining the surface after corrosion measurement has taken place, ignorin the fact that time-dependent processes may be obscured.

As the aim of the present study, both of these shortcomings were addressed throug the development and the evaluation of a specialized miniature electrochemical cell tha permits the real-time microscopic evaluation of a small (<1 mm^2), well-defined surface are during electrochemical measurement, with video-microscopic observations of phenomen such as corrosion product deposition and hydrogen evolution. The insights derived fror this method can be used to better understand and adapt conventional electrochemica measurement methods and corrosion models, by linking electrochemical measurement t physical surface phenomena, especially over time. This can improve their reproducibilit and predictive value, thus leading to the reduction of the in vitro/in vivo gap.

2. Materials and Methods

A graphical abstract of the Materials and Methods section is provided in Figure 1.

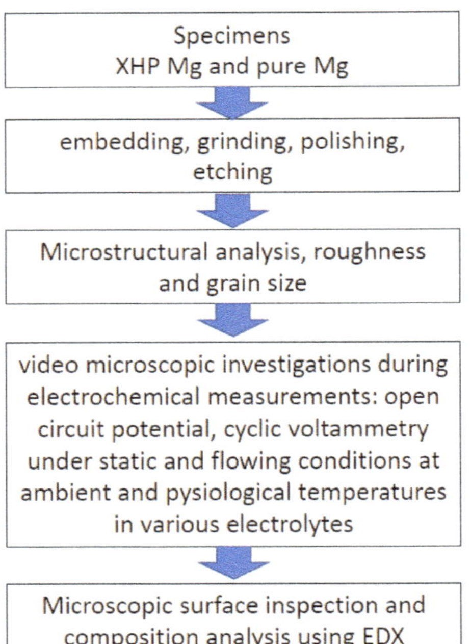

Figure 1. Graphical abstract of Material and Methods.

Specimens of two different magnesium purity grades, commercial pure (cp) an extra-high pure (XHP), were used.

The cp magnesium was cast and provided by Helmholtz Zentrum Geesthacht (Geesth Germany) and the XHP magnesium was cast and provided by ETH Zürich (Switzerlanc

The resulting rod-shaped bulk material was cut into discs of 10 mm in diameter and 1.5 mm in thickness. One such disc of each material was embedded in cold-curing poly (methyl methacrylate) resin (Technovit 4004, Heraeus Kulzer GmbH, Wehrheim, Germany) using cylindrically shaped molds, leaving only one circular surface of the disc exposed. The reverse side of the embedding material was opened up by drilling and the specimen connected with a small quantity of amalgam to a short length of copper wire, allowing the conductive connection of the specimen to the working electrode terminal of a potentiostat.

The specimens were ground on a rotary grinding and polishing machine (Struers Dap-V, Struers GmbH, Crinitz, Germany) with silica carbide sandpaper in steps of 500, 800, and 1200, and 2400 grit (WS Flex 18 C, Hermes Schleifmittel GmbH, Hamburg, Germany), using a rotation speed of 150 rpm; light, manual pressure; and water cooling. Each grinding step was continued for 5 min. Subsequently, the specimens were polished using a neoprene polishing plate (Mambo Chem, Industrieservice Siegmund Bigott, Kaarst, Germany) and a mixture of ethane diol-based colloidal silica dioxide suspension (OPS Extra, Industrieservice Siegmund Bigott, Kaarst, Germany) and alcohol-based 1 μ diamond spray (Diamantspray 1 μ, Industrieservice Siegmund Bigott, Kaarst, Germany) for 10 min, applying light, manual pressure. After polishing, the specimen surfaces were rinsed using pure ethanol and subsequently dried with pressurized air.

To visualize grain structure, the polished specimens were etched with a solution of picric and acetic acid for about 1 s, before being rinsed with ethanol and dried with pressurized air [44]. The specimens were stored in a vacuum desiccator.

Grain structure was recorded using a digital microscope (Keyence VHX-5000, Keyence Corporation, Osaka, Japan).

The specimens were polished once more according to the above protocol to prevent the remnants of the etching process from influencing the measurements.

The 3D areal surface texture of five randomly chosen rectangular areas of about 0.38 mm^2 per specimen was captured using an optical measurement system (Alicona InfiniteFocus, Alicona Imaging GmbH, Raaba, Austria) at 20-fold magnification. For surface roughness characterization, the value for Sa (absolute arithmetical mean height according to ISO 25178) was calculated from the recorded 3D images with the aid of the equipment manufacturer's software (Alicona IF-MeasureSuite 4.2).

For the electrochemical measurements and concurrent microscopic observation of the corroding surfaces, an opto-electrochemical cell was designed and constructed from Plexiglas (Evonik Röhm GmbH, Darmstadt, Germany). The miniaturized, electrochemical cell consisted of a chamber with a height of 7 mm and a contact area of about 0.8 mm^2 to the bare surface of the specimen. Furthermore, the chamber had two electrolyte ducts, one of which connected to an electrolyte reservoir for filling the cell with electrolyte at the beginning of the experiment and the other connected to the Mini-Cell System (MCS) (Figure 2). The Mini-Cell System [45] contained the platinum counter electrode (CE) and the saturated calomel reference electrode (SCE, E_0 = 0.241 V).

Connected to the MCS, a pump allowed for the aspiration and optional cycling of the electrolyte through the optical-electrochemical cell and MCS chamber. The MCS was connected to a mini-potentiostat (PalmSens Emstat3+, PalmSens BV, Houten, The Netherlands) and the associated PC hardware running the potentiostat manufacturer's software (PSTrace 5.3, PalmSens BV, Houten, The Netherlands). The top of the electrochemical cell consisted of a clear glass aperture of 2 mm in diameter, which allowed for direct observation of the corroding surface and associated phenomena with the aid of a microscope, in the present case, the digital microscope Keyence VHX-5000 (Keyence Corporation, Osaka, Japan).

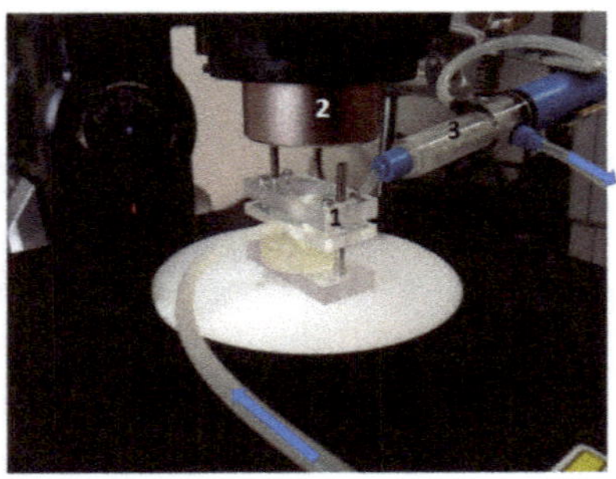

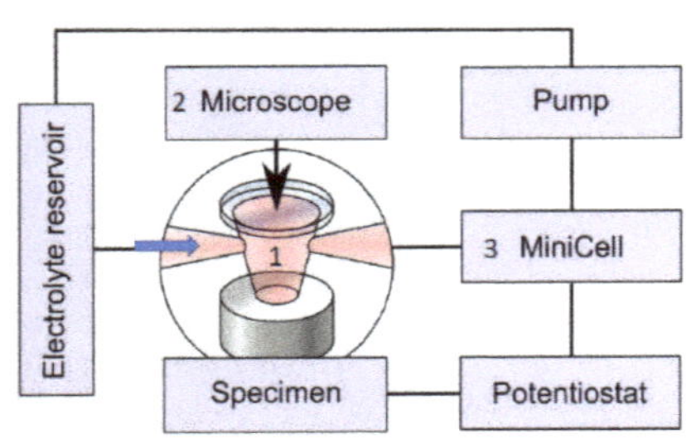

Figure 2. Video-optical-electrochemical cell with mounted specimen, connected to electrolyte inlet and Mini-Cell (background), beneath Keyence optical microscope. At left, a schematic view.

This setup allowed for an observation of a well-defined circular area of about 1 squa millimeter, undergoing corroding in contact with an electrolyte under a 200-fold micr scopic magnification and simultaneous measurement and/or manipulation of voltage a current flow over the corroding surface. An example of the appearance of the magnesiu surface as seen through the opto-electrochemical cell by microscope is shown in Figure

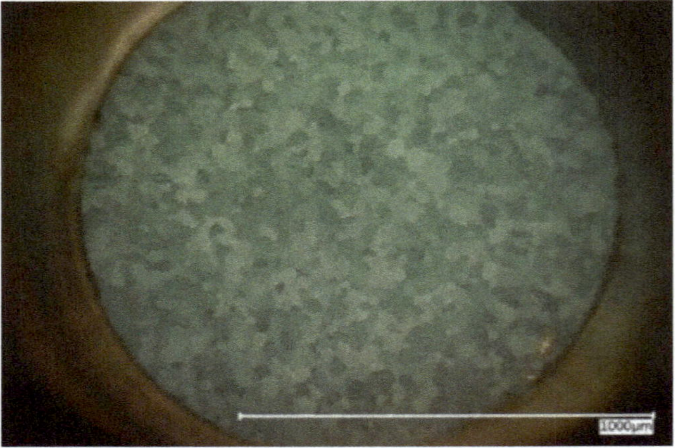

Figure 3. The cp Mg surface as seen through video-optical-electrochemical cell, polarized lightin

As electrolyte, MEM (minimum essential medium) Earle's w/o $NaHCO_3$, L-Glutami and Phenol red (Biochrom GmbH, Berlin, Germany) was chosen to achieve an appr imation of in vivo conditions [30] while establishing a baseline for the future additi of more complex components, such as fetal bovine serum (FBS) and additional buff systems [43,46]. This electrolyte is among those commonly used for magnesium biodeg dation measurements and results in relatively good correlations between in vitro a in vivo degradation rates [23]. It should be noted that electrochemical measurement resu differed considerably according to the chosen electrolyte [47]. The components of t selected electrolyte according to the manufacturer's data sheet [48] are listed in Table 1.

Table 1. Electrolyte composition [49].

Substance	Concentration [mg/L]
Amino acids	1696
Vitamins & growth factors	8
NaCl	6800
KCl	400
$NaH_2PO_4 \cdot H_2O$	140
$MgSO_4 \cdot 7H_20$	200
$CaCl_2$	200
d-Glucose	1000

After filling the chamber with electrolyte, OCP was measured for 30 min. The OCP was assumed to be at steady state at the end of the 30-minute measurement period. This was followed by six voltammetric cycles with an amplitude of ±0.5 volt starting at a value of –1.6 V and with a scan rate of 10 mV/s. Hereby, the first vertex was cathodic and the second vertex was anodic. Simultaneously, a video of the corroding surface was recorded with a digital microscope (Keyence VHX-5000, Keyence corporation, Osaka, Japan). The resulting video was cut into sequences, and relevant stills for analysis were extracted. The obtained electrochemical data (OCP vs. time and I vs. E curves) were exported to a corrosion data analysis software (CView 3,0, Scribner Associates Inc., Southern Pines, NC, USA) and the values for OCP, electrochemical corrosion potential (E_{corr}), linear polarization resistance (R_P), and total anodic current (I_{corr}) were extracted. Linear R_P was calculated from the reciprocal of the slope of a linear fit to the CV curve (defined as conductivity) over a segment of ±20 mV from E_{corr} multiplied by the exposed specimen surface and given in $\Omega \cdot cm^2$ (polarization resistance method). The corrosion current density (i_{corr}) was calculated from R_p and the Stern–Geary constant according to [49–51].

For the purpose of this study, a Stern–Geary constant of 12.8 mV was assumed for the calculation of i_{corr} in agreement with published data on cp magnesium [51]. This value corresponds to an anodic and cathodic Tafel slope of 60 mV/decade. When comparing the obtained values for i_{corr} to those published in literature, it is important to keep in mind that in the present case they were calculated from R_P derived from linear polarization and an estimated Stern–Geary constant. However, reported values for B differ considerably in the case of magnesium.

Experiments were conducted in triplicate. Tabular data are presented as mean value ± standard deviation.

3. Results

The etched surfaces of the uncorroded surfaces of both magnesium qualities can be seen in Figure 4. The corresponding values of surface roughness, expressed as S_a, are listed in Table 2.

Typical Open Circuit Potential (OCP) curves, obtained from cp Mg and XHP Mg measurement from corrosion onset up to 30 min after onset, are shown in Figure 5. As can be seen from these exemplary plots, for both magnesium grades, OCP generally started at potentials far catholically removed from the values at the 30-min mark. The cathodic OCP at the very beginning of the experiment, when the electrolyte first contacted the magnesium surface, corresponded with generalized production of hydrogen appearing as rapidly expanding and often detaching bubbles.

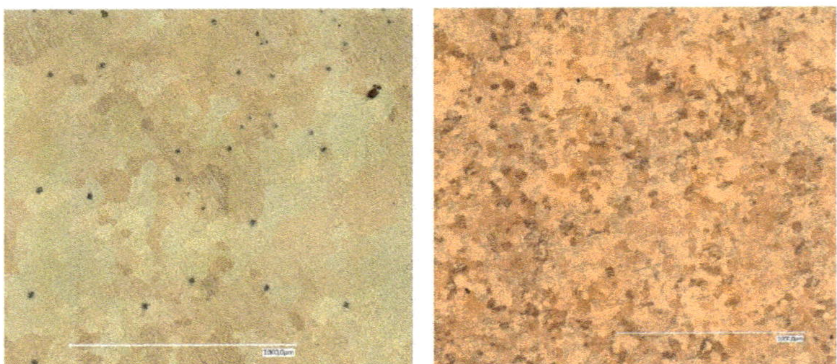

Figure 4. Microscopic appearance of etched surfaces of XHP (left) and cp (right) magnesium.

Table 2. Surface roughness.

Name	Sa-Value [μm]
XHP Mg	0.10 ± 0.010
CP Mg	0.13 ± 0.012

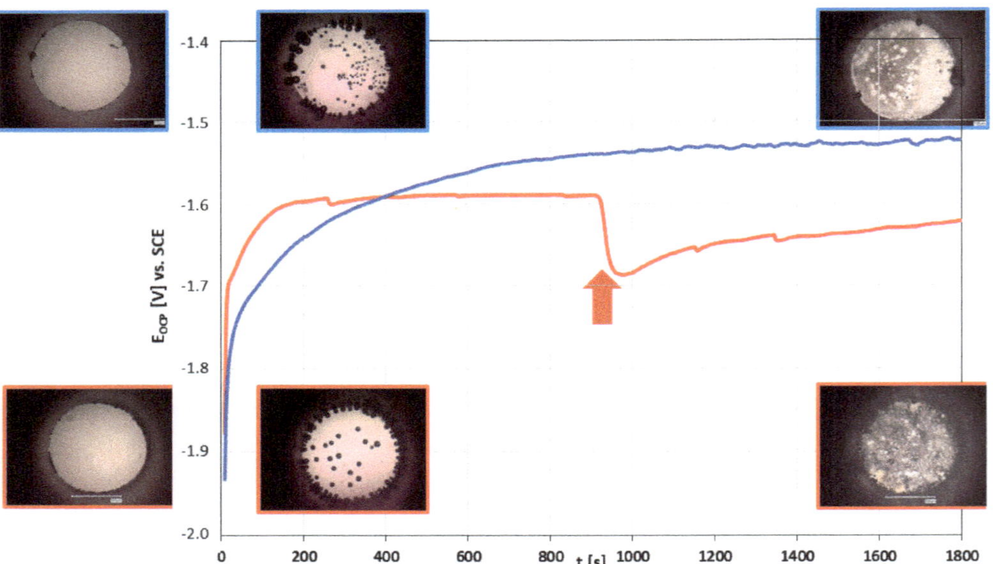

Figure 5. E_{OCP} vs. t plots for XHP (red) and cp (blue) magnesium aligned with microscopic pictures (prior to contact with electrolyte, as well as at t = 0 and 30 min, after removal of the hydrogen bubbles); arrow: spontaneous detachment of large hydrogen bubble exposing fresh magnesium surface.

Within minutes, the hydrogen evolution was observed as coming to a near standsti[l] while at the same time the formation of a thin corrosion layer of dark appearance wa[s] noticed. The corrosion layer appeared less dense in the vicinity of adhered hydroge[n] bubbles. Upon the occasional detachment or confluence of adhered bubbles at this stag[e] an instantaneous cathodic drop in OCP value was recorded, momentarily disturbing th[e] OCP equilibrium and initiating a renewed convergence toward the steady state OCP, a[s] exemplified in Figure 5. This corresponded with the exposure of previously uncorrode[d] areas where the particular bubble had been in direct contact with the specimen surface. Neither in the case of cp magnesium nor XHP magnesium was an onset of pitting corrosion observed within the duration of the measurement nor could a corresponding change in OCP value be measured. At the end of the 30-minute measuring period and after removal of adhered hydrogen bubbles the corrosion layer appeared relatively uniform with less dense areas in the vicinity of spots of hydrogen evolution, as can be seen in Figure 5. At the center of each film-free area of hydrogen evolution a dark spot of corrosion product formed. Evident from this observation was the fact that the locations of hydrogen evolution did not shift during the measurement period.

The obtained OCP values upon conclusion of the open circuit measurement after 30 min are given in Tables 3 and 4 and compared graphically in Figure 6. OCP values for static conditions at ambient temperature did not differ notably from those under dynamic conditions.

Table 3. Electrochemical measurement results, static conditions.

Name	E_{ocp} [V]	E_{corr} (Forward) [V]	E_{corr} (Back) [V]	i_{corr} (Forward) [A/cm^2]	i_{corr} (Back) [A/cm^2]
XHP Mg	-1.671 ± 0.035	-1.61 ± 0.13	-1.78 ± 0.07	$3.5210^{-6} \pm 1.8010^{-6}$	$2.2810^{-5} \pm 1.0610^{-5}$
CP Mg	-1.577 ± 0.049	-1.47 ± 0.02	-1.61 ± 0.01	$2.9710^{-6} \pm 1.78^{-6}$	$4.2610^{-5} \pm 9.5710^{-6}$

Table 4. Electrochemical measurement results, dynamic conditions

Name	E_{ocp} [V]	E_{corr} (Forward) [V]	E_{corr} (Back) [V]	i_{corr} (Forward) [A/cm^2]	i_{corr} (Back) [A/cm^2]
XHP Mg	-1.663 ± 0.037	-1.50 ± 0.05	-1.74 ± 0.04	$3.2610^{-6} \pm 1.0810^{-6}$	$2.1910^{-5} \pm 1.7010^{-6}$
CP Mg	-1.614 ± 0.032	-1.48 ± 0.01	-1.60 ± 0.00	$4.2410^{-6} \pm 1.37^{-6}$	$6.6910^{-5} \pm 1.9110^{-6}$

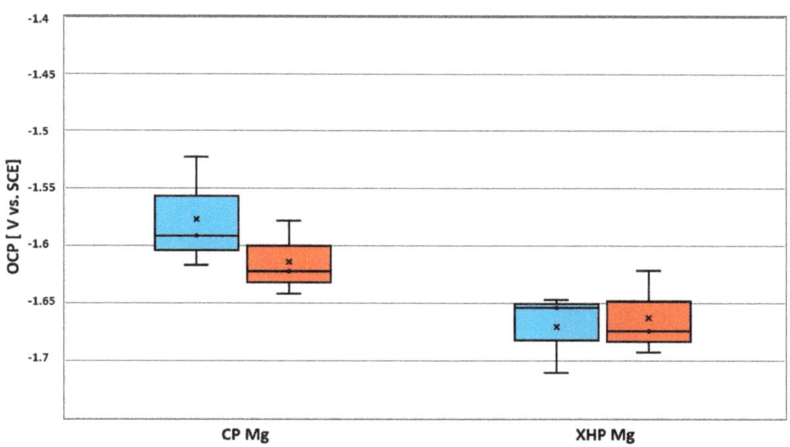

Figure 6. E_{OCP} after 30 min of XHP and cp magnesium in MEM at stationary and dynamic electrolyte flow conditions, at 25 °C. Left box relates to stationary and the right box to dynamic conditions.

A typical cyclic voltammetry (CV) plot is reproduced in Figure 7 using the example of cp Mg. In the interest of clarity, only one of the six cycles is reproduced. In general, the curve form and the corresponding key figures such as E_{corr} and R_p only varied slightly between cycles and adhered to the general characteristics presented hereafter.

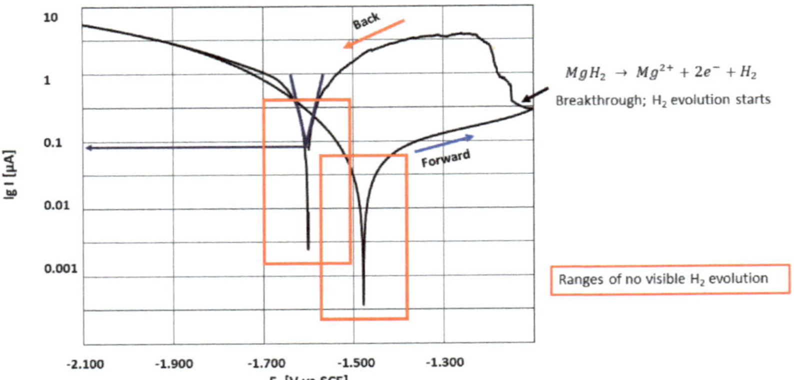

Figure 7. The principal shape of lgi vs. E curve. The red frames indicate the range where no hydrog evolution was observed.

The cathodic part of a cycle was marked by generalized hydrogen evolution, whi increased in intensity approaching the lower vertex, indicating the cathodic partial reacti of water electrolysis, and decreased as the potential approached the first $E_{corr,}$ in clo accordance with the measured electron flow.

Aside from the evolving hydrogen (Figure 5), no changes in surface appearance we observed during the cathodic cycle. Coming from the cathodic vertex, the hydrog evolution stopped approaching E_{corr} and remained that way for the beginning of t anodic part. This E_{corr} value, defined as the potential at the first zero current, is hencefor referred to as E_{corr}(forward). At the same time, the measured electron flow increased on slightly with increasing anodic potential, both phenomena indicating the formation o passive film. The anodic part of the cycle was characterized by the abrupt onset of pitti corrosion. The time of onset of pitting corrosion fluctuated considerably and no obvio relation to experiment parameters was evident. Pitting corrosion was extremely localiz often taking place at a single site. Video-optically, pitting corrosion was associated wi intense hydrogen evolution from the pitting site. The pitting site itself presented as a slow expanding dark front of deposited corrosion product.

In the case of cp magnesium, these spots expanded outward from the site of initiati along the surface preceded by a localized front of violent hydrogen evolution. In t case of XHP Mg, the hydrogen evolution was more strictly localized and associated wi the development of pits surrounded by volcano-like depositions of corrosion produ (Figure 8).

Localized corrosion coincided with a sharp increase in electron flow. Once initiate localized hydrogen evolution continued well into the following anodic cycle. Localiz anodic corrosion resulted in a cathodic shift of E_{corr} on the back scan (E_{corr} (back)) as w as increase of corrosion current density (i_{corr} (back)). In the subsequent cathodic pol ization, generalized hydrogen evolution could be once again observed; but, additional hydrogen evolution centered on the areas of anodic product deposition was preferenti without a change in appearance of these regions, indicating catalyzed cathodic hydrog evolution [52–54]. The areas where localized corrosion took place re-passivated during t following cathodic polarization. The onset of localized corrosion during the subseque anodic cycle was in every case at a different localization.

Apart from the described anodic localized corrosion (mode I), another anodic behavi could be observed, that of anodic passivation (mode II). In the case of the absence of anoc localized corrosion, no hydrogen evolution or product formation took place during t anodic part of the cycle. Correspondingly, no increase in current density took pla rather, the forward scan was retraced by the back scan. Accordingly, in these cases anoc polarization did not result in an increase in corrosion current density or a cathodic shift of E_{corr} (back).

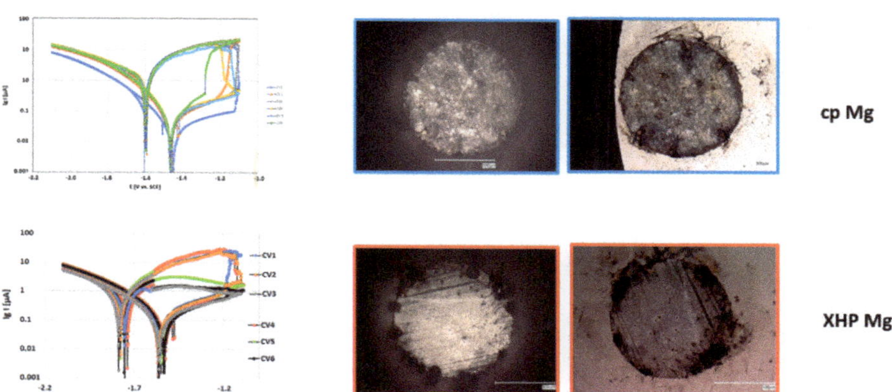

Figure 8. Left: lgi vs. E curves, cycle 1 and 6; right: microscopic view onto the surface after cyclic loading, (**left**) immediately after polarization and (**right**) after drying the surface. The upper row shows the CVs and microscopic pictures for cp Mg and the lower row for XHP Mg.

Both XHP and cp magnesium fluctuated between these two behaviors from cycle to cycle, independent of obvious experimental parameters. At times, hybrid forms could be detected.

The measured values for E_{corr} (forward) and E_{corr} (back), as well as i_{corr} (forward) and i_{corr} (back), are listed in Tables 3 and 4 and represented in Figures 9 and 10. E_{corr} (back) and i_{corr} (back) refer only to those cases when localized anodic corrosion took place (mode I). The values obtained in dynamic conditions at physiological temperature displayed less variation than those obtained in static conditions at ambient temperatures. This might be due to the fact that electrolyte depletion effects and surface obstruction through hydrogen bubble attachment introduced a random factor under static conditions that was ameliorated under dynamic conditions.

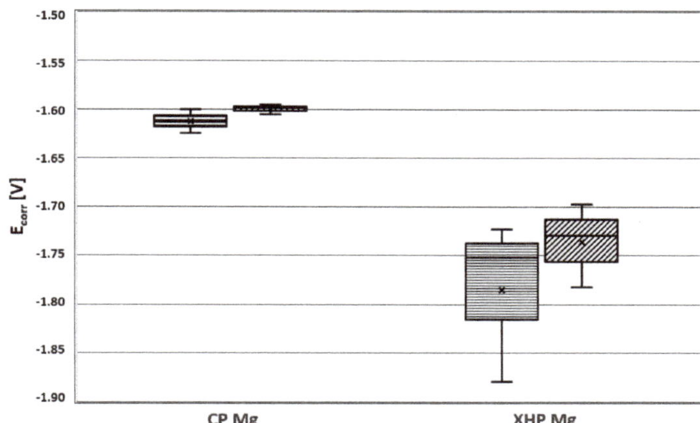

Figure 9. E_{corr} for static and dynamic conditions for XHP Mg and cp Mg in MEM.

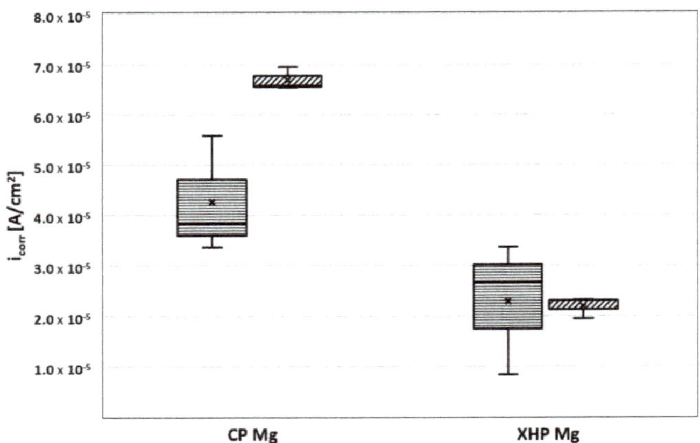

Figure 10. The i_{corr} for static and dynamic conditions for XHP Mg and cp Mg in MEM.

4. Discussion

The chosen measurement protocol utilizing the described optical-electrochemical ce revealed similar behavior across the tested XHP and cp magnesium specimens. These wer localized corrosion during anodic parts of a cycle and generalized hydrogen evolutio devoid of corrosion during cathodic parts of a cycle. Also, enhanced catalytic activit with anodic polarization could be observed; hereby, the corrosion products deposite during anodic part of a cycle became the preferred spots for hydrogen evolution [52–54 Furthermore, it was shown that the complete or partial passivation of the magnesiur surface through cathodic polarization led to a delay or prevention of anodic localize corrosion. The so-called negative difference effect (NDE) was also observed, this bein an increase in hydrogen formation as a function of increasing anodic potential at th breakthrough potential. In cases where this effect did not appear, no cathodic shift of th corrosion potential was observed. This can possibly be used as a proof of the hydroge accumulation realized by production of MgH_2, which was responsible for the anodic shi of the corrosion potential E_{corr} [43].

During corrosion of magnesium, the following reactions are possible:

$$Mg + 2H_2O \rightarrow Mg(OH)_2 + H_2 \quad (5)$$

$$2Mg + 2H_2O \rightarrow Mg(OH)_2 + MgH_2 \quad (6)$$

$$\text{Overall}: \ 3Mg + 4H_2O \rightarrow 2Mg(OH)_2 + H_2 + MgH_2 \quad (7)$$

With the help of these equations, the observations and results can be explained. Durin OCP measurement, here for a duration of 30 min, the initial hydrogen evolution was clearl visible. At the same time, the development of a surface layer, consisting of $Mg(OH)_2$, wa observed, which acted as a barrier for further fast degradation of Mg [55]. The OCP wa based on the galvanic coupling of anodic and cathodic reactions over the exposed surfac forming a mixed potential, initially described by reactions (1) and (2) and later on by (5 and (6), which caused an anodic shift of the OCP. The value of OCP depended on th relationship between the current densities of the abovementioned reactions and/or th relationship between the active surface areas of the anodic and cathodic reactions. Th can be used to explain the differences seen in OCP data between XHP Mg and cp Mg. Th more cathodic shift in OCP of cp Mg was caused by catalytic support of reaction (6) [52]

The subsequent cyclic voltammetry disturbed these equilibriums and produced ne data for the interpretation of the degradation XHP and cp Mg. The higher cathodic curren in the case of cp Mg revealed the higher catalytic activity of cp Mg in comparison to XH Mg, as described by Birbilis et al. [53]. In the voltammetric forward scan, the MgH_2 creatio in direct contact with $Mg(OH)_2$ was seen resulting in an anodic shift of E_{corr} of the forward scan. There were significant differences between XHP Mg and cp Mg, which were seen in the more anodic shift of E_{corr} forward, in the higher i_{corr} values, and its reproducibility. This demonstrated the higher activity, which was possibly caused by the impurities in cp Mg. An explanation for the scattering of the CVs of XHP Mg cannot be given at the moment, but the observations described in [55] are supported.

Utilizing potentiodynamic techniques to determine magnesium corrosion rate is not universally accepted in the science community [56]. This is due to incomplete knowledge and unsupported simplifications regarding the magnesium corrosion process such as the assumption of uniform corrosion. Because of parasitic chemical/electrochemical reactions and coverage effects of unknown magnitude occurring alongside the theoretical corrosion reaction, the calculated values for corrosion current density may not accurately reflect the actual corrosion rate [32,57], in particular, when using the Tafel equation for corrosion rate estimation [33]. Nevertheless, potentiodynamic methods can be employed to show the effects of experimental parameters and alloying on the cathodic and anodic branches [32].

The described measurement cell brings together a number of benefits. Among those is an easy, well-defined, and reproducible setup with minimal destruction to the sample surface, permitting repeated measurement on the same sample. Furthermore, the test conditions can be adjusted to approximate in vivo conditions.

Apart from the measurement protocols utilized in this study, further electrochemical protocols are possible without changing the setup. These include electrochemical impedance spectroscopy (EIS), linear sweep potentiometry, and long-term open circuit potentiometry. Depending on the research question, a number of parameters can be chosen or adjusted. These are perfusion rate, electrolyte composition and temperature, buffer system, electrolyte temperature, and specimen composition as well as measurement area. This way the often-unclear influence of the abovementioned factors on the measured results can be isolated. If a more pragmatic approach is needed, the parameters can be adjusted as far as possible to in vivo conditions.

The specimen composition in consideration was not limited to magnesium and its alloys. On the contrary, valuable insights can potentially be gleaned by utilizing the described measuring cell on other metallic biomaterials of interest, such as iron or zinc.

5. Conclusions

The described method consolidated versatility with reproducibility and brought together electrochemical data with the underlying optical surface phenomena in real time. Hydrogen evolution, corrosion product formation, and voltage/current can be correlated topographically and temporally.

These are the first investigations that show the complexity of such correlation, which are given here descriptively first.

Further measurements with different parameters (different alloys, electrolyte composition, temperature, pH, surface treatment, and measurement protocol) are possible and necessary to get more quantitative results where the relation of surfaces changes over time are involved into the assessment of the electrochemical data.

Author Contributions: Conceptualization, W.-D.M. and T.Z.; methodology, T.Z. and W.-D.M.; validation, W.-D.M. and A.S.; formal analysis, Y.Z.; investigation, T.Z. and Y.Z.; resources, N.H.; writing—original draft preparation, T.Z.; writing—review and editing, W.-D.M. and A.S.; visualization, T.Z., W.-D.M. and Y.Z.; supervision, W.-D.M. and A.S. All authors have read and agreed to the published version of the manuscript.

Funding: This research received no external funding.

Institutional Review Board Statement: Not applicable.

Informed Consent Statement: Informed consent was obtained from all subjects involved in the study.

Data Availability Statement: Not applicable.

Conflicts of Interest: The authors declare no conflict of interest.

References

1. Zhao, D.; Witte, F.; Lu, F.; Wang, J.; Li, J.; Qin, L. Current status on clinical applications of magnesium-based orthopaedic implar A review from clinical translational perspective. *Biomaterials* **2017**, *112*, 287–302. [CrossRef]
2. Walker, J.; Shadanbaz, S.; Woodfield, T.B.F.; Staiger, M.P.; Dias, G.J. Magnesium biomaterials for orthopedic application: A revi from a biological perspective. *J. Biomed. Mater. Res. Part B Appl. Biomater.* **2014**, *102*, 1316–1331. [CrossRef]
3. Staiger, M.P.; Pietak, A.M.; Huadmai, J.; Dias, G. Magnesium and its alloys as orthopedic biomaterials: A review. *Biomateri* **2006**, *27*, 1728–1734. [CrossRef] [PubMed]
4. Saris, N.E.L.; Mervaala, E.; Karppanen, H.; Khawaja, J.A.; Lewenstam, A. Magnesium: An update on physiological, clinical a analytical aspects. *Clin. Chim. Acta* **2000**, *294*, 1–26. [CrossRef]
5. Castellani, C.; Lindtner, R.A.; Hausbrandt, P.; Tschegg, E.; Stanzl-Tschegg, S.E.; Zanoni, G.; Beck, S.; Weinberg, A.-M. Bor implant interface strength and osseointegration: Biodegradable magnesium alloy versus standard titanium control. *Acta Bioma* **2011**, *7*, 432–440. [CrossRef]
6. Witte, F.; Kaese, V.; Haferkamp, H.; Switzer, E.; Meyer-Lindenberg, A.; Wirth, C.; Windhagen, H. In vivo corrosion of fo magnesium alloys and the associated bone response. *Biomaterials* **2005**, *26*, 3557–3563. [CrossRef]
7. Song, G. Control of biodegradation of biocompatable magnesium alloys. *Corros. Sci.* **2007**, *49*, 1696–1701. [CrossRef]
8. Yeung, K.W.; Wong, K.H. Biodegradable metallic materials for orthopaedic implantations: A review. *Technol. Health Care* **20** *20*, 345–362. [CrossRef]
9. Xin, Y.; Hu, T.; Chu, P. In vitro studies of biomedical magnesium alloys in a simulated physiological environment: A revi *Acta Biomater.* **2011**, *7*, 1452–1459. [CrossRef]
10. Wichelhaus, A.; Emmerich, J.; Mittlmeier, T. A case of implant failure in partial wrist fusion applying magnesium-based headl bone screws. *Case Rep. Orthoped.* **2016**, *2016*, 1–5. [CrossRef]
11. Kuhlmann, J.; Bartsch, I.; Willbold, E.; Schuchardt, S.; Holz, O.; Hort, N.; Höche, D.; Heineman, W.R.; Witte, F. Fast escape hydrogen from gas cavities around corroding magnesium implants. *Acta Biomater.* **2013**, *9*, 8714–8721. [CrossRef]
12. Pogorielov, M.; Husak, E.; Solodivnik, A.; Zhdanov, S. Magnesium-based biodegradable alloys: Degradation, application, a alloying elements. *Interv. Med. Appl. Sci.* **2017**, *9*, 27–38. [CrossRef] [PubMed]
13. Li, N.; Zheng, Y. Novel magnesium alloys developed for biomedical application: A review. *J. Mater. Sci. Technol.* **2013**, *29*, 489–5 [CrossRef]
14. Chen, Y.; Xu, Z.; Smith, C.; Sankar, J. Recent advances on the development of magnesium alloys for biodegradable implar *Acta Biomater.* **2014**, *10*, 4561–4573. [CrossRef] [PubMed]
15. Hornberger, H.; Virtanen, S.; Boccaccini, A. Biomedical coatings on magnesium alloys—A review. *Acta Biomater.* **2012**, *8*, 2442–24 [CrossRef] [PubMed]
16. Wu, G.; Ibrahim, J.M.; Chu, P.K. Surface design of biodegradable magnesium alloys—A review. *Surf. Coat. Technol.* **2013**, *233*, 2– [CrossRef]
17. Gray-Munro, J.E.; Seguin, C.; Strong, M. Influence of surface modification on the in vitro corrosion rate of magnesium alloy AZ *J. Biomed. Mater. Res. Part A* **2009**, *91*, 221–230. [CrossRef]
18. Cifuentes, S.; Gavilán, R.; Lieblich, M.; Benavente, R.; González-Carrasco, J. In vitro degradation of biodegradable polylac acid/magnesium composites: Relevance of Mg particle shape. *Acta Biomater.* **2016**, *32*, 348–357. [CrossRef] [PubMed]
19. Wong, H.M.; Yeung, K.W.; Lam, K.O.; Tam, V.; Chu, P.K.; Luk, K.D.; Cheung, K.M. A biodegradable polymer-based coating control the performance of magnesium alloy orthopaedic implants. *Biomaterials* **2010**, *31*, 2084–2096. [CrossRef] [PubMed]
20. Gray, J.; Luan, B. Protective coatings on magnesium and its alloys—A critical review. *J. Alloy. Compd.* **2002**, *336*, 88–113. [CrossR
21. Gu, X.; Zheng, Y.; Cheng, Y.; Zhong, S.; Xi, T. In vitro corrosion and biocompatibility of binary magnesium alloys. *Biomateri* **2009**, *30*, 484–498. [CrossRef] [PubMed]
22. Witte, F.; Fischer, J.; Nellesen, J.; Crostack, H.-A.; Kaese, V.; Pisch, A.; Beckmann, F.; Windhagen, H. In vitro and in vivo corrosi measurements of magnesium alloys. *Biomaterials* **2006**, *27*, 1013–1018. [CrossRef] [PubMed]
23. Sanchez, A.H.M.; Luthringer, B.J.; Feyerabend, F.; Willumeit, R. Mg and Mg alloys: How comparable are in vitro and in vi corrosion rates? A review. *Acta Biomater.* **2015**, *13*, 16–31. [CrossRef] [PubMed]
24. Song, G.-L.L.; Atrens, A. Corrosion mechanisms of magnesium alloys. *Adv. Eng. Mater.* **1999**, *1*, 11–33. [CrossRef]
25. Song, G.-L. Corrosion electrochemistry of magnesium (Mg) and its alloys. In *Corrosion of Magnesium Alloys*; Elsevier: Amsterda The Netherlands, 2011; pp. 3–65.
26. Song, G.; Atrens, A. Understanding magnesium corrosion—A framework for improved alloy performance. *Adv. Eng. Mater.* **20** *5*, 837–858. [CrossRef]
27. Witte, F.; Hort, N.; Vogt, C.; Cohen, S.; Kainer, K.U.; Willumeit, R.; Feyerabend, F. Degradable biomaterials based on magnesi corrosion. *Curr. Opin. Solid State Mater. Sci.* **2008**, *12*, 63–72. [CrossRef]
28. Shaw, B.A. Corrosion resistance of magnesium alloys. In *Corrosion: Fundamentals, Testing, and Protection*; ASM Internatio Geauga, OH, USA, 2003; Volume 13, pp. 693–696.
29. Hanawalt, J. Corrosion studies of magnesium and its alloys. *Trans AIME* **1942**, *147*, 273–299.

Kirkland, N.; Lespagnol, J.; Birbilis, N.; Staiger, M. A survey of bio-corrosion rates of magnesium alloys. *Corros. Sci.* **2010**, *52*, 287–291. [CrossRef]

Kirkland, N.; Birbilis, N.; Staiger, M. Assessing the corrosion of biodegradable magnesium implants: A critical review of current methodologies and their limitations. *Acta Biomater.* **2012**, *8*, 925–936. [CrossRef] [PubMed]

Esmaily, M.; Svensson, J.; Fajardo, S.; Birbilis, N.; Frankel, G.; Virtanen, S.; Arrabal, R.; Thomas, S.; Johansson, L. Fundamentals and advances in magnesium alloy corrosion. *Prog. Mater. Sci.* **2017**, *89*, 92–193. [CrossRef]

Atrens, A.; Song, G.-L.; Cao, F.; Shi, Z.; Bowen, P.K. Advances in Mg corrosion and research suggestions. *J. Magnes. Alloy.* **2013**, *1*, 177–200. [CrossRef]

Bard, A.J.; Faulkner, L.R. *Electrochemical Methods Fundamentals and Applications*, 2nd ed.; Allen, J., Bard, L.R.F., Eds.; John Wiley & Sons Inc.: Hoboken, NJ, USA, 2001.

Staiger, M.; Feyerabend, F.; Willumeit, R.; Sfeir, C.; Zheng, Y.; Virtanen, S.; Mueller, W.; Atrens, A.; Peuster, M.; Kumta, P.; et al. Summary of the panel discussions at the 2nd Symposium on Biodegradable Metals, Maratea, Italy, 2010. *Mater. Sci. Eng. B* **2011**, *176*, 1596–1599. [CrossRef]

Mueller, M.D. Electrochemical techniques for assessment of corrosion behaviour of Mg and Mg-alloys. *BioNanoMaterials* **2015**, *16*, 31–39. [CrossRef]

King, A.; Birbilis, N.; Scully, J. Accurate electrochemical measurement of magnesium corrosion rates—A combined impedance, mass-loss and hydrogen collection study. *Electrochim. Acta* **2014**, *121*, 394–406. [CrossRef]

Müller, W.D.; Nascimento, M.L.; Zeddies, M.; Córsico, M.; Gassa, L.M.; De Mele, M.A.F.L. Magnesium and its alloys as degradable biomaterials: Corrosion studies using potentiodynamic and EIS electrochemical techniques. *Mater. Res.* **2007**, *10*, 5–10. [CrossRef]

Nascimento, M.L.; Fleck, C.; Müller, W.-D.; Löhe, D. Electrochemical characterisation of magnesium and wrought magnesium alloys. *Int. J. Mater. Res.* **2006**, *97*, 1586–1593. [CrossRef]

Orazem, M.E.; Tribollet, B. *Electrochemical Impedance Spectroscopy*; John Wiley & Sons Inc.: Hoboken, NJ, USA, 2017; Volume 2.

Gu, X.N.; Zheng, Y.; Chen, L.J. Influence of artificial biological fluid composition on the biocorrosion of potential orthopedic Mg–Ca, AZ31, AZ91 alloys. *Biomed. Mater.* **2009**, *4*, 065011. [CrossRef]

Mueller, W.D.; Nascimento, M.L.; De Mele, M.F. Critical discussion of the results from different corrosion studies of Mg and Mg alloys for biomaterial applications. *Acta Biomater.* **2010**, *6*, 1749–1755. [CrossRef]

Willumeit, R.; Fischer, J.; Feyerabend, F.; Hort, N.; Bismayer, U.; Heidrich, S.; Mihailova, B. Chemical surface alteration of biodegradable magnesium exposed to corrosion media. *Acta Biomater.* **2011**, *7*, 2704–2715. [CrossRef]

Kree, V.; Bohlen, J.; Letzig, K.U.K. Practical metallography. *Prakt. Metallogr.* **2004**, *41*, 233.

Mueller, W.D.; Schoepf, C.; Nascimento, M.L.; Carvalho, A.C.; Moisel, M.; Schenk, A. Electrochemical characterization of dental alloys: Its possibilities and limitations. *Anal. Bioanal. Chem.* **2005**, *381*, 1520–1525. [CrossRef] [PubMed]

Yamamoto, A.; Hiromoto, S. Effect of inorganic salts, amino acids and proteins on the degradation of pure magnesium in vitro. *Mater. Sci. Eng. C* **2009**, *29*, 1559–1568. [CrossRef]

Xin, Y.; Hu, T.; Chu, P.K. Influence of test solutions on in vitro studies of biomedical magnesium alloys. *J. Electrochem. Soc.* **2010**, *157*, C238–C243. [CrossRef]

American Society for Testing and Materials. *G 102-89 Standard Practice for Calculation of Corrosion Rates and Related Information*; American Society for Testing and Materials: Conshohocken, PA, USA, 1999.

Biochrom GmbH. MEM (Earle) Liquid Medium Product Information. Available online: http://www.biochrom.de/fileadmin/user_upload/service/produktinformation/deutsch/BC_Katalog_58_59_60_MEM.pdf (accessed on 5 February 2021).

Stern, M.; Geary, A. Electrochemical polarization—I. A theoretical analysis of the shape of polarization curves. *J. Electrochem. Soc.* **1957**, *104*, 559. [CrossRef]

Bland, L.; King, A.; Birbilis, N.; Scully, J. Assessing the corrosion of commercially pure magnesium and commercial AZ31B by electrochemical impedance, mass-loss, hydrogen collection, and inductively coupled plasma optical emission spectrometry solution analysis. *Corrosion* **2015**, *71*, 128–145. [CrossRef]

Birbilis, N.; King, A.; Thomas, S.; Frankel, G.; Scully, J. Evidence for enhanced catalytic activity of magnesium arising from anodic dissolution. *Electrochim. Acta* **2014**, *132*, 277–283. [CrossRef]

Taheri, M.; Kish, J.; Birbilis, N.; Danaie, M.; McNally, E.; McDermid, J. Towards a physical description for the origin of enhanced catalytic activity of corroding Magnesium surfaces. *Electrochim. Acta* **2014**, *116*, 396–403. [CrossRef]

Cano, Z.P.; McDermid, J.R.; Kish, J.R. Cathodic activity of corrosion filaments formed on Mg alloy AM30. *J. Electrochem. Soc.* **2015**, *162*, C732–C740. [CrossRef]

Mueller, W.D.; Hornberger, H. The Influence of MgH_2 on the assessment of electrochemical data to predict the degradation Rate of Mg and Mg alloys. *Int. J. Mol. Sci.* **2014**, *15*, 11456–11472. [CrossRef]

Shi, Z.; Liu, M.; Atrens, A. Measurement of the corrosion rate of magnesium alloys using Tafel extrapolation. *Corros. Sci.* **2010**, *52*, 579–588. [CrossRef]

Cao, F.; Shi, Z.; Hofstetter, J.; Uggowitzer, P.J.; Song, G.; Liu, M.; Atrens, A. Corrosion of ultra-high-purity Mg in 3.5% NaCl solution saturated with $Mg(OH)_2$. *Corros. Sci.* **2013**, *75*, 78–99. [CrossRef]

ticle

1 Vitro Studies on Mg-Zn-Sn-Based Alloys Developed as a Jew Kind of Biodegradable Metal

feng Wen [1], Qingshan Liu [2], Weikang Zhao [1], Qiming Yang [1], Jingfeng Wang [2] and Dianming Jiang [1,3,*]

1 Department of Orthopaedics, The First Affiliated Hospital of Chongqing Medical University, No. 1 Youyi Road, Yuzhong District, Chongqing 400016, China; wenyafeng@stu.cqmu.edu.cn (Y.W.); weikang-zhao@cqu.edu.cn (W.Z.); 2018010024@stu.cqmu.edu.cn (Q.Y.)

2 National Engineering Research Center for Magnesium Alloys, College of Materials Science and Engineering, Chongqing University, No. 174, Shapingba Main Street, Shapingba District, Chongqing 400044, China; 201709021005@cqu.edu.cn (Q.L.); jfwang@cqu.edu.cn (J.W.)

3 Department of Orthopaedics, The Third Affiliated Hospital of Chongqing Medical University, No.1 Shuanghu Road, Yubei District, Chongqing 401120, China

* Correspondence: 201296@hospital.cqmu.edu.cn

Abstract: Mg-Zn-Sn-based alloys are widely used in the industrial field because of their low-cost, high-strength and heat-resistant characteristics. However, their application in the biomedical field has been rarely reported. In the present study, biodegradable Mg-1Zn-1Sn and Mg-1Zn-1Sn-0.2Sr alloys were fabricated. Their microstructure, surface characteristics, mechanical properties and bio-corrosion properties were carried out using an optical microscope (OM), X-ray diffraction (XRD), electron microscopy (SEM), mechanical testing, electrochemical and immersion test. The cell viability and morphology were studied by cell counting kit-8 (CCK-8) assay, live/dead cell assay, confocal laser scanning microscopy (CLSM) and SEM. The osteogenic activity was systematically investigated by alkaline phosphatase (ALP) assay, Alizarin Red S (ARS) staining, immunofluorescence staining and quantitative real time-polymerase chain reaction (qRT-PCR). The results showed that a small amount of strontium (Sr) (0.2 wt.%) significantly enhanced the corrosion resistance of the Mg-1Zn-1Sn alloy by grain refinement and decreasing the corrosion current density. Meanwhile, the mechanical properties were also improved via the second phase strengthening. Both Mg-1Zn-1Sn and Mg-1Zn-1Sn-0.2Sr alloys showed excellent biocompatibility, significantly promoted cell proliferation, adhesion and spreading. Particularly, significant increases in ALP activity, ARS staining, type I collagen (COL-I) expression as well as the expressions of three osteogenesis-related genes (runt-related transcription factor 2 (Runx2), osteopontin (OPN), and osteocalcin (Bglap)) were observed for the Mg-1Zn-1Sn-0.2Sr group. In summary, this study demonstrated that Mg-Zn-Sn-based alloy has great application potential in orthopedics and Sr is an ideal alloying element of Mg-Zn-Sn-based alloy, which optimizes its corrosion resistance, mechanical properties and osteoinductive activity.

Keywords: Mg-Zn-Sn alloy; corrosion behavior; mechanical properties; biocompatibility; osteoinductive activity

ation: Wen, Y.; Liu, Q.; Zhao, W.; g, Q.; Wang, J.; Jiang, D. In Vitro dies on Mg-Zn-Sn-Based Alloys reloped as a New Kind of degradable Metal. *Materials* **2021**, 1606. https://doi.org/10.3390/ 14071606

demic Editors: Hendra Hermawan, ndi Razavi and Irina Hussainova

eived: 9 February 2021
epted: 19 March 2021
lished: 25 March 2021

1. Introduction

Biodegradable implants represented by magnesium (Mg) alloys have attracted increasing interest in the last few years. Compared with traditional metal materials, the biggest advantage of Mg alloys is the ability to be completely degraded gradually after exerting biological functions in the body, thereby avoiding subsequent surgical removal procedures [1,2]. Consequently, lifelong problems caused by permanent implants such as tissue dysfunction, long-term foreign body stimulation and local chronic inflammatory reactions can be effectively alleviated or eliminated [3]. In addition, the density and elastic modulus of Mg alloys are close to those of human cortical bone [4], which are more suitable for orthopedic implant applications and can effectively eliminate/decrease the stress shielding effect and resulting osteoporosis induced by traditional metal materials [Moreover, Mg implants have been proven to stimulate the formation of new bone [6].

However, the rapid degradation rate of Mg alloys in the physiological environment the main reason hindering its clinical application. Furthermore, during the degradati of Mg alloys, it will result in increasing of local pH and accumulation of hydrogen (H causing inflammation and destruction of surrounding tissues [7]. This also means th the Mg alloy may lose sufficient mechanical support strength before the expected task completed. Therefore, the development of new biodegradable Mg alloy by adding ne alloy elements is a current research focus.

From a clinical viewpoint, an ideal Mg alloy orthopedic implant should meet t following standards: First, the degradation rate should be less than or equal to 0.5 mm Second, the mechanical strength must be higher than 200 MPa and the elongation preferably greater than 10%. Simultaneously, the mechanical integrity should be maintain for at least 90–180 days in vivo [6,8]. Third, the limit of acceptable H_2 evolution rate f humans was reported as 0.01 $mL/cm^2/day$ [6,9]. Moreover, biosafety and bioactivi etc. must also be considered. To date, Mg-based alloys that have been developed a intensively studied include Mg-Zn [10], Mg-Ca [11], Mg-Zr [12], Mg-Sr [13] et al., a ternary or multicomponent alloys developed on those bases. In 2013, magnesium-bas alloy compression screws received the CE mark and became the first Class III medic device made of Mg alloy approved for clinical use. However, due to the presence rare-earth (RE) metal elements, their long-term biosafety remains controversial.

Mg-Zn-Sn-based alloys are widely used in the industrial field due to their low-co high-strength and heat-resistant characteristics. However, there are few reports on appli tion of Mg-Zn-Sn-based alloys in the biomedical field. Tin (Sn) is an ultra-trace element th does not exceed 1 mg per kilogram of body weight in the human body. The task of Sn as trace element in the body is not really known, but a deficiency of Sn may disrupt kidn function. Studies have reported that Sn may be a good alloying element for biodegradab magnesium alloys [14]. In addition, Sn is an element with high H_2 evolution overpotenti which can control H_2 release of Mg alloys [15]. Strontium (Sr), as an essential trace eleme for the human body, has been in good repute to endorse the proliferation of osetoblasts a restrain the activity of osteoclasts [16]. Moreover, microalloying Sr in different Mg-bas systems can indeed improve the performance of Mg alloys [17], but in the Mg-Zn-Sn-bas alloy, no relevant literature has been reported.

In the present study, the application possibility and potential in the biomedical fie of Mg-1Zn-1Sn and Mg-1Zn-1Sn-0.2Sr alloys were systematically studied, focusing the surface characteristics, mechanical properties, corrosion performance and in vit biocompatibility and bioactivity.

2. Materials and Methods

2.1. Alloys Preparation

The ingots of Mg-1Zn-1Sn and Mg-1Zn-1Sn-0.2Sr alloys were prepared by casti from melting Pure Mg (purity > 99.98), pure Zn (purity > 99.99), pure Sn (purity > 99.9 and Mg-15 wt.% Sr master alloys in an electric resistance furnace under the protection an SF_6 and CO_2 gas mixture in a graphite crucible.

The actual compositions of the alloys are shown in Table 1. After homogenizati treatment at 500 °C for 6 h, the ingot casting was extruded at 280 °C with an extrusi ratio of 28:1. Subsequently, the Mg–1Zn–1Sn and Mg–1Zn–1Sn-0.2Sr ingots were cut in sheets, and commercial pure-Mg (p-Mg) was used as a reference. The sheets were cut in φ15 mm × 1 mm wafers for follow-up experiments.

Table 1. Actual composition of the Mg-1Zn-1Sn and Mg-1Zn-1Sn-0.2Sr alloys (wt.%).

Nominal Composition	Actual Composition						
	Zn	Sn	Sr	Fe	Si	Ni	Mg
Mg-1Zn-1Sn	1.04	1.13	0	<0.01	<0.01	<0.001	Balance
Mg-1Zn-1Sn-0.2Sr	1.02	1.12	0.21	<0.01	<0.01	<0.001	Balance

2.2. Microstructure Analysis and Mechanical Testing

The microstructure and surface morphology of the samples were examined by using an OM (OLYMPUS, Fuji, Japan) and a SEM (Vega III LMH, TESCAN, Shanghai, China) equipped with energy-dispersive X-ray spectroscopy (EDS). Moreover, the phase compositions were detected by XRD (Rigaku, Tokyo, Japan) with CuKα radiation at a scan rate of 5°/min. Compression tests were conducted in accordance with the procedures listed in ASTM standard E9-09 [18] by using a universal testing machine (AOUDE, Beijing, China), and the initial strain rate is 10^{-3}/s. According to the specifications of American Society for Testing Materials (ASTM) standard B557-15 [19], tensile specimens were tested at room temperature, and the tensile speed is 2 mm/min. Three parallel samples were tested for each group of materials.

2.3. In Vitro Corrosion Evaluation

Potentiodynamic polarization (PDP) curves were obtained using an electrochemical workstation (CHI600C, CHINSTRUMENTS, Shanghai, China) in simulated body fluid (SBF). Specific details are given elsewhere [20]. Briefly, The Mg-1Zn-1Sn and Mg-1Zn-1Sn-0.2Sr alloys were embedded in epoxy resin as the working electrode (only 1 cm^2 exposed). Platinum foil was used as the counter electrode, and a saturated calomel electrode was employed as reference. Afterward, a PDP curve was performed at a scanning speed of 1 $mV{\cdot}s^{-1}$ for all the measurements. Taking ASTM G102-89 [21] as the guide standard, the Tafel method was used to calculate the values of corrosion current density (i_{corr}) and corrosion potential (E_{corr}).

Immersion tests were carried out in SBF. The chemical compositions of SBF were shown in Table 2. The ratio of SBF volume to the surface area of the material was 50 mL: 1 cm^2. The H_2 evolution and pH value were monitored during a period of 336 h. The mass loss was also recorded after 3, 7, 15 and 35 day-immersion, respectively. A mixed solution composed of chromic acid (180 g/L) and silver nitrate (10 g/L) was used to eliminate the corrosion products on the surface of samples. The corrosion rate (CR) was calculated by the following formula:

$$CR = (K \times W) \div (D \times T \times A)$$

where the coefficient K = 8.76×10^4, W is the mass loss (g), D is the density of the material ($g.cm^{-3}$), T is the exposure time (h) and A is the sample area exposed to the solution (cm^2).

Table 2. The chemical compositions of SBF.

Components	Concentration
NaCl	7.996 g/L
KCl	0.224 g/L
$CaCl_2$	0.278 g/L
$NaHCO_3$	0.350 g/L
$MgCl{\cdot}6H_2O$	0.305 g/L
$K_2HPO_4{\cdot}3H_2O$	0.228 g/L
Na_2SO_4	0.071 g/L
HCl (1 mol/L)	40 mL
$(CH_2OH)_2CNH_2$	6.051 g/L

2.4. *In Vitro Cell Test*

2.4.1. Cell Culture and Preparation of Extraction

The murine calvarial preosteoblasts (MC3T3-E1) were utilized to realize the evaluatio of in vitro experiments. P-Mg extracts and normal culture medium were used as a contr and reference, respectively. The extraction was prepared according to a reference [22]. I short, samples were immersed in Dulbecco's modified Eagle's medium (DMEM, Hyclon containing 10% (v/v) fetal bovine serum (FBS, Gibco), 100 U/mL penicillin and 100 mg/m streptomycin for 72 h under a standard cell culture environment (95% humidity, 5% CC and 37 °C). The extraction ratio of medium volume to material weight was 5 mL/g.

2.4.2. Cytocompatibility and Cell Morphology

MC3T3-E1 cells were seeded in 96-well plates with a density of 1×10^4 cells per we and incubated for 1 day to allow the cells to adhere completely, and then replace the cultu medium with the extracts. After incubating for 1, 2 and 3 days, 20 μL CCK-8 assay reager was added to each well and incubated for another 1 h in an incubator. The optical densit (OD) was measured by using a microplate reader (Thermo Scientific, Waltham, MA, USA at the wavelength of 450 nm.

The live/dead cell assay was performed according to the protocol from the manufa turer (BestBio, Shanghai, China). In brief, cells treated with extracts for 3 days were staine with 200 μL (1:10,000) of calcein-AM solution for 30 min and 200 μL (1:5000) of PI solutio for 5 min. Images were collected by fluorescence microscopy (Zeiss, Jenoptik, Germany the viable cells were stained green, while dead cells were stained red.

The morphology of cells co-cultured with extracts and materials were observed b CLSM. Briefly, cells were fixed and permeabilized with 4% paraformaldehyde (PFA and 0.1% Triton X-100 successively, and then dyed the nucleus with 4′,6-diamidino- phenylindole dilactate (DAPI). F-actins were stained with Actin-Tracker Green (Beyotim Biotechnology, Shanghai, China) and rhodamine-phalloidin (Sigma-Aldrich Co., Shangha China). Fluorescence images were captured under the same exposure condition.

SEM was used to observe the cell adhesion and spreading morphology on the surfac of the samples. In brief, samples were immersed in 2.5% glutaraldehyde for 15 mi Next, samples were dehydrated in graded ethanol (10, 30, 50, 70, 90, and 100% ethanc sequentially; 10 min each) and then immersed in graded tertiary butanol (50, 70, 90, 95, 10 and 100% tertiary butanol sequentially, 5 min each). Finally, samples were air-dried an sprayed with gold. SEM was used to observe the morphology of cells.

2.4.3. In Vitro Osteogenic Differentiation

The alkaline phosphatase (ALP) assay was used to analyze the intracellular AL activity. Briefly, cells seeded in 12-well plates were incubated with extracts (containin 100 nM dexamethasone, 10 mM β-glycerophosphate, 50 mM ascorbate and glutamine for 14 days and the extracts were replaced every other day. The cellular ALP expressio was detected by BCIP/NBT ALP Color Development Kit (Beyotime, Shanghai, China) an corresponding quantitative analysis was measured by the Alkaline Phosphatase Assay K (Beyotime, Shanghai, China) according to manufacturer's instructions.

Calcium nodules, an important sign of extracellular matrix mineralization, was qua itatively tested by Alizarin Red S (ARS) (Sigma-Aldrich Co., Shanghai, China) stainin In brief, after 14 days of osteogenesis induction, samples were collected and fixed wit 4% PFA for 30 min and then stained with 1% AR solution for 45 min. Then, the calciu nodules were observed under the microscope. In order to semi-quantitatively analyze th mineralization of extracellular matrix, cetylpyridinium chloride (10%, Sigma-Aldrich Cc Shanghai, China) solution was used to dissolve the mineralized nodules, and then th absorbance value of the dissolving solution was detected at 620 nm by using a micropla reader (Thermo Scientific, Waltham, MA, USA).

The expression of COL-I protein after 14 days osteogenic induction was determine by immunofluorescence staining. Cells were washed gently with PBS three times an then fixed in 4% PFA for 30 min. Then, Triton X-100 (0.1 *v/v*%) and bovine serum albumin (10%) were used to permeabilize and block cells, respectively. Next, incubated the cells overnight with anti-COL-I (Abcam) primary antibody. Afterward, cells were incubated with fluorescent secondary antibody for 1 h. Finally, cells were stained with DAPI for 5 min and observed by CLSM.

The expression of osteogenic genes such as Runx2, OPN, and Bglap were determined by qRT-PCR. Briefly, MC3T3-E1 cells were harvested for RNA extraction using Trizol (Sigma-Aldrich Co., Shanghai, China) method. Nanodrop 2000 (Thermo Scientific, Waltham, MA, USA) was used to detect the purity and concentration of RNA. The First Strand cDNA Kit (Takara, Dalian, China) was employed to reverse transcribe the total mRNA into cDNA. Next, 1 μL of synthesized cDNA was taken from each group and added into a 10 μL reaction system containing SYBR Green Mastermix and primers (Table 3). The expression of Runx2, CON and Bglap was quantified by the ABI 7900HT real-time PCR system (Applied Biosystems, Foster City, California, USA), β-actin was employed as a housekeeping gene. The expression of relative genes was calculated by using the $2^{-\triangle\triangle CT}$ formula.

Table 3. Primer sequences used for qRT-PCR.

Gene	Primers (F = Forward; R = Reverse)
Runx2	F: 5′-ATCCAGCCACCTTCACTTACAAA-3′ R: 5′-GGGACCATTGGGAACTGATAGG-3′
OPN	F: 5′-CCAAGCGTGGAAACACACAGCC-3′ R: 5′-GGCTTTGGAACTCGCCTGACTG-3′
Bglap	F: 5′-GAGCTGCCCTGCACTGGGTG-3′ R: 5′-TGGCCCCAGACCTCTTCCCG-3′
β-actin	F: 5′-ATCGTGGGCCGCCCTAGGCA-3′ R: 5′-TGGCCTTAGGGTTCAGAGGGG-3′

2.5. Statistical Analysis

Statistical analysis was performed using a Student's *t*-test for two groups comparison and one-way ANOVA followed by post-hoc tests for multiple-group comparisons via SPSS 18.0 software. Results that were statistically significant were determined by *p*-values < 0.05.

3. Results and Discussion

3.1. Microstructures and Electrochemical Evaluations

Figure 1A shows the OM, SEM images and corresponding EDS diagram of the Mg-1Zn-1Sn and Mg-1Zn-1Sn-0.2Sr alloys. From the OM images, the grain size was slightly reduced from 26 ± 3 μm to 20 ± 2 μm with the incorporation of Sr, which is consistent with a previous study indicating that Sr has the effect of grain refinement [23–25]. With the addition of Sr, the second phase (bright spot) in the alloy matrix increases, the volume fraction of the second phase increased from 0.5% (Mg-1Zn-1Sn) to 1.1% (Mg-1Zn-1Sn-0.2Sr). To determine the composition of the bright spot, EDS was used to analyze the two marked points (A and B) in the alloys. The results (Figure 1C) showed that the elemental composition of point A included Mg, Zn and Sn, while those of point B included Mg, Zn, Sn and Sr, which was the same as that of the alloy matrix.

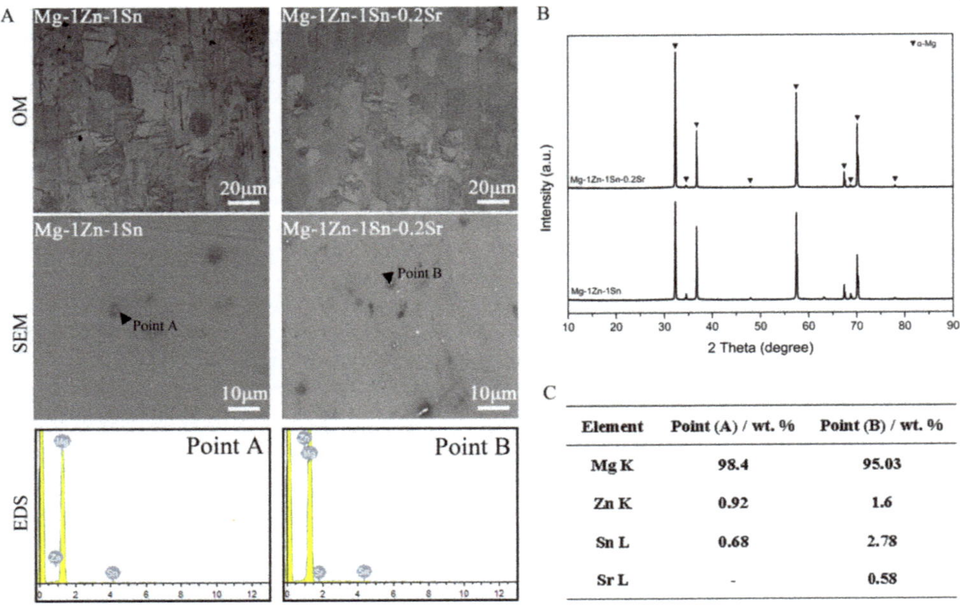

Element	Point (A) / wt. %	Point (B) / wt. %
Mg K	98.4	95.03
Zn K	0.92	1.6
Sn L	0.68	2.78
Sr L	-	0.58

Figure 1. (**A**) Optical microstructure, SEM images, and corresponding EDS results of as-extruded Mg-1Zn-1Sn and Mg-1Zn-1Sn-0.2Sr alloys. The black triangle indicates the second phase. (**B**) XRD patterns of Mg-1Zn-1Sn and Mg-1Zn-1Sn-0.2Sr. (**C**) The chemical composition of point A and point B in Figure 1A.

In order to further analyze the phase composition of the alloy microstructure, t XRD patterns of the Mg-1Zn-1Sn and Mg-1Zn-1Sn-0.2Sr alloys are shown in Figure The microstructure of both alloys is mainly composed of α-Mg; although the second pha can be clearly observed in the SEM images, it is difficult to detect by XRD due to t low content and high solid solubility of Zn and Sn in Mg. Moreover, the content of is as low as 0.2 wt.%, which may even exceed the detection accuracy of XRD. Figure shows the representative dynamic polarization (PDP) curves of p-Mg, Mg-1Zn-1Sn a Mg-1Zn-1Sn-0.2Sr samples, and the values of E_{corr} and I_{corr} are also obtained (Figure 2 Significant differences in I_{corr} values were detected between the mean value for Mg-1Zn-1 (7.36 ± 0.86 μA cm^{-2}) and Mg-1Zn-1Sn-0.2Sr (6.55 ± 0.41 μA cm^{-2}), both of which we all significantly lower than p-Mg (14.05 ± 1.5). Similarly, the E_{corr} of p-Mg was significan greater than Mg-1Zn-1Sn and Mg-1Zn-1Sn-0.2Sr. Electrochemical test results confirm that the corrosion resistance of Mg-1Zn-1Sn and Mg-1Zn-1Sn-0.2Sr was significantly bet than p-Mg. Moreover, the addition of Sr can further reduce the I_{corr} of Mg-1Zn-1Sn alloy enhance corrosion resistance.

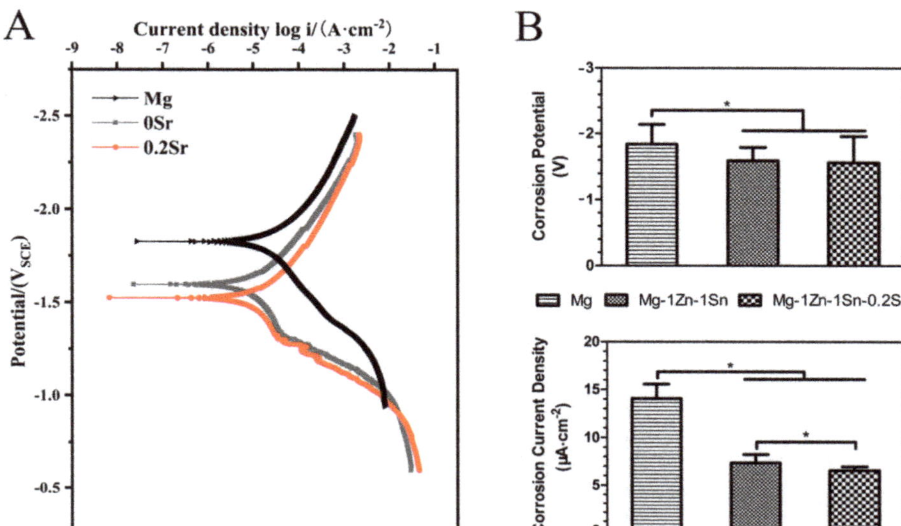

Figure 2. (**A**) Representative PDP curves of as-extruded Mg-1Zn-1Sn and Mg-1Zn-1Sn-0.2Sr alloys, (**B**) corrosion potential (E_{corr}) and corrosion current density (i_{corr}) obtained from Tafel extrapolation of PDP curves; values are mean ± SD, n = 3, * $p < 0.05$.

3.2. Mechanical Properties

The mechanical properties of Mg-1Zn-1Sn and Mg-1Zn-1Sn-0.2Sr alloys are shown in Figure 3. The compressive yield strength, ultimate compressive strength, and compressive strain of p-Mg were 64 ± 6 MPa, 261 ± 18 MPa and 26.4 ± 1.5%, those of Mg-1Zn-1Sn alloy were 88 ± 3 MPa, 404 ± 14 MPa and 20.5 ± 1.1%, and those of Mg-1Zn-1Sn-0.2Sr alloy were 93 ± 5 MPa, 410 ± 12 MPa and 19.9 ± 0.8%, respectively. The tensile yield strength, ultimate tensile strength, and tensile strain of p-Mg were 121 ± 3 MPa, 178 ± 10 MPa and 15.4 ± 1%, those of Mg-1Zn-1Sn alloy were 151 ± 2 MPa, 229 ± 1 MPa and 9.0 ± 0.9%, and those of Mg-1Zn-1Sn-0.2Sr alloy were 168 ± 8 MPa, 245 ± 8 MPa and 9.0 ± 0.4%, respectively. The mechanical strength of Mg-1Zn-1Sn and Mg-1Zn-1Sn-0.2Sr were significantly stronger than that of p-Mg. It also showed that the addition of Sr further improved the mechanical properties of Mg-1Zn-1Sn alloy. We speculate that the reasons for the improved mechanical properties may be due to the second phase strengthening [26,27] or slight texture variations, and the specific mechanism still needs further study.

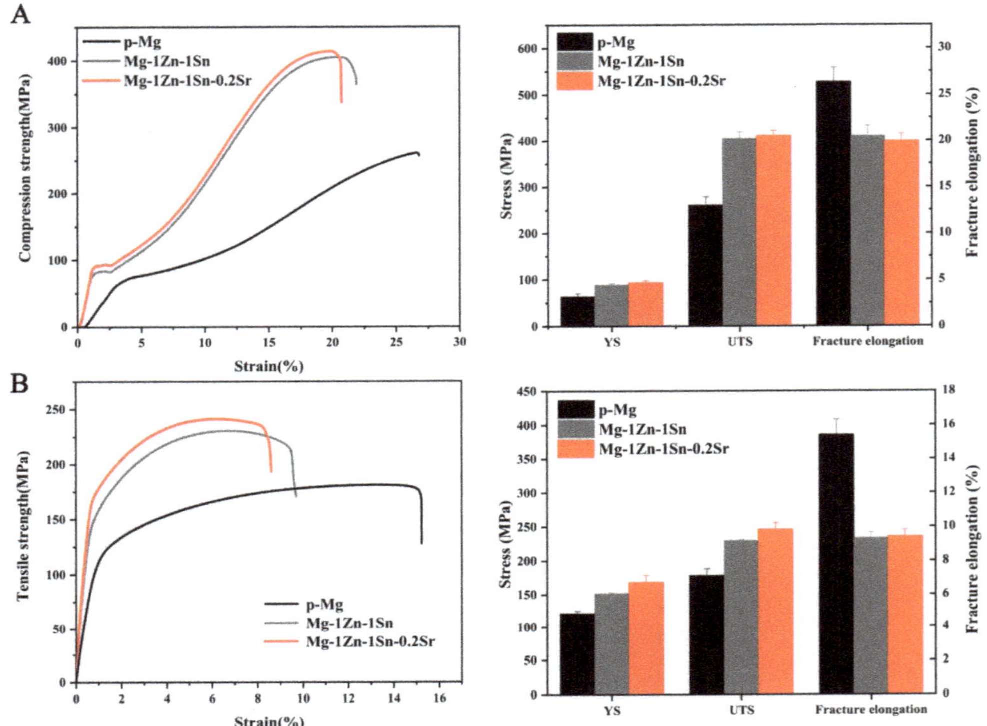

Figure 3. Mechanical properties of the p-Mg, Mg-1Zn-1Sn and Mg-1Zn-1Sn-0.2Sr alloys: (**A**) Compression stress–strain curves and graphs; (**B**) tensile stress–strain curves and graphs.

3.3. Immersion Tests

The results of immersion tests about general observation, H_2 evolution, pH change, mass loss and corrosion rate calculated by mass loss are shown in Figure 4. Obviously, the degradation products were deposited on the surface of all samples as the extension of the immersion time. Specifically, corrosion products deposited on the surface of p-Mg samples were more and uneven, while those of that on the Mg-1Zn-1Sn and Mg-1Zn-1Sn-0.2Sr samples are less and evenly distributed. After removing the corrosion products, there are obvious corrosion holes on p-Mg and Mg-1Zn-1Sn samples, while the Mg-1Zn-1Sn-0.2Sr sample tends to be uniformly corroded without obvious corrosion pits.

Mg degrades in physiological solutions according to the following reaction.

$$Mg_{(s)} + 2H_2O_{(aq)} \rightarrow Mg(OH)_{2(s)}\downarrow + H_{2(g)}\uparrow$$

$$Mg(OH)_{2(s)} + 2Cl^-_{(aq)} \rightarrow MgCl_{2(s)} + 2OH^-$$

During the first 24 h, the volume of H_2 in Mg-1Zn-1Sn and Mg-1Zn-1Sn-0.2Sr samples increased rapidly, and then leveled off. It is obvious that the H_2 release of Mg-1Zn-1Sn was much less than that of p-Mg, and when 0.2 wt.% Sr was added to the alloy, the amount of H_2 evolution was further reduced. The total H_2 released from Mg-1Zn-1Sn and Mg-1Zn-1Sn-0.2Sr samples over 336 h were 0.42 ± 0.02 mL·cm^{-2} and 0.13 ± 0.01 mL·cm^{-2}, respectively, both of which were significantly lower than that of p-Mg samples (1.45 ± 0.41 mL·cm^{-2}). We infer that the main reasons contain the following aspects. First, the Sn addition will decrease the H_2 evolution of Mg alloys substantially for its high H_2 evolution overpotential [16,28,29], which can effectively capture the H atom than the matrix and inhibit the

H_2 evolution rate [30]. Secondly, Sr can further moderate the corrosion rate of Mg alloys by refining the grains and decreasing the corrosion current density, thereby inhibiting H_2 evolution [17,31].

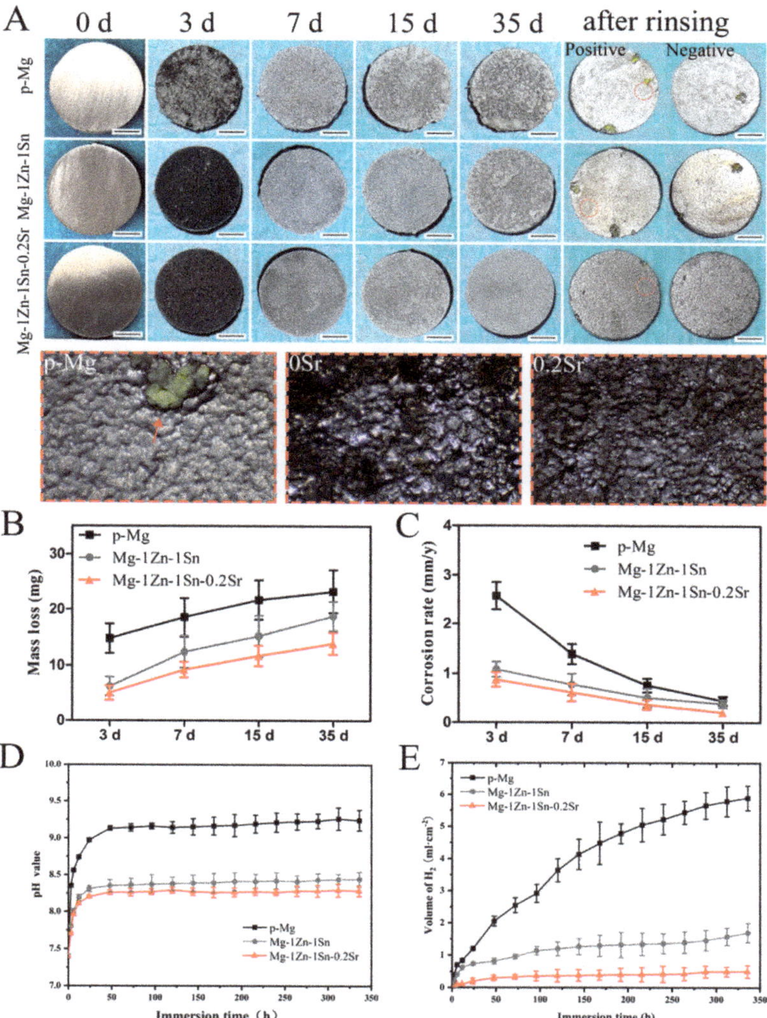

Figure 4. (**A**) Macroscopic images of p-Mg, Mg-1Zn-1Sn and Mg-1Zn-1Sn-0.2Sr at each prescribed time point during 35 days of immersion degradation in SBF. All wafers had a starting dimension of 15 mm in diameter and 1 mm in thickness. The red arrow indicates the typical pitting characteristics of p-Mg. (**B**) Mass loss, (**C**) corrosion rate calculated by mass loss, (**D**) pH value and (**E**) H_2 release of p-Mg, Mg-1Zn-1Sn and Mg-1Zn-1Sn-0.2Sr at various immersion times in SBF. Scale bar = 5.

The pH values of Mg-1Zn-1Sn group are comparable to that of Mg-1Zn-1Sn-0.2Sr group. In the first 24 h, the pH value of Mg-1Zn-1Sn group increased from 7.4 to 8.31, those of Mg-1Zn-1Sn-0.2Sr group increased from 7.4 to 8.25, and then maintained a relatively gentle upward trend throughout the inspection process. However, the pH value of Mg group increased rapidly from 7.4 to 8.9, which indicated that Mg-1Zn-1Sn and Mg-1Zn-1Sn-0.2Sr have better corrosion resistance than p-Mg.

Next, the corrosion rate based on mass loss during the immersion process was c culated, the mass loss of p-Mg, Mg-1Zn-1Sn and Mg-1Zn-1Sn-0.2Sr was 23.31 ± 3.91 m 18.78 ± 2.75 mg and 13.85 ± 1.93 mg after 35 days of immersion, and the corrosion rate w 0.44 ± 0.09, 0.31 ± 0.05 and 0.20 ± 0.03 mm/y, respectively. The average corrosion rate c culated from H_2 evolution, mass loss and I_{corr} of p-Mg, Mg-1Zn-1Sn and Mg-1Zn-1Sn-0.2 alloys in SBF solution was shown in Table 4.

Table 4. The average corrosion rate calculated from H_2 evolution, mass loss and I_{corr} of p-Mg, Mg-1Zn-1Sn and Mg-1Zn-1Sn-0.2Sr alloys in SBF solution.

Materials	Corrosion Rate (mm/y) Calculated by H_2	Corrosion Rate (mm/y) Calculated by Mass Loss	Corrosion Rate (mm/y) Calculated by I_{corr}
Mg	0.72 ± 0.28	0.44 ± 0.09	0.61 ± 0.16
Mg-1Zn-1Sn	0.24 ± 0.09	0.31 ± 0.05	0.29 ± 0.08
Mg-1Zn-1Sn-0.2Sr	0.18 ± 0.02	0.20 ± 0.03	0.23 ± 0.05

Figure 5 shows the surface topographies, elemental compositions and morphologi of a cross-section of p-Mg, Mg-1Zn-1Sn and Mg-1Zn-1Sn-0.2Sr after immersion in SBF f 7 days. The corrosion products on the p-Mg samples had an uneven thickness and were the form of crystal clusters, while those of that on the Mg-1Zn-1Sn and Mg-1Zn-1Sn-0.2 samples were uniform and dense with deposited white clusters/particles. The EDS resu demonstrate that the corrosion products mainly compose of oxygen (O), magnesium (M phosphorous (P) and calcium (Ca) on Mg-1Zn-1Sn and Mg-1Zn-1Sn-0.2Sr samples, wh the corrosion products on p-Mg are mainly carbon (C), oxygen (O) and magnesium (M indicating that Ca and P are more likely to be deposited on the surface of Mg-1Zn-1Sn a Mg-1Zn-1Sn-0.2Sr alloys, which are more conducive to the osteogenic differentiation cells [32]. The formation of a large number of corrosion products was observed from t cross-sectional image of the p-Mg sample, indicating that the surface corrosion was serio while the Mg-1Zn-1Sn sample showed fewer corrosion products, and the cross-section w clear and uniform. Remarkably, the cross-section of Mg-1Zn-1Sn-0.2Sr sample had t shallowest corrosion beneath the surface; no obvious corrosion products were observed

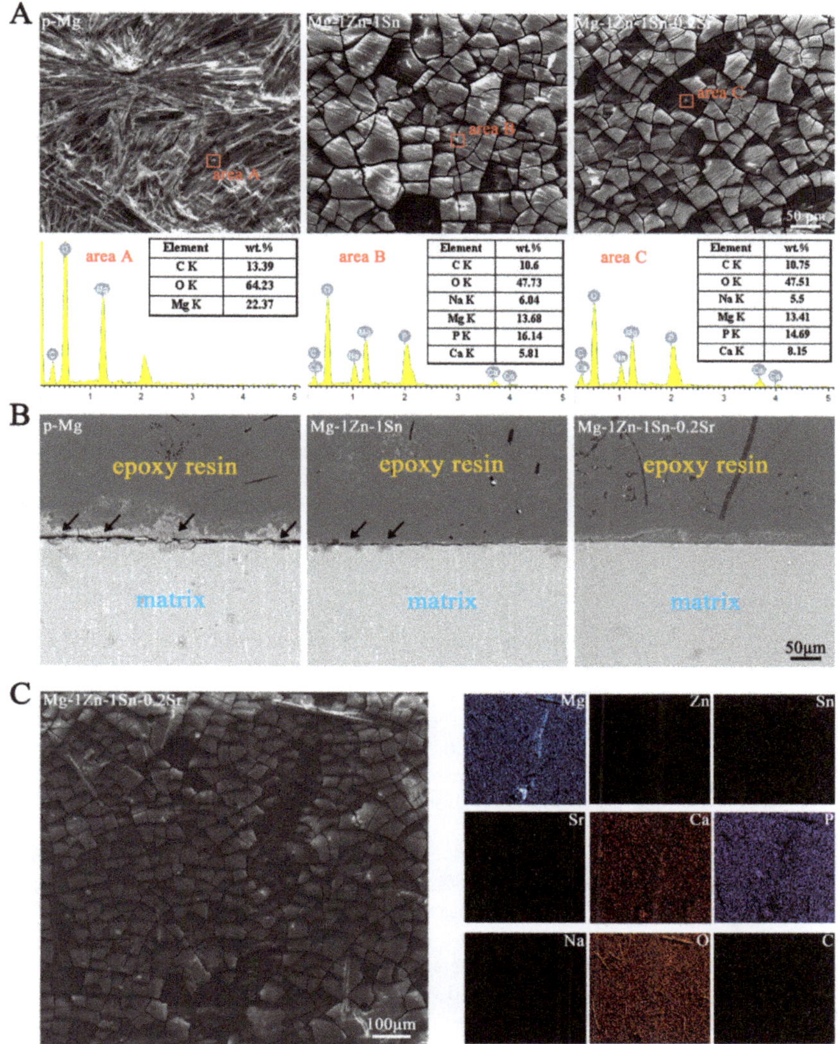

Figure 5. (**A**) Surface topographies and EDS results of p-Mg, Mg-1Zn-1Sn and Mg-1Zn-1Sn-0.2Sr alloys after 7 days of immersion in SBF; scale bar = 50 μm. (**B**) The morphologies of the cross-section of p-Mg, Mg-1Zn-1Sn and Mg-1Zn-1Sn-0.2Sr alloys after 7 days of immersion in SBF; scale bar = 50 μm. The black arrow indicates corrosion products. (**C**) SEM-EDS composite image of the surface of Mg-1Zn-1Sn-0.2Sr sample immersed in SBF at 37 °C for 7 days, scale bar = 100 μm.

3.4. Cell Viability, Cytocompatibility and Cell Morphology

The viability of different extracts towards MC3T3-E1 is shown in Figure 6. The cell viability of Mg-1Zn-1Sn and Mg-1Zn-1Sn-0.2Sr reached the highest on the second day, which were 108 ± 3%, 125 ± 8% and 139 ± 6%, respectively, of the control group. This suggested that Mg-Zn-Sn-based alloys have excellent cell compatibility and can significantly promote cell proliferation.

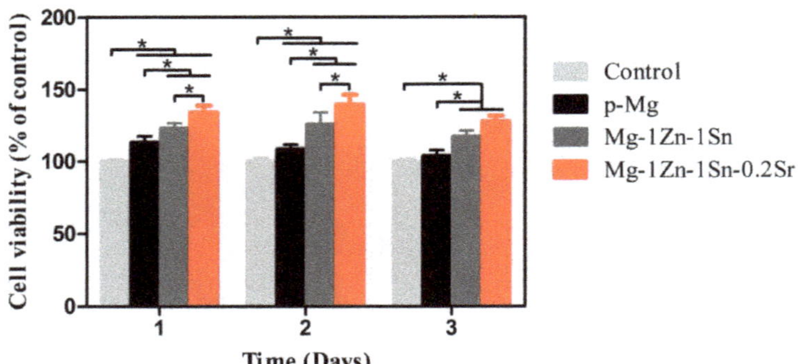

Figure 6. Cell viability of MC3T3-E1 cells cocultured with extracts for 1, 2 and 3 days. Values a mean ± SD, n = 3, * $p < 0.05$.

Figure 7A shows the live/dead staining images of MCET3-E1 co-cultured with eac extract for 3 days. Live cells with green fluorescence almost filled the entire field of vie while dead cells with red fluorescence also existed. The number of dead cells in the p-M group was obviously more than in other groups. However, no significant difference wa detected among the control, Mg-1Zn-1Sn, and Mg-1Zn-1Sn-0.2Sr groups. These result also demonstrated the excellent cytocompatibility of the Mg-1Zn-1Sn and Mg-1Zn-1S 0.2Sr alloys.

Fluorescence images of MC3T3-E1 cultured in different extracts for 3 days are summ rized in Figure 7B. Cells showed satisfactory adhesion state in multiple directions as we as intercellular connections and visibly stained cytoskeleton filaments in Mg-1Zn-1Sn an Mg-1Zn-1Sn-0.2Sr groups. Obviously, the F-actin area of cells adhered to the p-Mg sample was significantly smaller compared with the control, Mg-1Zn-1Sn and Mg-1Zn-1Sn-0.2S groups, which may be attributed to the rapid degradation of p-Mg in the early stage c immersion, resulting in drastic changes in the metal ion concentration, pH value an osmotic pressure, etc. [33].

Fluorescence and SEM images from a direct culture on the surfaces of the p-Mg, M 1Zn-1Sn and Mg-1Zn-1Sn-0.2Sr samples for one day are shown in Figure 7C,D. Attache cells were observed on all samples. The number of MC3T3-E1 on the surface of p-M was less and cannot spread well, which perform round or long fusiform morphologie In contrast, cells on the surface of Mg-1Zn-1Sn and Mg-1Zn-1Sn-0.2Sr samples increase significantly and expanded well, which perform spindle or polygonal morphologies an have more pseudopods.

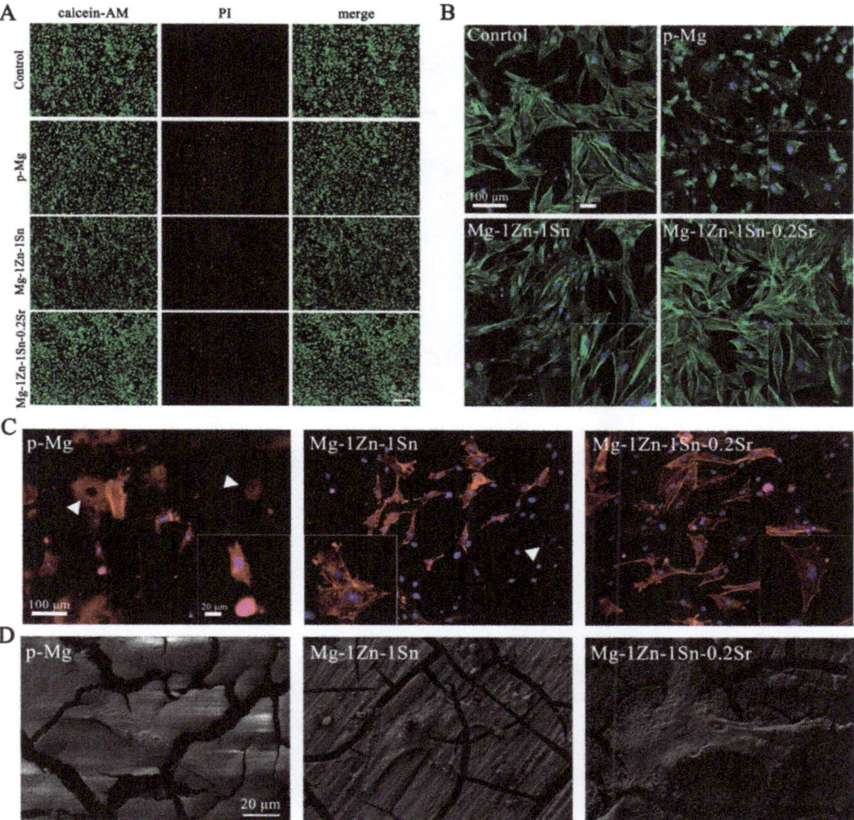

Figure 7. (**A**) Live/dead staining of MC3T3-E1 cells coculture with p-Mg, Mg-1Zn-1Sn and Mg-1Zn-1Sn-0.2Sr alloy extracts and DMEM control for 3 days; scale bar = 200 μm. (**B**) Actin-nucleus co-staining of MC3T3-E1 morphologies coculture with p-Mg, Mg-1Zn-1Sn and Mg-1Zn-1Sn-0.2Sr alloy extracts and DMEM control for 3 days. Scale bar = 100 μm. Insets were taken at 200 original magnification with scale bar = 20 μm. (**C**) Fluorescence images of MC3T3-E1 adhered to the surface of p-Mg, Mg-1Zn-1Sn and Mg-1Zn-1Sn-0.2Sr alloys after coculture for 3 days. Blue indicates nuclei, and red indicates cytoskeleton. White triangle indicates corrosion pit on the sample surface. The scale bar = 100 μm. Insets were taken at 200× original magnification with scale bar = 20 μm. (**D**) Morphologies of MC3T3-E1 cells adhered to the surface of p-Mg, Mg-1Zn-1Sn and Mg-1Zn-1Sn-0.2Sr alloys after coculture for 1 day. Scale bar = 20 μm.

In addition, corrosion holes were found on the surface of p-Mg and Mg-1Zn-1Sn samples—yet were not found on Mg-1Zn-1Sn-0.2Sr sample—indicating that the corrosion of the Mg-1Zn-1Sn-0.2Sr sample was more uniform and slower. Ion concentration in the extracts is also determined by inductively coupled plasma optical emission spectrometry (ICP-OES) (Figure 8). The Mg ions in the Mg-1Zn-1Sn and Mg-1Zn-1Sn-0.2Sr groups were lower than p-Mg group, implying their better corrosion resistance in vitro. Moreover, the solubilized Mg, Zn, Sn and Sr ions in the media were all within the average daily intake range of the human body [34,35], indicating their excellent biosafety.

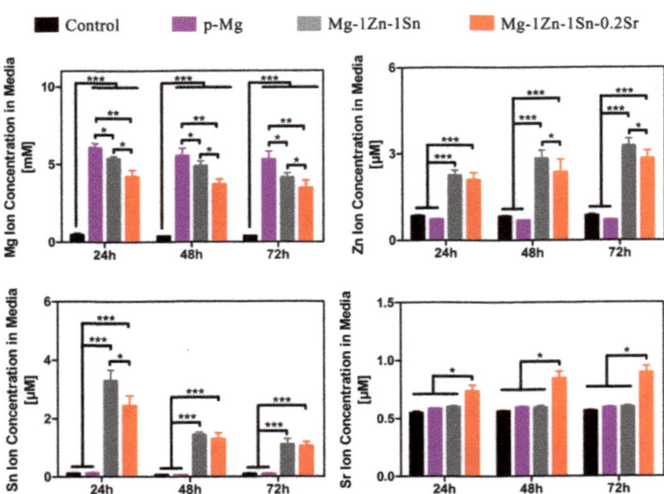

Figure 8. Mg, Zn, Sn and Sr ion concentrations in culture medium incubated with the samples duri a 72 h period. n = 3 for all groups and time points. Values are mean ± SD. * $p < 0.05$, ** $p < 0.$ *** $p < 0.001$.

Cell adhesion to the surface of the material is the most important step and is cruc for subsequent cell proliferation, long-term functions and organization of tissues [36]. the present study, the adhesion of cells may be affected by many factors, such as ion cc centration, pH value, material surface morphology and H_2 evolution, etc. Romani A. et reported that the active transport of Mg^{2+} ions across the cell membrane was strictly cc trolled to keep the intracellular concentration within a normal range, regardless of t extracellular concentration [37]. In addition, the bicarbonate buffer system in the DME medium can effectively reduce the rapid increase in pH caused by the degradation of t material [20]. This indicates that the Mg^{2+} concentration and pH value in the mediu are unlikely to be the main factors affecting cell adhesion. Therefore, we believe th changes in surface morphology and H_2 evolution caused by substrate degradation m play important roles in regulating cell adhesion.

As shown in Figures 4 and 5, the faster the material corroded, the more severe t surface morphology changed, the more the H_2 produced. The CLSM results (Figure 7 indicated that cells on the surface of Mg-1Zn-1Sn and Mg-1Zn-1Sn-0.2Sr samples show more adherence and better adhesion morphology (significantly larger F-actin area ar more pseudopods), which was consistent with the results observed by SEM (Figure 7 The above-mentioned results indicated that Mg-1Zn-1Sn and Mg-1Zn-1Sn-0.2Sr allo were more suitable for cell adhesion and spreading. We think that the reasons may be follows: As an element with high H_2 evolution overpotential, Sn can effectively inhibit t H_2 evolution of Mg alloys [28], thereby minimizing the adverse effect of H_2 release on c adhesion. Moreover, the incorporation of Sr further improves the corrosion resistance ar enhances cell viability and proliferation.

3.5. In Vitro Osteogenesis Ability

Figure 9A,B represents the images of ALP activity and ARS staining and correspondi quantitative analysis results. The ALP activity in p-Mg, Mg-1Zn-1Sn and Mg-1Zn-1S 0.2Sr groups was significantly enhanced. Quantitative analysis also showed that the AI activities of p-Mg, Mg-1Zn-1Sn and Mg-1Zn-1Sn-0.2Sr groups were 1.8, 2.0 and 2.7 tim that of the control group, respectively. Similarly, the ARS staining area of p-Mg, M 1Zn-1Sn and Mg-1Zn-1Sn-0.2Sr groups were significantly improved, and the absorban of the extracellular matrix mineralization was 0.42 ± 0.06, 0.45 ± 0.03 and 0.60 ± 0.

respectively, which were all significantly higher than that of the control group (0.25 ± 0.01). These results indicated that the Mg-1Zn-1Sn-0.2Sr group has the strongest osteoinductive activity. Figure 9C shows the expression of COL-I protein in cells after being cultured in various extracts for 14 days. Obviously, the Mg-1Zn-1Sn-0.2Sr group had the strongest fluorescence intensity among all groups, the order of fluorescence intensity was Mg-1Zn-1Sn-0.2Sr > Mg-1Zn-1Sn > p-Mg > Control. Our experimental results on COL-I expression were consistent with previous studies, which demonstrated that strontium has the ability to enhance the synthesis of COL-I [38,39].

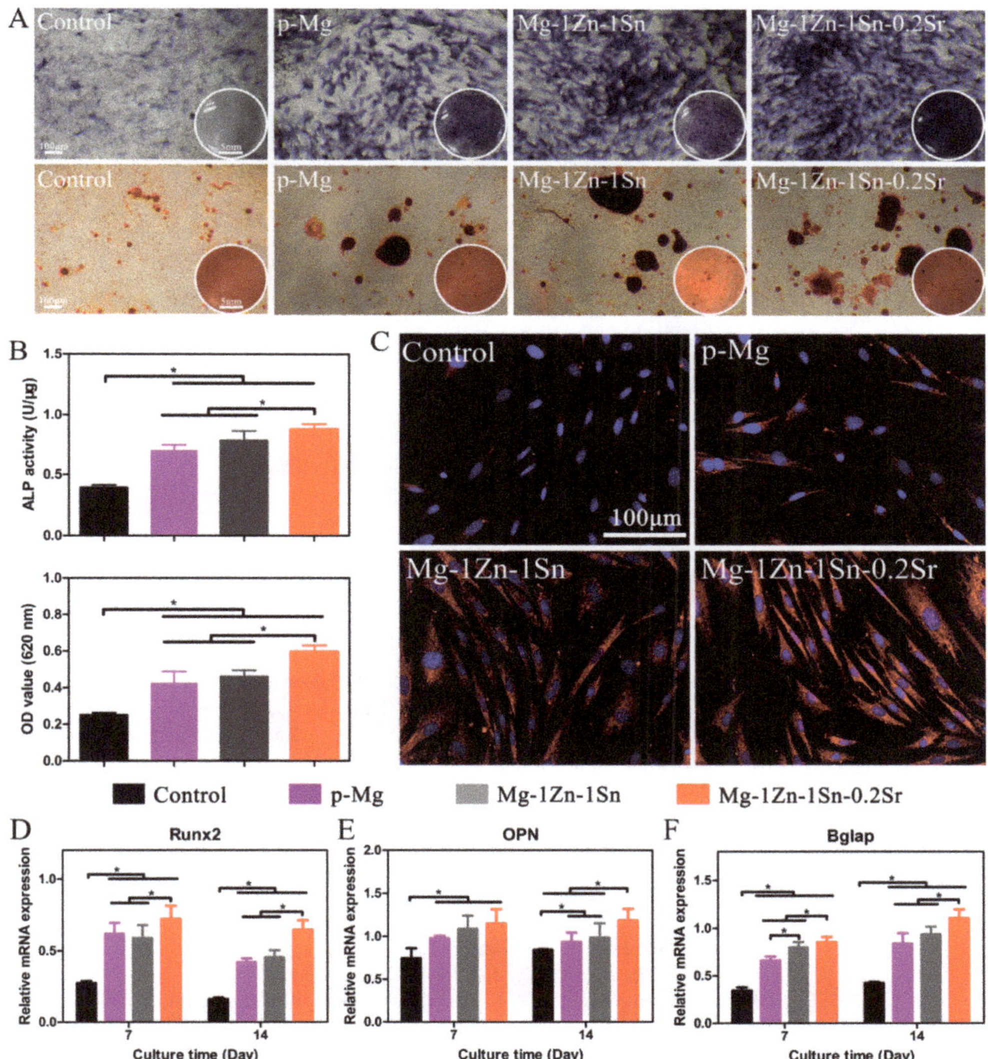

Figure 9. (**A**) Images of ALP activity and matrix mineralization, and (**B**) corresponding quantitative analysis. (**C**) Immunofluorescent staining of the expression of COL-I protein after osteogenic induction for 14 days. Osteogenesis-related genes Runx2 (**D**), OPN (**E**) and Bglap (**F**) expression in MC3T3-E1 cells after osteogenic induction for 7 and 14 days. Values are mean ± SD. * $p < 0.05$.

We further explored the effect of various extracts on osteogenic differentiation MC3T3-E1 cells on the molecular level. Runx2 is identified as a key transcription fact at the early stage of bone development [40]. OPN, a negatively charged non-collagenou bone matrix glycoprotein, is closely related to the formation of the bone matrix. Bglap a vitamin K-dependent calcium-binding protein synthesized and secreted by osteoblas which is associated with the maturation of osteoblasts and matrix mineralization [41]. A shown in Figure 9D, the expression of these three genes in p-Mg, Mg-1Zn-1Sn and M 1Zn-1Sn-0.2Sr groups was significantly higher than the control group. In accordance wit the previous results, the expression of the aforementioned genes related to osteogenes was much stronger in the Mg-1Zn-1Sn-0.2Sr group than the other groups, both at da 7 and day 14. Sr has received widespread attention because of its not only stimulatin osteoblast differentiation but also inhibiting osteoclast differentiation [42]. Specificall Sr has the function of promoting the proliferation of osteoblasts by regulating calciur sensing receptors and phosphorylation of extracellular regulated protein kinases1/2, a well as inhibiting bone resorption by reducing receptor activator of NF-κB ligand (RANK or enhancing the expression of osteoprotegerin [43]. In addition, Sr can also positivel affect the interaction between osteocytes and osteoblasts by regulating the paracrine sign transduction [44].

Recently, Geoffroy et al. [45] have confirmed that Sr is related to the regulation selective osteoinductive genes/their induction products. In this study, though the co centration of Sr is lower than 1.75 μg/mL (the lower limit strontium ranelate stimulate the proliferation and differentiation of osteoblasts) [46,47], the improvement of trace S in osteogenic activity of Mg-1Zn-1Sn-based alloy cannot be ignored. Park et al. [48] als reported that Sr ion concentrations as low as 103–135 ng/mL can still enhance osteogen differentiation. We infer that the co-release of various metal ions such as Mg, Zn, Sn and S during the material degradation may have a synergistic effect, which greatly decreases th limit value, but the specific mechanism needs further study.

4. Conclusions

In this work, the possibility of using Mg-1Zn-1Sn and Mg-1Zn-1Sn-0.2Sr alloys a biomedical materials was explored for the first time; we systematically investigated the microstructure, surface characteristics, mechanical properties, bio-corrosion behavior biocompatibility and biological activity. The major conclusions are as follows:

(1) Mg-1Zn-1Sn and Mg-1Zn-1Sn-0.2Sr alloys have excellent corrosion resistanc (0.31 ± 0.05 and 0.20 ± 0.03 mm/y), low H_2 evolution (0.42 ± 0.02 and 0.13 ± 0.01 mL·cm^{-} and suitable mechanical strength (229 ± 1 and 245 ± 8 MPa) as well as better biocompa ibility compared with p-Mg, showing their significant application potential for using a orthopedic implants.

(2) Mg-1Zn-1Sn and Mg-1Zn-1Sn-0.2Sr alloys have osteoinductive activity comparab to or even significantly better than p-Mg, which may benefit from the contribution of Sr.

(3) By incorporating 0.2 wt.% of Sr into the Mg-1Zn-1Sn-based alloy, the corrosion r sistance, mechanical properties, biocompatibility and biological activity of the material ar all enhanced, demonstrating that Sr is an ideal alloy element for Mg-1Zn-1Sn-based alloy

Author Contributions: Y.W. carried out the molecular lab work, participated in data analysis, parti ipated in the design of the study and drafted the manuscript. Q.L. carried out the statistical analyse and drafted the manuscript. W.Z. carried out the statistical analyses, Q.Y. collected field data. J.W critically revised the manuscript. D.J. conceived of the study, designed the study, coordinated th study and helped draft the manuscript. All authors have read and agreed to the published version c the manuscript.

Funding: This work was funded by [the National Natural Science Foundation of China] grar number [51874062]; and [the Chongqing Technology Innovation and Application Developmer Special Project] grant number [cstc2020jsxc-msxmX0219]. The APC was funded by [the Chongqin Technology Innovation and Application Development Special Project].

Institutional Review Board Statement: Not applicable.

Informed Consent Statement: Not applicable.

Data Availability Statement: The data presented in this study are available on request from the corresponding author.

Conflicts of Interest: The authors declare no conflict of interest.

ferences

Staiger, M.P.; Pietak, A.M.; Huadmai, J.; Dias, G. Magnesium and its alloys as orthopedic biomaterials: A review. *Biomaterials* **2006**, *27*, 1728–1734. [CrossRef]

Atrens, A.; Liu, M.; Abidin, N.I.Z. Corrosion mechanism applicable to biodegradable magnesium implants. *Mater. Sci. Eng. B* **2011**, *176*, 1609–1636. [CrossRef]

Moravej, M.; Mantovani, D. Biodegradable Metals for Cardiovascular Stent Application: Interests and New Opportunities. *Int. J. Mol. Sci.* **2011**, *12*, 4250–4270. [CrossRef]

Chen, Y.; Xu, Z.; Smith, C.; Sankar, J. Recent advances on the development of magnesium alloys for biodegradable implants. *Acta Biomater.* **2014**, *10*, 4561–4573. [CrossRef]

Li, Z.; Gu, X.; Lou, S.; Zheng, Y. The development of binary Mg–Ca alloys for use as biodegradable materials within bone. *Biomaterials* **2008**, *29*, 1329–1344. [CrossRef]

Witte, F.; Fischer, J.; Nellesen, J.; Crostack, H.A.; Kaese, V.; Pisch, A.; Beckmann, F.; Windhagen, H. In vitro and in vivo corro-sion measurements of magnesium alloys. *Biomaterials* **2006**, *27*, 1013–1018. [CrossRef]

Song, G. Control of biodegradation of biocompatable magnesium alloys. *Corros. Sci.* **2007**, *49*, 1696–1701. [CrossRef]

Ding, W. Opportunities and challenges for the biodegradable magnesium alloys as next-generation biomaterials. *Regen. Biomater.* **2016**, *3*, 79–86. [CrossRef]

Munir, K.; Lin, J.; Wen, C.; Wright, P.F.A.; Li, Y. Mechanical, corrosion, and biocompatibility properties of Mg-Zr-Sr-Sc alloys for biodegradable implant applications. *Acta Biomater.* **2020**, *102*, 493–507. [CrossRef]

Gu, X.; Zheng, Y.; Cheng, Y.; Zhong, S.; Xi, T. In vitro corrosion and biocompatibility of binary magnesium alloys. *Biomaterials* **2009**, *30*, 484–498. [CrossRef]

Salahshoor, M.; Guo, Y. Biodegradable Orthopedic Magnesium-Calcium (MgCa) Alloys, Processing, and Corrosion Performance. *Materials* **2012**, *5*, 135–155. [CrossRef]

Pande, G.; Ragamouni, S.; Li, Y.; Kumar, M.J.; Harishankar, N.; Hodgson, P.D.; Wen, C.; Mushahary, D.; Sravanthi, R. Zirconium, calcium, and strontium contents in magnesium based biodegradable alloys modulate the efficiency of implant-induced osseointegration. *Int. J. Nanomed.* **2013**, *8*, 2887–2902. [CrossRef]

Bornapour, M.; Muja, N.; Shum-Tim, D.; Cerruti, M.; Pekguleryuz, M. Biocompatibility and biodegradability of Mg–Sr alloys: The formation of Sr-substituted hydroxyapatite. *Acta Biomater.* **2013**, *9*, 5319–5330. [CrossRef]

Kubasek, J.; Vojtech, D.; Lipov, J.; Ruml, T. Structure, mechanical properties, corrosion behavior and cytotoxicity of bio-degradable Mg–X (X=Sn, Ga, In) alloys. *Mater. Sci. Eng. C Mater. Biol. Appl.* **2013**, *33*, 2421–2432. [CrossRef]

Liu, X.; Shan, D.; Song, Y.; Chen, R.; Han, E. Influences of the quantity ofMg2Sn phase on the corrosion behavior of Mg–7Sn magnesium alloy. *Electrochimica Acta.* **2011**, *56*, 2582–2590. [CrossRef]

Marie, P.J.; Felsenberg, D.; Brandi, M.L. How strontium ranelate, via opposite effects on bone resorption and formation, prevents osteoporosis. *Osteoporos. Int.* **2011**, *22*, 1659–1667. [CrossRef] [PubMed]

Cipriano, A.F.; Lin, J.; Lin, A.; Sallee, A.; Celene, C.A.M.; Alcaraz, M.C.C.; Guan, R.-G.; Botimer, G.; Inceoğlu, S.; Liu, H. Degradation of Bioresorbable Mg–4Zn–1Sr Intramedullary Pins and Associated Biological Responses in Vitro and in Vivo. *ACS Appl. Mater. Interfaces* **2017**, *9*, 44332–44355. [CrossRef]

ASTM International. *A Standard, E9-09, Standard Test Methods of Compression Testing of Metallic Materials at Room Temperature*; ASTM International: West Conshohocken, PA, USA, 2009.

ASTM International. *A Standard, ASTM E8/E8M-16a, Standard Test Methods for Tension Testing of Metallic Materials*; ASTM International: West Conshohocken, PA, USA, 2016.

Cipriano, A.F.; Sallee, A.; Guan, R.-G.; Zhao, Z.-Y.; Tayoba, M.; Sanchez, J.; Liu, H. Investigation of magnesium–zinc–calcium alloys and bone marrow derived mesenchymal stem cell response in direct culture. *Acta Biomater.* **2015**, *12*, 298–321. [CrossRef]

ASTM International. *A Standard, G102−89 (Reapproved 2015) e1 Standard Practice for Calculation of Corrosion Rates and Related Information from Electrochemical Measurements*; ASTM International: West Conshohocken, PA, USA, 2015.

Wang, J.; Witte, F.; Xi, T.; Zheng, Y.; Yang, K.; Yang, Y.; Zhao, D.; Meng, J.; Li, Y.; Li, W.; et al. Recommendation for modifying current cytotoxicity testing standards for biodegradable magnesium-based materials. *Acta Biomater.* **2015**, *21*, 237–249. [CrossRef]

Lai, H.; Li, J.; Li, J.; Zhang, Y.; Xu, Y. Effects of Sr on the microstructure, mechanical properties and corrosion behavior of Mg-2Zn-xSr alloys. *J. Mater. Sci. Mater. Med.* **2018**, *29*, 87. [CrossRef]

Guan, R.G.; Cipriano, A.F.; Zhao, Z.Y.; Lock, J.; Tie, D.; Zhao, T.; Cui, T.; Liu, H. Development and evaluation of a magnesium–zinc–strontium alloy for biomedical applications—Alloy processing, microstructure, mechanical properties, and biodegradation. *Mater. Sci. Eng. C Mater. Biol. Appl.* **2013**, *33*, 3661–3669. [CrossRef]

25. Ding, Y.; Li, Y.; Wen, C. Effects of Mg17Sr2Phase on the Bio-Corrosion Behavior of Mg-Zr-Sr Alloys. *Adv. Eng. Mater.* **2015**, 259–268. [CrossRef]
26. Li, H.; Peng, Q.; Li, X.; Li, K.; Han, Z.; Fang, D. Microstructures, mechanical and cytocompatibility of degradable Mg–Zn bas orthopedic biomaterials. *Mater. Des.* **2014**, *58*, 43–51. [CrossRef]
27. Chu, P.-W.; Le Mire, E.; Marquis, E.A. Microstructure of localized corrosion front on Mg alloys and the relationship with hydrog evolution. *Corros. Sci.* **2017**, *128*, 253–264. [CrossRef]
28. Jiang, W.; Wang, J.; Liu, Q.; Zhao, W.; Jiang, D.; Guo, S. Low hydrogen release behavior and antibacterial property of Mg-4Zn-x alloys. *Mater. Lett.* **2019**, *241*, 88–91. [CrossRef]
29. Zhao, C.; Pan, F.; Zhao, S.; Pan, H.; Song, K.; Tang, A. Microstructure, corrosion behavior and cytotoxicity of biodegradal Mg–Sn implant alloys prepared by sub-rapid solidification. *Mater. Sci. Eng. C* **2015**, *54*, 245–251. [CrossRef] [PubMed]
30. Ha, H.-Y.; Kang, J.-Y.; Yang, J.; Yim, C.D.; You, B.S. Role of Sn in corrosion and passive behavior of extruded Mg-5 wt%Sn al *Corros. Sci.* **2016**, *102*, 355–362. [CrossRef]
31. Cheng, M.; Chen, J.; Yan, H.; Su, B.; Yu, Z.; Xia, W.; Gong, X. Effects of minor Sr addition on microstructure, mechanical a bio-corrosion properties of the Mg-5Zn based alloy system. *J. Alloy. Compd.* **2017**, *691*, 95–102. [CrossRef]
32. Xu, Y.; Li, H.; Wu, J.; Yang, Q.; Jiang, D.; Qiao, B. Polydopamine-induced hydroxyapatite coating facilitates hydroxy atite/polyamide 66 implant osteogenesis: An in vitro and in vivo evaluation. *Int. J. Nanomed.* **2018**, *13*, 8179–8193. [CrossRef]
33. Cipriano, A.F.; Sallee, A.; Tayoba, M.; Alcaraz, M.C.C.; Lin, A.; Guan, R.-G.; Zhao, Z.-Y.; Liu, H. Cytocompatibility and ea inflammatory response of human endothelial cells in direct culture with Mg-Zn-Sr alloys. *Acta Biomater.* **2017**, *48*, 499–5 [CrossRef]
34. Bian, D.; Zhou, W.; Deng, J.; Liu, Y.; Li, W.; Chu, X.; Xiu, P.; Cai, H.; Kou, Y.; Jiang, B.; et al. Development of magnesium-bas biodegradable metals with dietary trace element germanium as orthopaedic implant applications. *Acta Biomater.* **2017**, *64*, 421–4 [CrossRef] [PubMed]
35. Cipriano, A.F.; Zhao, T.; Johnson, I.; Guan, R.-G.; Garcia, S.; Liu, H. In vitro degradation of four magnesium–zinc–strontiu alloys and their cytocompatibility with human embryonic stem cells. *J. Mater. Sci. Mater. Med.* **2013**, *24*, 989–1003. [CrossR [PubMed]
36. Györgyey, Á.; Ungvári, K.; Kecskeméti, G.; Kopniczky, J.; Hopp, B.; Oszkó, A.; Pelsöczi, I.; Rakonczay, Z.; Nagy, K.; Turzó, Attachment and proliferation of human osteoblast-like cells (MG-63) on laser-ablated titanium implant material. *Mater. Sci. E C* **2013**, *33*, 4251–4259. [CrossRef] [PubMed]
37. Romani, A. Regulation of magnesium homeostasis and transport in mammalian cells. *Arch. Biochem. Biophys.* **2007**, *458*, 90–1 [CrossRef]
38. Marie, P. Strontium ranelate: A physiological approach for optimizing bone formation and resorption. *Bone* **2006**, *38*, 10– [CrossRef]
39. Capuccini, C.; Torricelli, P.; Sima, F.; Boanini, E.; Ristoscu, C.; Bracci, B.; Socol, G.; Fini, M.; Mihailescu, I.N.; Bigi, A. Stronti-u substituted hydroxyapatite coatings synthesized by pulsed-laser deposition: In vitro osteoblast and osteoclast response. *A Biomater.* **2008**, *4*, 1885–1893. [CrossRef]
40. Mao, L.; Liu, J.; Zhao, J.; Chang, J.; Xia, L.; Jiang, L.; Wang, X.; Lin, K.; Fang, B. Effect of micro-nano-hybrid structured hydr yapatite bioceramics on osteogenic and cementogenic differentiation of human periodontal ligament stem cell via Wnt sig-nali pathway. *Int. J. Nanomed.* **2015**, *10*, 7031–7044. [CrossRef]
41. Lin, K.; Xia, L.; Gan, J.; Zhang, Z.; Chen, H.; Jiang, X.; Chang, J. Tailoring the nanostructured surfaces of hydroxyapati bi-oceramics to promote protein adsorption, osteoblast growth, and osteogenic differentiation. *ACS Appl. Mater. Interfaces* **2013**, 8008–8017. [CrossRef]
42. Marie, P.J. Strontium ranelate: New insights into its dual mode of action. *Bone* **2007**, *40*, S5–S8. [CrossRef]
43. Pilmane, M.; Salma-Ancane, K.; Loca, D.; Locs, J.; Berzina-Cimdina, L. Strontium and strontium ranelate: Historical review some of their functions. *Mater. Sci. Eng. C* **2017**, *78*, 1222–1230. [CrossRef]
44. Bakker, A.D.; Zandieh-Doulabi, B.; Klein-Nulend, J. Strontium Ranelate affects signaling from mechanically-stimulated osteocy towards osteoclasts and osteoblasts. *Bone* **2013**, *53*, 112–119. [CrossRef] [PubMed]
45. Geoffroy, V.; Chappard, D.; Marty, C.; Libouban, H.; Ostertag, A.; Lalande, A.; de Vernejoul, M.C. Strontium ranelate de-creas the incidence of new caudal vertebral fractures in a growing mouse model with spontaneous fractures by improving bo microarchitecture. *Osteoporos Int.* **2011**, *22*, 289–297. [CrossRef]
46. Grynpas, M.; Hamilton, E.; Cheung, R.; Tsouderos, Y.; Deloffre, P.; Hott, M.; Marie, P. Strontium increases vertebral bone volur in rats at a low dose that does not induce detectable mineralization defect. *Bone* **1996**, *18*, 253–259. [CrossRef]
47. Grynpas, M.; Marie, P. Effects of low doses of strontium on bone quality and quantity in rats. *Bone* **1990**, *11*, 313–319. [CrossR
48. Park, J.W.; Kim, H.K.; Kim, Y.J.; Jang, J.H.; Song, H.; Hanawa, T. Osteoblast response and osseointegration of a Ti-6Al-4V al-l implant incorporating strontium. *Acta Biomater.* **2010**, *6*, 2843–2851. [CrossRef]

Review

Is There a Role for Absorbable Metals in Surgery? A Systematic Review and Meta-Analysis of Mg/Mg Alloy Based Implants

Cortino Sukotjo [1,*], Tiburtino J. Lima-Neto [2], Joel Fereira Santiago Júnior [3], Leonardo P. Faverani [4] and Michael Miloro [5]

1 Department of Restorative Dentistry, College of Dentistry, University of Illinois at Chicago, Chicago, IL 60612, USA
2 Oral and Maxillofacial Surgery, Department of Diagnosis and Surgery, Division of Oral and Maxillofacial Surgery, School of Dentistry, São Paulo State University—Unesp, Araçatuba, São Paulo 16015-050, Brazil; tiburtinoneto@hotmail.com
3 Department of Health Sciences, Centro Universitário Sagrado Coração-UNISAGRADO, Bauru, São Paulo 16011-160, Brazil; jf.santiagojunior@gmail.com
4 Department of Diagnosis and Surgery, Division of Oral and Maxillofacial Surgery and Implantology, School of Dentistry, São Paulo State University—Unesp, Araçatuba, São Paulo 16015-050, Brazil; leonardo.faverani@unesp.br
5 Department of Oral and Maxillofacial Surgery, College of Dentistry, University of Illinois at Chicago, Chicago, IL 60612, USA; mmiloro@uic.edu
* Correspondence: csukotjo@uic.edu; Tel.: +1-617-272-5512

Received: 21 July 2020; Accepted: 1 September 2020; Published: 4 September 2020

Abstract: Magnesium (Mg) alloys have received attention in the literature as potential biomaterials for use as absorbable implants in oral and maxillofacial and orthopedic surgery applications. This study aimed to evaluate the available clinical studies related to patients who underwent bone fixation (patients), and received conventional fixation (intervention), in comparison to absorbable metals (comparison), in terms of follow-up and complications (outcomes). A systematic review and meta-analysis were performed in accordance with the PRISMA statement and PROSPERO (CRD42020188654), PICO question, ROBINS-I, and ROB scales. The relative risk (RR) of complications and failures were calculated considering a confidence interval (CI) of 95%. Eight studies (three randomized clinical trial (RCT), one retrospective studies, two case-control studies, and two prospective studies) involving 468 patients, including 230 Mg screws and 213 Titanium (Ti) screws, were analyzed. The meta-analysis did not show any significant differences when comparing the use of Mg and Ti screws for complications ($p = 0.868$). The estimated complication rate was 13.3% (95% CI: 8.3% to 20.6%) for the comparison group who received an absorbable Mg screw. The use of absorbable metals is feasible for clinical applications in bone surgery with equivalent outcomes to standard metal fixation devices.

Keywords: bone surgery; absorbable implants; magnesium (Mg); oral and maxillofacial; orthopedic; titanium (Ti)

---

1. Introduction

One of the most significant public health concerns is the high incidence of traumatic accidents resulting in skeletal injuries with the need for bone reduction and fixation [1–5]. These traumatic events significantly affect the quality of life of these accident victims. Elderly patients and those suffering from chronic systemic conditions, such as diabetes, osteoporosis, and other bone metabolic disorders, have an increased potential for poor outcomes and worse complications from the management of these injuries. In orthopedic and oral and maxillofacial surgical specialties, plates and screws are used to stabilize the fractured bone fragments [4,6–9], and the most commonly used material is commercially pure Ti (CPTi) and its alloys, especially Ti grade 5 (Ti-6Al-4V). Other metals have also been used, for example, cobalt-chromium-molybdenum alloys, as well as stainless steel. These are considered biocompatible materials and possess the mechanical resistance necessary to prevent bony segment mobility and allow for primary bone healing and revascularization between the bone fracture segments. However, investigators may have misunderstood the concept of inertia of the metals implanted into the human body over the past few decades, since some metals used for bone fixation purposes are nobler than others. However, after implantation, these metals are subjected regularly to mechanical, electrochemical, and temperature alterations, which have resulted in complications, such as infection, metal hypersensitivity, and foreign body reactions [10–16].

Furthermore, metal plates and screws used for permanent implantation for bone fixation may have issues when placed in growing children with disturbances in normal growth patterns. Therefore, to resolve these problems, absorbable materials were developed, with commercially-available polymers and co-polymer materials(polyglycolic and polylactic acids) manufactured into bone plates and screws. After fracture repair and completion of the bone healing process (at approximately six months), the resorbable fixation devices begin to degrade into carbon dioxide and water; therefore, a second stage surgery is not needed to remove the plates and screws after healing is complete. However, many questions still remain regarding the use of resorbable fixation devices since they possess lower mechanical resistance than conventional metal devices. In addition, there is difficulty in bending (molding) the resorbable plates during surgery using heated water or ultrasonic methods, the absence of radiopaque implants on post-surgical radiologic evaluation, and unpredictable tissue responses with possible bone resorption due to the process of acidic degradation of the co-polymer materials, such as poly-l-lactic acid [17–20].

The concept of absorbable metals has been developed recently to reduce these possible complications [21–23]. Mg and zinc-based degradable metal alloys were recently developed since these metals possess desirable characteristics such as adequate strength (tensile, bending, and torsional) for bone fracture fixation. The final degradation product is not acidic as with poly-l-lactic acid materials(PLLA). In vitro and in vivo investigations have been performed and have led to improvements in the biocompatibility, bone healing properties, and corrosion resistance of the absorbable metals [24–26]. The Zn-based alloys have been investigated, primarily due to their excellent electrochemical process [21,27], which does not result in the accumulation of gas cavities such as hydrogen [28]. Both metals have shown excellent biocompatibility during the degradation process. They were both safely metabolized, including simulation for osteoblastogenesis in bone surgeries [28,29]. However, pure Zn does not have enough mechanical properties for osteosynthesis materials for use in bone fracture fixation. The primary element used to increase its strength is copper (Cu), which results in a suitable alloy for use (ZnCu) [18,28]. However, it is no longer an absorbable metal [30]. Further, Zn-based alloys have only been used for cardiovascular stents thus far [30,31]. Therefore, for this review, we focus on the clinical outcomes of the Mg and Mg-based alloy implants only.

Regarding the clinical applications, recently, the literature has shown exemplary behavior of Mg alloys used for bone fracture fixation. In Germany, the first report was published using screws from Mg alloys to fixate hallux fractures [11,32,33]. Any decision-making related to a clinical situation should be performed after obtaining an acceptable level of scientific evidence. Based on this principle, this study aims to answer the research question: "Is there evidence to support the clinical application of absorbable metals for bone fixation, particularly Mg/Mg alloy based implants?" This systematic review aims to evaluate clinical studies related to patients who underwent bone fixation (patients). Clinical data on patients who received conventional non-resorbable metal plates or screws (intervention, e.g., Ti) will be compared to data on absorbable plates and screws (comparison). Biological outcomes and clinical follow-up, as well as complications (outcomes) will also be investigated [10,34]. Therefore, to understand the clinical behavior of these materials, this systematic review and meta-analysis will gather information from clinical studies regarding these issues. Additionally, in vivo studies are included to demonstrate the biological responses of these materials.

2. Materials and Methods

2.1. Standard Criteria and Type of Study

This systematic review followed the Cochrane criteria [35,36], and the PRISMA-P and PRISMA Statement [37,38] on systematic review and meta-analysis.

2.1.1. Protocol and Registration

The researchers registered this systematic review in the PROSPERO database, under submission: CRD42020188654 entitle "Is there viability in the use of absorbable metals in bone surgery? Systematic Review and Meta-analysis". The authors followed the PRISMA-P protocol for planning a systematic review [37].

2.1.2. Eligibility Criteria

The researchers performed the analyses based on the PICO index:

1. Population: Patients undergoing surgical treatment of bone fractures or deformities with fixation devices.
2. Intervention: Fixation using conventional metal plates and/or screws, such as Ti alloys.
3. Comparison: Fixation using absorbable metal plates and/or screws (Mg).
4. Outcome: Survival rates of the fixation systems, systemic complications, pain scale, quality of life, and functional analysis.

2.2. Inclusion/Exclusion Criteria

2.2.1. Inclusion Criteria

The studies were selected according to the search strategy with the following inclusion criteria: (1) English language; (2) clinical follow-up studies of at least 6 months including the following study types: retrospective, prospective, and controlled and randomized clinical trial (RCTs); (3) publication period analysis until 20 June 2020; (4) adults and children with no upper or lower age limit; (5) consecutive cases including over 5 patients.

2.2.2. Exclusion Criteria

The exclusion criteria included: studies related to in vitro methodology; animal studies; studies with less than 5 patients or with incomplete data; studies only related to the absorbable plates and/or screws from PLLA; review papers; studies that did not allow the collection of the required information.

2.3. Study Search Strategy

The databases used were: Medline/PubMed; Cochrane Library; EMBASE. These searches were carried out for articles published until 20 June 2020. Additional contact was made with the authors when it was not possible to locate the article via the national online system or COMUT.

2.4. Searches

The keywords based on MeSH/PubMed were: Surgery, Bone Plates, Absorbable implants. The articles were selected on the following bases (Cochrane, 46 articles; Embase, 517 articles; Pubmed, 556 articles; total = 1119 articles).

A manual search was also carried out in the specific journals in the area: Biomedical Engineering, Foot Ankle Surgurgery, Musculoskelet Disorders, Journal of Orthopaedic Science, Journal of Orthopaedic Research, Biomaterials and Journal of Oral and Maxillofacial Surgery, totalizing 9 articles and open grey.

2.5. Data Collection Process

This research was carried out by methods used by previously calibrated researchers. The selection of articles and data collection was performed by a calibrated reviewers (T.J.L.-N.and L.P.F.). All titles and abstracts evaluated as eligible were separated and analyzed thoroughly to assess the titles and abstracts found, to obtain a concordance thesis value for the articles selected in both databases, and to reduce the possibility of bias in selecting articles. A meeting was required to reach consensus, and discrepancies were discussed and resolved by the third reviewer (J.F.S.Jr.).

2.6. Items to Be Extracted

The extracted data from each study were analyzed in an orderly manner, and the needed information was obtained in a standardized fashion. The following data were collected from the articles: authors, type of study, number of patients, age, sex, operated region of the body, type of screw, number of screws, surgery time, follow up, radiologic measures, functional recovery, laboratory values (mertal ion release), and complications. All data were collected by one reviewer (T.J.L.-N.) was then verified by another reviewer (L.P.F.). The data collection was entered in Excel spreadsheet (Excel, Microsoft, Washington, DC, USA).

2.7. Assessment of Study Quality and Risk of Bias

The ROBINS-I scale was applied for Non-Randomized studies of the effects of interventions [39]. This scale was developed by members of the Cochrane Bias Methods Group and the Cochrane Non-Randomised Studies Methods Group. For RCTs studies, the risk of bias in randomized trials was applied [35,40,41]. The online Robvis website (https://mcguinlu.shinyapps.io/robvis/) (accessed: 29 June 2020), was used to prepare responses for the seven areas presented in ROBINS-I and for the 5 domains presented on the ROB scale [40].

2.8. Types of Outcomes

2.8.1. Primary Outcome

Evaluation of the clinical complications and failure rates of absorbable metal (Mg) plates/screws compared to Ti plates/screws used for bone surgeries

2.8.2. Secondary Outcome

Analyses of radiologic measures, functional recovery, and laboratory results of the metal ions released.

2.8.3. Additional Analysis

Sensitivity tests for subgroup analysis were performed in order to avoid the potential for heterogeneity considering, for example, possible differences in the different bone regions rehabilitated [42,43].

2.9. Meta-Analysis

Summary Measures

Quantitative data were grouped for some variables: the number of complications in patients who received absorbable Mg screws compared to the control group (Ti screws or bone grafting), the prevalence of severe complications, and failure in the Mg screw group was also calculated. This grouped information was evaluated for the event rate considering 95% CI. The number of treated patients who received surgical treatment was considered for data analysis (dichotomous data), which was used as a risk ratio (RR) [44,45].

A p value < 0.05 was considered significance. For event rate analyzes, the total number of patients who received absorbable Mg screws, and the total number of complications and failures were considered. The contribution weight of each study was also assessed. The Comprehensive Meta-Analysis software (Software version 3.0—Biostat, Englewood, NJ, USA) was used to construct the Forest plot [46].

3. Results

3.1. Qualitative Analysis

In the initial search, 1199 articles were found according to the flowchart represented in Figure 1. After analysis of the inclusion criteria, eight articles were eligible, which 468 patients were included (control group or experimental group), with 230 screws of Mg. The age of patients included in all studies was 48.14 years [7,10,11,21,22,33,34,47]. The main results are summarized in Table 1.

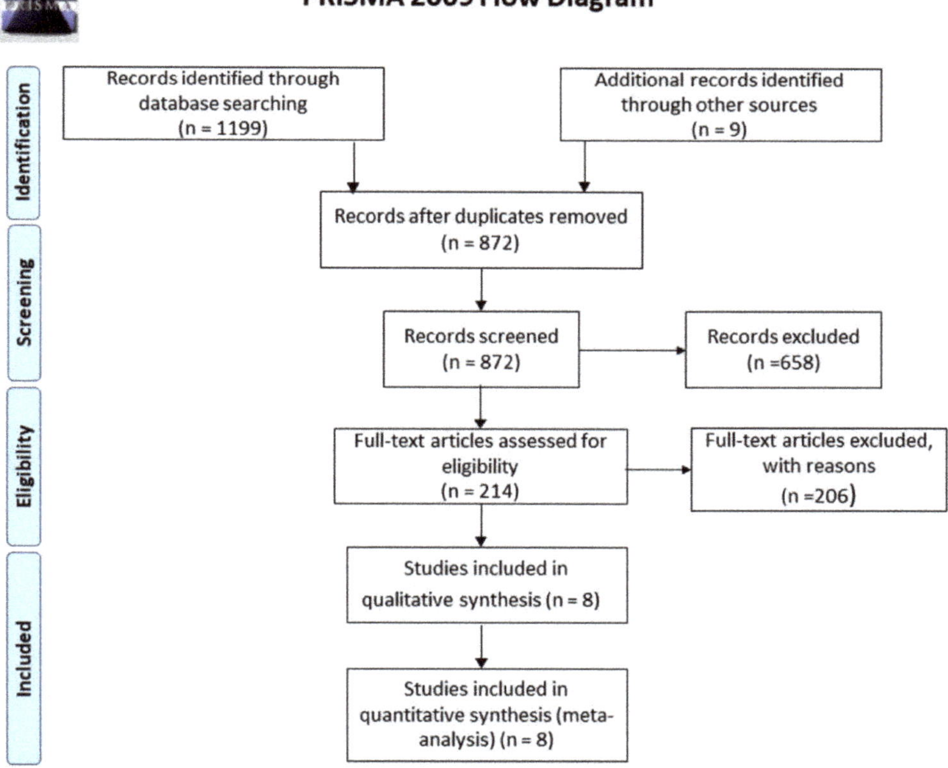

Figure 1. Flowchart of the studies selected for the systematic review.

Table 1. Quantitative data from selected studies.

Article	Type of Study	Number of Patients	Age (Mean in Years)	Sex (M or F)	Operated Region	Type of Screw	Number of Screws	Surgery Time (Min)	Follow Up (Months)
Windhagen et al.	Randomized Clinical Trial	26 (13 Mg; 13 Ti)	57.2 ± 7.2 Mg; 49.9 ± 16.5 Ti	11 m. 2 f. (Mg); 13 m. 0 f. (Ti)	Hallux	Mg; Ti	26 (13 (Mg); 13 (Ti))	40 ± 9.1 (Mg); 34 ± 3.3 (Ti)	6
Klauser	Retrospective	200 (100 Mg; 100 Ti)	52.34 (Mg); 50.87 (Ti)	NR	Hallux	Mg; Ti	200 (100 (Mg); 100 (Ti))	60.6 (Mg); 55.6 (Ti)	12.2 Mg; 11.7 Ti
Atkinson et al.	Case control study	36 (11 Mg; 25 Ti)	38 (Mg); 41 (Ti)	2 M., 9 F. (Mg); 2 M., 23 F. (Ti)	Hallux	Mg; Ti	36 (11 (Mg); 25 (Ti))	35 (Mg); 34 (Ti)	19 (12–30)
Choo et al.	Case control study	93 (24 Mg; 69 Ti)	54.5 ± 12 (Mg-Ti)	M.:1, F.:23 (Mg)	Hallux	Mg; Ti	93 (24 (Mg); 69 (Ti))	NR	12
Plaass et al.	Prospective case series	45 (Mg)	45.5 ± 10.6: 19.6–68.2	45 (2 m., 43 fe.)	Hallux	Mg	45	NR	12
Plaass et al.	Randomized Clinical Trial	14 (8 Mg; 6 Ti)	56 ± 8.9 (Mg); 52 ± 9.0 (Ti)	14 f. (Mg-Ti)	Hallux	Mg; Ti	14 (8 (Mg); 6 (Ti))	NR	36
Zhao et al.	Randomized Clinical Trial	48 (23 Mg; 25 C.)	30 ± 7 (Mg); 33 ± 8 (C.)	9 f./14 m. (Mg); 10 f./15 m. (c.)	femoral head	Mg	23 Mg	NR	12
Leonhardt et al.	Prospective case series	6 (Mg)	43.2: 30–66	4 m.; 2 f.	Mandibular condyle	Mg	6 Mg	NR	12

C: Control; Mg: Mg; Ti: Ti; M: Male; F:Female; Min: Minute.

3.2. Experimental Design

From eight studies selected, three were RCT studies [10,11,33], one were retrospective studies [34], two were case-control studies [11,22], and lastly, two were prospective case series [21,47]. All of these studies were published between 2013–2020. These studies are unclear about the specific location of the surgeries, but the report stated that surgeons were well-trained in their respective fields (Table 1).

3.3. Patient Selection

The studies analyzed reported various inclusion criteria for patient selection. Inclusion criteria were: age between 18 and 79 years, no medical contraindications, and surgical procedures that require fixation using screws [10,22]. The exclusion criteria included: patients with neurological diseases, surgeries in the same body region operated previously, allergies against the materials used for testing components of the screws [33,34] (Table 1).

3.4. Operated Region

Of the eight studies selected, six studies were related to the hallux region [7,10,21,22,33,34], one related to the treatment of necrosis of the femoral head [11] and last one related to mandibular condyle fracture [47]. All studies used screws for fixation. For each region operated, the most appropriate surgical procedure had a follow-up to evaluate the success of the treatment (Table 1).

3.5. Type of Screws

Five studies compared Mg and Ti screws related to the treatment of hallux fracture [7,10,22,33,34], one study used Mg screw compared to a control group with no graft fixation [11], and two studies had no control or comparison group [21,47] (Table 1).

3.6. Surgery Time

Only three studies determined the surgical time 40 min ± 9.1 (Mg) vs. 34 min ± 3.3 (Ti) [33]; 60.6 min (Mg) vs. 55.6 min (Ti) [34] and 35 min (Mg) vs. 34 min (Ti) [22]. The surgical times were closed for both groups (Table 1).

3.7. Radiologic Measures

All studies did some method of radiologic evaluation, and only one study described significant changes between the Mg and Ti group. In the Mg group, the authors classified 60% of the radiographs as satisfactory and 40% with some alteration, but these alterations were not specified. They only described areas of radiolucency, lytic areas, signs of plate or screw loosening, and bone resorption areas of demineralization, but no patients demonstrated any painful symptoms [34].

In the other studies, the radiographic evaluation showed no difference between the Mg and Ti group [7,10,22,33] (Table 2).

Table 2. Qualitative data from the included studies.

Article	Radiologic Measures	Functional Recovery	Laboratory	Complications
Windhagen et al.	Correct placement of the implants and early signs of union and bone healing	all healed patients	No *	MgG(two patients had problems in healing) TiG (one patient had problems in healing; one patient had exposure of screw head)
Klauser	TiG. (All postoperative radiographs were satisfied.); Mg group: 60% of the radiographs as satisfy. and 40% with some alteration	There was no difference between groups (Mg group 3% vs. Ti group 4%)	NR	MgG (one broken screw; three patients with superficial infection; two patients with deep infection) TiG (one patient had prominence of the screw; four patients with superficial infections; one patient deep infection)
Atkinson et al.	No radiographic changes.	Mg—Improvement in postoperative results	NR	MgG (There were no post-operative complications of intraoperative technical)
Choo et al.	No radiographic changes of the screws in any group	The Ti group shows better results compared to the Mg group	NR	MgG (three cases of infection; one case of local pain) TiG (three cases of cellulite; one case of regional pain; one patient had implant removed)
Plaass et al.	The x-rays showed a significant improvement of all; Radiographic signs of bony healing	Improvement in postoperative results was observed	NR	Five patients (early implant disintegration, dislocation, radiolucency's, or pain); two patients of early disintegration; seven patients showed functional problems after surgery)

Table 2. *Cont.*

Article	Radiologic Measures	Functional Recovery	Laboratory	Complications
Plaass et al.	There was no difference between the study groups regarding fracture repair	No difference regarding the rehabilitation of patients	NR	MgG no complications TiG (two patients' pain during running; three patients had residual pain)
Zhao et al.	The tom. shows an increase in bone density compared to the control group	Favorable results for the Mg group compared to the CG	No *	There were no complications associated with the Mg group
Leonhardt et al.	Adequate repair of fractures was observed at 6-months postoperative tomography	All patients had experienced excellent restoration of their occlusion, and no revisions were required.	NR	No postoperative complications were reported

* NO: no change in the level of Mg in the blood; CG: Control Group; TiG: Ti Group; MgG: Mg Group. NR: Not is reported.

3.8. Follow-Up

The follow-up period ranged from 6 months to 36 months [10,33]. The mean postoperative follow-up was 12 months [7,11,21,22,34,47]. Despite the variation in follow-up time, the authors reported that it was enough to notice possible complications (Table 2).

3.9. Functional Recovery

All studies did some type of postoperative recovery assessment. Only one study reported positive results for Ti compared to the Mg group [7], whereas other studies showed similar results between the Mg group and the Ti group [10,22,33,34] (Table 2).

3.10. Laboratory Results

Only two studies reported Mg blood level assessment. Regarding the lab alterations, there was no difference between the groups [11,33] (Table 2).

3.11. Complications

The studies described the number of complications, one patient for Ti group with a screw head displayed, but the patient declined a re-operation surgery [34]. One patient in another study had a prominence of the screw, but no re-treatment was performed [22].

Regarding the evaluation of infections and postoperative healing, two studies reported patients with postoperative infection [7,34]. Klauser reported three patients had healing problems without signs of infection, two patients in the Mg group and one patient in the Ti group, and complications healed during the follow-up [34]. Thirteen patients had some type of infection of the surgical site, six from the Mg group and seven from the Ti group. The treatment for those infections ranged from the prescription of systemic antibiotic therapy to surgical treatment [7,22].

Regarding postoperative pain, one study [7] was very clear to discuss the pain evaluation between groups. In the Mg group, there were three cases of infection and one case of local pain. In the Ti group there were three cases of cellulitis; one case of regional pain; and one patient required implant removal. One study [5] reported that three patients for the Ti group had residual postoperative pain, while no patient in the Mg group reported any type of pain. In two other studies [11,47], no complications were related to the Mg group (Table 2).

4. Quantitative Analysis (Meta-Analysis)

4.1. Primary Outcomes

4.1.1. Complications in the Absorbable Mg Screws vs. Control Group

Five studies [7,10,11,33,34] involving a total of 156 patients who received Mg screws identified 15 complications, and 213 patients received Ti screws or bone grafting, with 18 complications. The meta-analysis did not indicate a significant difference in this comparison (RR 1.071; 95% CI 0.475 to 2.417, $p = 0.868$, Figure 2). The heterogeneity was Q-value: 5.442, $p = 0.245$, $I^2 = 26.499$.

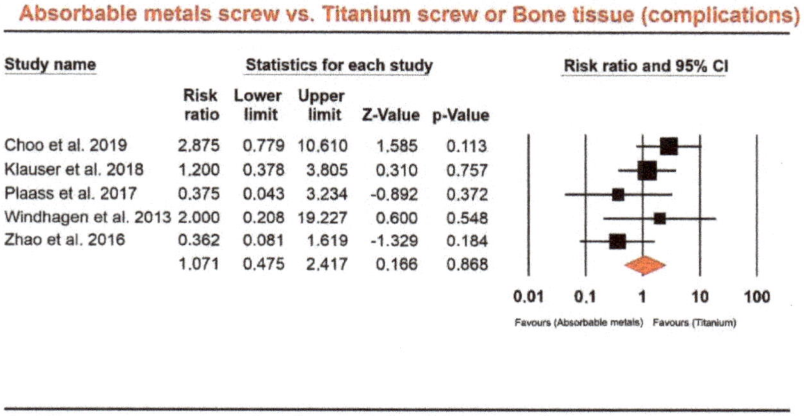

Absorbable metals screw vs. Titanium screw or Bone tissue (complications)

Study name	Risk ratio	Lower limit	Upper limit	Z-Value	p-Value
Choo et al. 2019	2.875	0.779	10.610	1.585	0.113
Klauser et al. 2018	1.200	0.378	3.805	0.310	0.757
Plaass et al. 2017	0.375	0.043	3.234	-0.892	0.372
Windhagen et al. 2013	2.000	0.208	19.227	0.600	0.548
Zhao et al. 2016	0.362	0.081	1.619	-1.329	0.184
	1.071	0.475	2.417	0.166	0.868

Figure 2. Forest plot for absorbable Mg-based screw vs. other materials.

4.1.2. Complications in the Absorbable Mg Screw vs. Ti Screw Group and Hallux Valgus Deformity Surgery

Four studies [7,10,33,34] involving a total of 145 patients who received absorbable Mg screws identified 13 complications and 188 patients received Ti screws in addition to being specifically for the region: Hallux valgus deformity, with 12 complications. The meta-analysis did not indicate a significant difference in this comparison (RR 1.476; 95% CI 0.693 to 3.144, $p = 0.313$, Figure 3). The heterogeneity was Q-value: 2.748, $p = 0.432$, $I^2 = 00.000$.

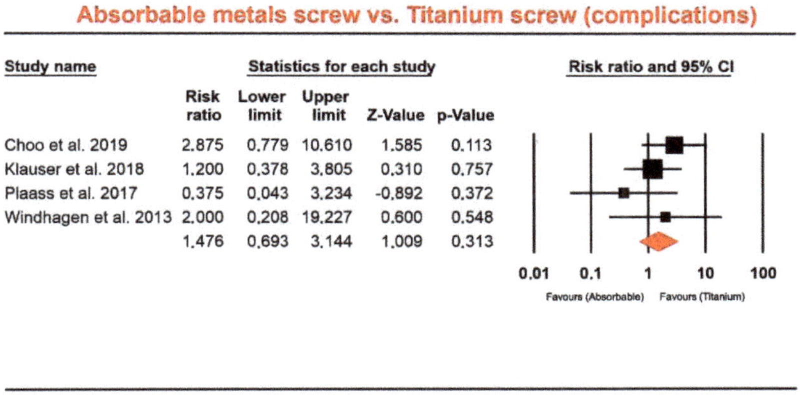

Absorbable metals screw vs. Titanium screw (complications)

Study name	Risk ratio	Lower limit	Upper limit	Z-Value	p-Value
Choo et al. 2019	2.875	0.779	10.610	1.585	0.113
Klauser et al. 2018	1.200	0.378	3.805	0.310	0.757
Plaass et al. 2017	0.375	0.043	3.234	-0.892	0.372
Windhagen et al. 2013	2.000	0.208	19.227	0.600	0.548
	1.476	0.693	3.144	1.009	0.313

Figure 3. Forest plot for absorbable Mg-based screw vs. Ti screw (complication).

4.1.3. Failure in the Mg vs. Absorbable Screw Group Control (Ti Screw and Region: Hallux Valgus Deformity)

Four studies [7,10,33,34] involving a total of 145 patients who received absorbable Mg screws identified 1 failure and 188 patients received Ti screw or bone grafting, showing four failures. The meta-analysis did not indicate a significant difference in this comparison (RR 0.548; 95% CI 0.122 to 2.463, $p = 0.433$, Figure 4). The heterogeneity was de Q-value: 0.622, $p = 0.891$, $I^2 = 00.000$.

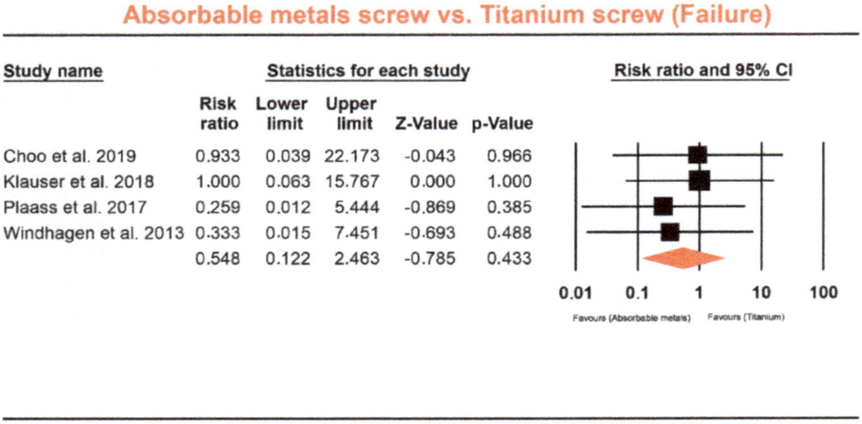

Absorbable metals screw vs. Titanium screw (Failure)

Study name	Risk ratio	Lower limit	Upper limit	Z-Value	p-Value
Choo et al. 2019	0.933	0.039	22.173	-0.043	0.966
Klauser et al. 2018	1.000	0.063	15.767	0.000	1.000
Plaass et al. 2017	0.259	0.012	5.444	-0.869	0.385
Windhagen et al. 2013	0.333	0.015	7.451	-0.693	0.488
	0.548	0.122	2.463	-0.785	0.433

Risk ratio and 95% CI: 0.01 0.1 1 10 100; Favours (Absorbable metals) Favours (Titanium)

Figure 4. Forest plot for absorbable Mg-based screw vs. Ti screw (failure).

4.1.4. The Event Rate for Complications in Absorbable Mg Screw-In Operated Patients

Eight studies [7,10,11,21,22,33,34,47] involving a total of 230 patients who received absorbable Mg screws identified 25 complications. Event rate data ranged from 8.3% to 20.6%. The overall pooled for event rate was 13.3% (random; 95% CI: 8.3% to 20.6%; Figure 5). Regions considered: Hallux valgus deformity, osteosynthesis of the mandibular condyle, osteonecrosis of the femoral head. The heterogeneity of the event rate for complications was considered to be Q-value: 9.448, $p = 0.222$, $I^2 = 25.907$.

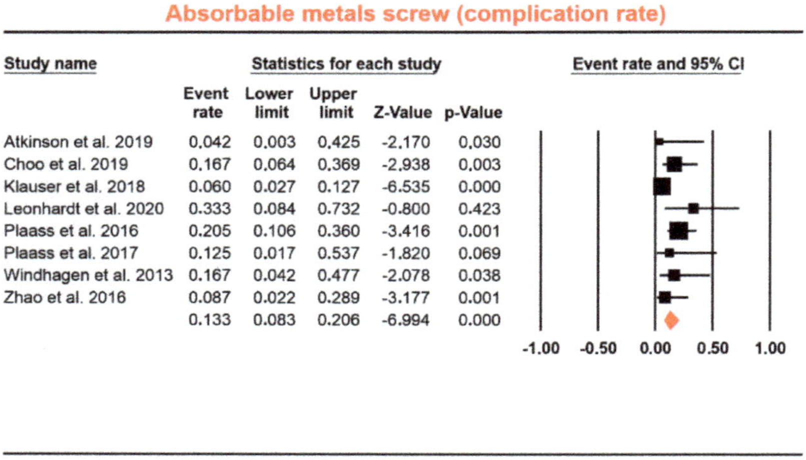

Absorbable metals screw (complication rate)

Study name	Event rate	Lower limit	Upper limit	Z-Value	p-Value
Atkinson et al. 2019	0.042	0.003	0.425	-2.170	0.030
Choo et al. 2019	0.167	0.064	0.369	-2.938	0.003
Klauser et al. 2018	0.060	0.027	0.127	-6.535	0.000
Leonhardt et al. 2020	0.333	0.084	0.732	-0.800	0.423
Plaass et al. 2016	0.205	0.106	0.360	-3.416	0.001
Plaass et al. 2017	0.125	0.017	0.537	-1.820	0.069
Windhagen et al. 2013	0.167	0.042	0.477	-2.078	0.038
Zhao et al. 2016	0.087	0.022	0.289	-3.177	0.001
	0.133	0.083	0.206	-6.994	0.000

Event rate and 95% CI: -1.00 -0.50 0.00 0.50 1.00

Figure 5. Forest plot for absorbable Mg-based screw (complication rate).

4.1.5. Event Rate for Absorbable Mg Screw Failure in Operated Patients

Seven studies(7, 10, 21, 22, 33, 34, 47) involving a total of 207 patients who received absorbable Mg screws identified 3 failures. Event rate data ranged from 1.5% to 7.7%. The overall pooled for event rate was 3.4% (random; 95% CI: 1.5% to 7.7%; Figure 6). The heterogeneity of the event rate for complications was considered to be *Q*-value: 2.474, $p = 0.871$, $I^2 = 0.000$.

Absorbable metals screw (complication rate)

Study name	Statistics for each study					Event rate and 95% CI
	Event rate	Lower limit	Upper limit	Z-Value	p-Value	
Atkinson et al. 2019	0.042	0.003	0.425	-2.170	0.030	
Choo et al. 2019	0.167	0.064	0.369	-2.938	0.003	
Klauser et al. 2018	0.060	0.027	0.127	-6.535	0.000	
Leonhardt et al. 2020	0.333	0.084	0.732	-0.800	0.423	
Plaass et al. 2016	0.205	0.106	0.360	-3.416	0.001	
Plaass et al. 2017	0.125	0.017	0.537	-1.820	0.069	
Windhagen et al. 2013	0.167	0.042	0.477	-2.078	0.038	
Zhao et al. 2016	0.087	0.022	0.289	-3.177	0.001	
	0.133	0.083	0.206	-6.994	0.000	

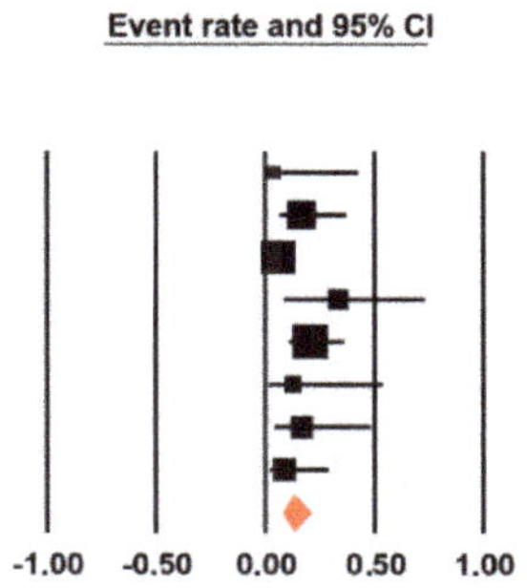

-1.00 -0.50 0.00 0.50 1.00

Figure 6. Forest plot for the absorbable Mg-based screw (failure rate).

4.2. *Risk of Bias in the Studies*

Heterogeneity was used using the Q method and the value of I^2 was analyzed [45,48] heterogeneity above 75 (0–100) may reflect greater significance [45,49], we adopted analysis random for all meta-analyzes in order to reduce the potential for heterogeneity [50]. Particularities of the sample designs of each study were also evaluated and particularities of each forest plot were considered, considering, for example, specific analysis for hallux valgus deformity disregarding other regions, a control group containing only Ti screw was also considered, disregarding other materials.

4.3. *Study Quality and Risk of Bias*

Non-Randomized Studies

For non-randomized clinical studies [7,10,22,34,47], some studies either lacked the sample design, data were not proportional, or used retrospective data for test or control groups, which may all have influenced the outcomes. Limitations on the follow-up and lacking tomographic analysis for all the groups were also noted in some studies. Lastly, a lack of identification of the size of the screws in all studies, and only one study showed a sample size calculation. Figures 7 and 8 show the main data on the risk of bias scale.

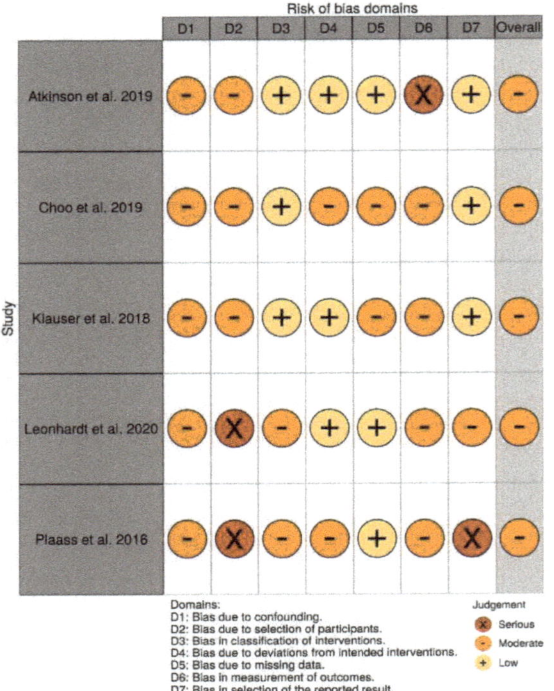

Figure 7. Risk of bias domains—ROBINS-I—Individual studies.

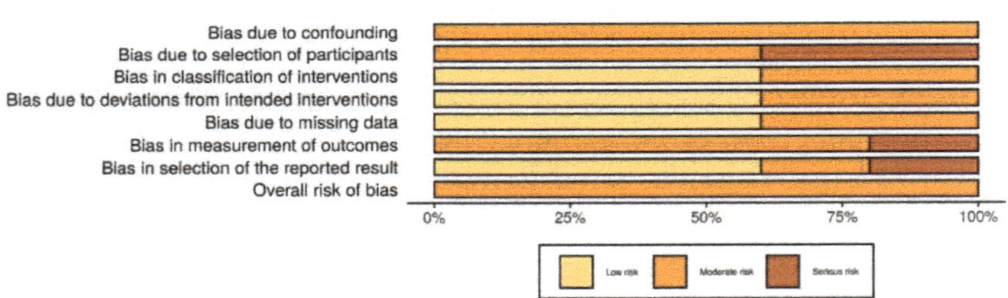

Figure 8. Risk of bias—Risk of bias domains—ROBINS-I—General information.

4.4. Randomized Studies

For randomized clinical studies [10,11,33], there was a limitation in the randomization methods. There was a lack of organization of the failures differently from the complications on each step of the evaluation. A short period of follow-up, or lack of information about systemic disorders or etiological factors also were noted. The main results for the evaluated domains are shown in Figures 9 and 10, related to individual and general valuation, respectively.

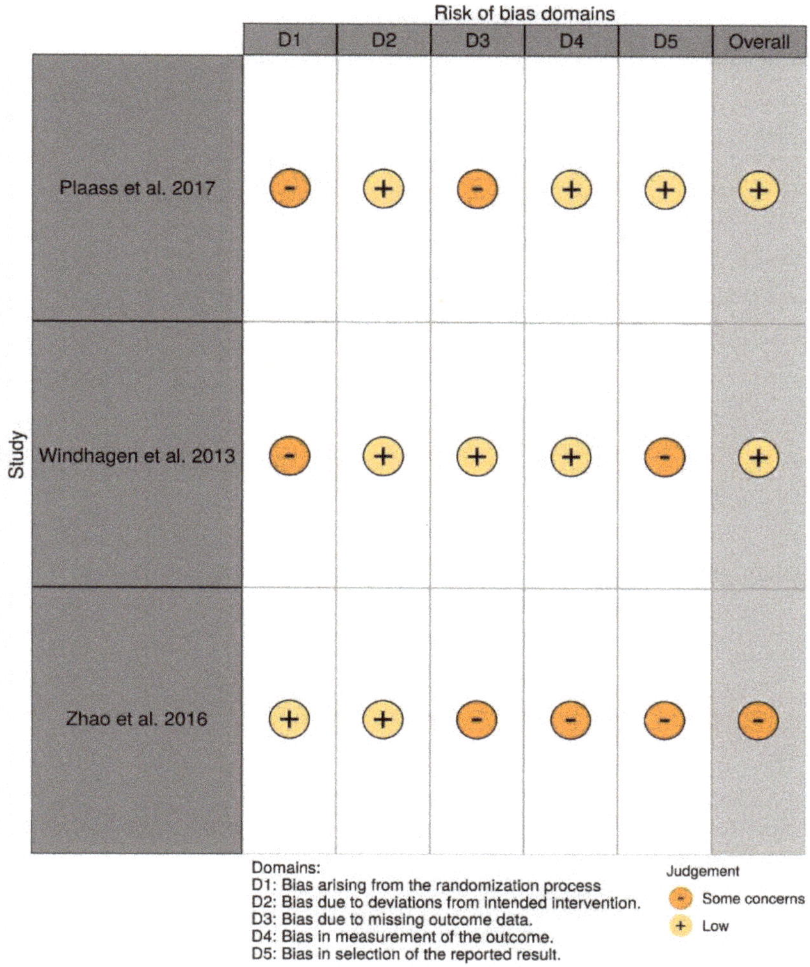

Risk of bias domains

Study	D1	D2	D3	D4	D5	Overall
Plaass et al. 2017	-	+	-	+	+	+
Windhagen et al. 2013	-	+	+	+	-	+
Zhao et al. 2016	+	+	-	-	-	-

Domains:
D1: Bias arising from the randomization process
D2: Bias due to deviations from intended intervention.
D3: Bias due to missing outcome data.
D4: Bias in measurement of the outcome.
D5: Bias in selection of the reported result.

Judgement
- Some concerns
+ Low

Figure 9. Risk of bias domains—ROB—Individual studies.

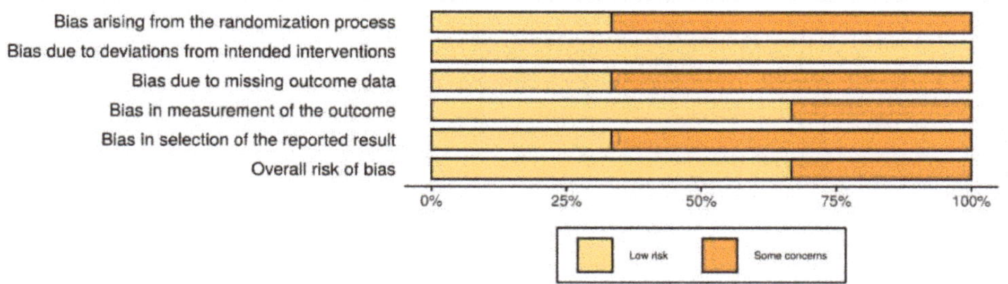

Figure 10. Risk of bias domains—ROB—General studies.

Other methodologies information was also assessed, and it was noticed limitations related to data organization and identification of the sample size calculation (Table 3).

Table 3. Additional data verified.

Studies	Randomization	Sample Size Calculation	Suggestions	Limitation
Atkinson et al., 2019	No	No	Learning curve and multicentric studies	Sample not standardized before experiment
Choo et al., 2019	No	Yes	Higher sample	Largest sample; different screw sizes
Klauser et al., 2018	No	No	Higher sample	Short follow-up
Leonhardt et al., 2020	No	No	Higher sample and control group	Sample and comparison group
Plaass et al., 2016	No	No	Higher sample and follow-up	Reduced sample, short follow-up, absent of a control group
Plaass et al., 2017	Yes, but there was no description of the technique.	No	Learning curve	Reduced sample, data making some analyzes impossible
Windhagen et al., 2013	Yes. There was no description of the technique, but there was extern monitoring.	No	NR	Short follow-up, and some considerations related to the assessment of the screws and radiological images
Zhao et al., 2016	Sim	No	More multicentric studies	Consider etiological and other systemic factors

5. Discussion

This systematic review with a meta-analysis yielded that absorbable metals used for bone surgery, especially Mg alloys, were clinically useful and biologically acceptable compared to standard Ti implants. Of the 468 patients assessed in the studies selected, 230 Mg-alloys screws were used to stabilize a bone fracture [47], bone graft fixation [11], or correction of bone deformity [7,10,21,22,33,34]. Similar statistical data were noted for comparison between Ti and Mg screws ($p = 0.433$), as shown in Figure 1, showing acceptable biological responses for these applications.

Previous animal studies have found interesting results for absorbable metals in bone surgeries [34,51,52]. Most studies have suggested that Mg increases osteoblastic activity and has anti-inflammatory properties during the degradation process [26,53,54]. The vascular endothelial grown factor (VEGF), one of the most important factors to vascular proliferation, was significantly increased around Mg implants placed in bone marrow defects of rats [54]. The osteogenic and angiogenic properties with its degradation were achieved because of Mg ion and a co-enzyme of more than 200 enzymes from the organism which can be responsible to improve bone healing [55]. A few parts of Mg were distributed in the adjacent muscles, and other parts were metabolized by the kidney and liver without any postoperative blood chemistry alteration, as noticed by Windhagen et al., 2013 [33] and Choo et al., 2019 [7] (Table 2). The serum levels of calcium, Mg, and phosphorus were normal [11,26,33,56–60].

One of the concerns during Mg degradation is the hydrogen gas release. Some studies have speculated that the corrosion occurs with non-absorbable and absorbable metals. The particles around the bone and soft tissues could cause inflammation and bone resorption, including hypersensitivity reactions. However, none of the investigations found any evidence of allergenic effects in the presence of Mg-alloys [10,61,62]. One possible complication could be edema and emphysema during hydrogen gas release. Some cases indicated some radiographic signals of peri-implant gas radiolucencies.

However, no clinical symptomatology existed, and the radiographic image change was solved at three months postoperatively [33,47,63,64].

In regards to the biomechanical properties, Mg screws have Young' modulus very similar to bone tissue. However, over time, Mg could fatigue, especially when it requires a higher insertion torque. Recent studies have added other ions or metal particles, such as iron and zinc, to increase the mechanical resistance features [26]. Regardless, most of the clinical studies have found similar fatigue complications in the comparison between Mg and Ti materials for bone fixation [65,66]. One clinical strategy to decrease that complication is to use a countersink drill before screw insertion, since Mg screws are not self-drilling or self-tapping. Although this process may increase the time of surgery, the studies did not show a significant difference between Ti or Mg fixation, with only five-six minutes more time required for Mg surgeries [22,33,34] (Table 1).

The use of absorbable metals is very applicable in oral and maxillofacial surgery. For instance, in mandibular condyle fractures, when there is any instability of the temporomandibular joint complex during the postoperative period, an absorbable fixation system may avoid the development of some TMJ disorders or pathologies. Mg-MgHA/collagen-based scaffolds have been successfully used for sinus augmentation procedures. However, the result should be interpreted carefully as controls were not used for comparison [67]. The use of conventional Ti fixation in children is still controversial. Many studies suggested fixation removal after at least six months postoperatively, leading to more indirect costs related to the surgery and additional days for recovery of the patients. Therefore, an absorbable metal fixation becomes useful in children, with no necessity for a second surgery only to remove the fixation system. Although this study does not focus on an analysis of children, further research needs to be performed in the area [68–72].

In the orthopedic field, no significant differences were found between Ti and Mg screws used to bone fixation in all studies that assessed postoperative clinical parameters such as pain, walking/standing, and social interaction through standard scales demonstrated. In regards to better postoperative function, both treatments were effective [22].

Another important finding in the studies is the use of absorbable polymers, especially from the PLLA and poly-glycolic acid (PGA). Although the studies did not demonstrate significant differences in postoperative complications, absorbable polymers are mechanically weaker, and their degradation process through hydrolysis resulted in acidic elements, which can increase infections and bone resorption osteoclasts activation. The findings indicated that the use of PLLA and PGA might not be applicable for clinical applications, and the search for other degradable materials is warranted [14,73–75].

Several limitations were noticed in this systematic review as described above in the results section. These are fundamental to achieve an acceptable level of evidence for a clinical problem. The first one is the number of well-designed clinical investigations, including lack of clear methods of randomization, equality in the size of the sample for each experimental group (absence of sample calculations), limited follow-up, and different periods of follow-up among studies, limited organization of the failure's assessment. Future studies should be designed following those standard parameters, especially through RCTs, involving control groups within the same study, and finally increasing the level of evidence of the results. Numerous studies have investigated Mg screws for fracture fixation of the maxillofacial and orthopedic bone surgeries. Future studies evaluating Mg plates for bioactivity properties should be warranted.

6. Conclusions

The use of Mg-based implants as absorbable metals for osteosynthesis show feasible applications in bone surgery procedures. There are no differences in a comparison between Mg-implants and conventional implants (Ti) with regards to biocompatibility or complication rates. Therefore, Mg-based implants should be considered for clinical applications in oral and orthopedic reconstructive surgery.

Author Contributions: C.S.: methodology, supervision; T.J.L.-N.: investigation, data curation; J.F.S.J.: formal analysis, data curation; L.P.F.: investigation, data curation, writing original draft preparation, supervision; M.M.: supervision, project administration. All authors have read and agreed to the published version of the manuscript.

Funding: Study financed with own resources.

Acknowledgments: The authors would like to express gratitude for support and scholarship from the Coordination for the Improvement of Higher Education Personnel (CAPES) in the PrINT Program (Fellowship Research #88887.373422/2019-00).

Conflicts of Interest: The authors state that there is no conflict of interest.

References

1. Dupaix, J.P.; Opanova, M.I.; Elston, M.J.; Lee, L.S. A Comparison of Skeletal Injuries Arising from Moped and Motorcycle Collisions. *Hawaii J. Health Soc. Welf.* **2019**, *78*, 311–315. [PubMed]
2. Kudo, D.; Miyakoshi, N.; Hongo, M.; Kasukawa, Y.; Ishikawa, Y.; Ishikawa, N.; Shimada, Y. An epidemiological study of traumatic spinal cord injuries in the fastest aging area in Japan. *Spinal Cord* **2019**, *57*, 509–515. [CrossRef] [PubMed]
3. Getzmann, J.M.; Slankamenac, K.; Sprengel, K.; Mannil, L.; Giovanoli, P.; Plock, J.A. The impact of non-thermal injuries in combined burn trauma: A retrospective analysis over the past 35 years. *J. Plast. Reconstr. Aesthetic Surg.* **2019**, *72*, 438–446. [CrossRef] [PubMed]
4. Burkhard, J.P.M.; Pitteloud, C.; Klukowska-Rötzler, J.; Exadaktylos, A.K.; Iizuka, T.; Schaller, B. Changing trends in epidemiology and management of facial trauma in a Swiss geriatric population. *Gerodontology* **2019**, *36*, 358–364. [CrossRef] [PubMed]
5. Stathopoulos, P.; Igoumenakis, D.; Mezitis, M.; Rallis, G. Blindness after facial trauma: Epidemiology, incidence and risk factors: A 27-year cohort study of 5708 patients. *Oral Surgery Oral Med. Oral Pathol. Oral Radiol.* **2018**, *126*, 129–133. [CrossRef]
6. Chagnon, M.; Guy, L.-G.; Jackson, N. Evaluation of Magnesium-based Medical Devices in Preclinical Studies: Challenges and Points to Consider. *Toxicol. Pathol.* **2019**, *47*, 390–400. [CrossRef]
7. Choo, J.T.; Lai, S.H.S.; Ying, C.T.Q.; Thevendran, G. Magnesium-based bioabsorbable screw fixation for hallux valgus surgery—A suitable alternative to metallic implants. *Foot Ankle Surg.* **2019**, *25*, 727–732. [CrossRef]
8. Brucoli, M.; Boffano, P.; Romeo, I.; Corio, C.; Benech, A.; Ruslin, M.; Forouzanfar, T.; Jensen, T.S.; Rodríguez-Santamarta, T.; De Vicente, J.C.; et al. Epidemiology of maxillofacial trauma in the elderly: A European multicenter study. *J. Stomatol. Oral Maxillofac. Surg.* **2019**. [CrossRef]
9. Neto, I.C.P.; Franco, J.M.P.L.; Junior, J.L.D.A.; Santana, M.D.R.; De Abreu, L.C.; Bezerra, I.M.P.; Soares, E.C.S.; Gondim, D.G.D.A.; Rodrigues, L.M.R. Factors Associated With the Complexity of Facial Trauma. *J. Craniofacial Surg.* **2018**, *29*, e562–e566. [CrossRef]
10. Plaass, C.; Von Falck, C.; Ettinger, S.; Sonnow, L.; Calderone, F.; Weizbauer, A.; Reifenrath, J.; Claassen, L.; Waizy, H.; Daniilidis, K.; et al. Bioabsorbable magnesium versus standard titanium compression screws for fixation of distal metatarsal osteotomies—3 year results of a randomized clinical trial. *J. Orthop. Sci.* **2018**, *23*, 321–327. [CrossRef]
11. Zhao, D.W.; Huang, S.; Lu, F.; Wang, B.; Yang, L.; Qin, L.; Yang, K.; Li, Y.; Li, W.; Wang, W.; et al. Vascularized bone grafting fixed by biodegradable magnesium screw for treating osteonecrosis of the femoral head. *Biomaterials* **2016**, *81*, 84–92. [CrossRef] [PubMed]
12. Kasai, T.; Matsumoto, T.; Iga, T.; Tanaka, S. Complications of implant removal in ankle fractures. *J. Orthop.* **2019**, *16*, 191–194. [CrossRef] [PubMed]
13. Schepers, T.; Van Lieshout, E.M.; De Vries, M.R.; Van Der Elst, M. Complications of syndesmotic screw removal. *Foot Ankle Int.* **2011**, *32*, 1040–1044. [CrossRef] [PubMed]
14. Sukegawa, S.; Kanno, T.; Matsumoto, K.; Sukegawa-Takahashi, Y.; Masui, M.; Furuki, Y. Complications of a poly-l-lactic acid and polyglycolic acid osteosynthesis device for internal fixation in maxillofacial surgery. *Odontology* **2018**, *106*, 360–368. [CrossRef] [PubMed]
15. Fage, S.W.; Muris, J.; Jakobsen, S.S.; Thyssen, J.P. Titanium: A review on exposure, release, penetration, allergy, epidemiology, and clinical reactivity. *Contact Dermat.* **2016**, *74*, 323–345. [CrossRef]
16. Sun, Y.; Hu, Y.; Yuan, Q.; Yu, J.; Wu, X.; Du, Z.; Wu, X.; Hu, J. Association between metal hypersensitivity and implant failure in patients who underwent titanium cranioplasty. *J. Neurosurg.* **2019**, *131*, 40–46. [CrossRef]

17. Chen, X.; Giambini, H.; Ben-Abraham, E.; An, K.-N.; Nassr, A.; Zhao, C. Effect of Bone Mineral Density on Rotator Cuff Tear: An Osteoporotic Rabbit Model. *PLoS ONE* **2015**, *10*, e0139384. [CrossRef]
18. Seitz, J.-M.; Durisin, M.; Goldman, J.; Drelich, J.W. Recent Advances in Biodegradable Metals for Medical Sutures: A Critical Review. *Adv. Healthc. Mater.* **2015**, *4*, 1915–1936. [CrossRef]
19. Pelto-Vasenius, K.; Hirvensalo, E.; Vasenius, J.; Rokkanen, P. Osteolytic Changes After Polyglycolide Pin Fixation in Chevron Osteotomy. *Foot Ankle Int.* **1997**, *18*, 21–25. [CrossRef]
20. Clanton, T.O.; Betech, A.A.; Bott, A.M.; Matheny, L.M.; Hartline, B.; Hanson, T.W.; McGarvey, W.C. Complications After Tendon Transfers in the Foot and Ankle Using Bioabsorbable Screws. *Foot Ankle Int.* **2013**, *34*, 486–490. [CrossRef]
21. Plaass, C.; Ettinger, S.; Sonnow, L.; Koenneker, S.; Noll, Y.; Weizbauer, A.; Reifenrath, J.; Claassen, L.; Daniilidis, K.; Stukenborg-Colsman, C.; et al. Early results using a biodegradable magnesium screw for modified chevron osteotomies. *J. Orthop. Res.* **2016**, *34*, 2207–2214. [CrossRef] [PubMed]
22. E Atkinson, H.D.; Khan, S.; Lashgari, Y.; Ziegler, A. Hallux valgus correction utilising a modified short scarf osteotomy with a magnesium biodegradable or titanium compression screws—A comparative study of clinical outcomes. *BMC Musculoskelet. Disord.* **2019**, *20*, 334. [CrossRef] [PubMed]
23. Willbold, E.; Weizbauer, A.; Loos, A.; Seitz, J.-M.; Angrisani, N.; Windhagen, H.; Reifenrath, J. Magnesium alloys: A stony pathway from intensive research to clinical reality. Different test methods and approval-related considerations. *J. Biomed. Mater. Res. Part A* **2016**, *105*, 329–347. [CrossRef] [PubMed]
24. Kim, B.J.; Piao, Y.; Wufuer, M.; Son, W.-C.; Choi, T.H. Biocompatibility and Efficiency of Biodegradable Magnesium-Based Plates and Screws in the Facial Fracture Model of Beagles. *J. Oral Maxillofac. Surg.* **2018**, *76*, 1055.e1–1055.e9. [CrossRef] [PubMed]
25. Naujokat, H.; Seitz, J.-M.; Açil, Y.; Damm, T.; Möller, I.; Gülses, A.; Wiltfang, J. Osteosynthesis of a cranio-osteoplasty with a biodegradable magnesium plate system in miniature pigs. *Acta Biomater.* **2017**, *62*, 434–445. [CrossRef]
26. Waizy, H.; Diekmann, J.; Weizbauer, A.; Reifenrath, J.; Bartsch, I.; Neubert, V.; Schavan, R.; Windhagen, H. In vivo study of a biodegradable orthopedic screw (MgYREZr-alloy) in a rabbit model for up to 12 months. *J. Biomater. Appl.* **2013**, *28*, 667–675. [CrossRef]
27. Guo, Y.; Zhang, S.; Wei, B.; Legut, D.; Germann, T.C.; Zhang, H.; Zhang, R. A generalized solid strengthening rule for biocompatible Zn-based alloys, a comparison with Mg-based alloys. *Phys. Chem. Chem. Phys.* **2019**, *21*, 22629–22638. [CrossRef]
28. Yang, H.; Jia, B.; Zhang, Z.; Qu, X.; Li, G.; Lin, W.; Zhu, D.; Dai, K.; Zheng, Y. Alloying design of biodegradable zinc as promising bone implants for load-bearing applications. *Nat. Commun.* **2020**, *11*, 1–16. [CrossRef]
29. Chen, Y.; Dou, J.; Yu, H.; Chen, C. Degradable magnesium-based alloys for biomedical applications: The role of critical alloying elements. *J. Biomater. Appl.* **2019**, *33*, 1348–1372. [CrossRef]
30. Li, P.; Zhang, W.; Dai, J.; Xepapadeas, A.B.; Schweizer, E.; Alexander, D.; Scheideler, L.; Zhou, C.; Zhang, H.; Wan, G.; et al. Investigation of zinc-copper alloys as potential materials for craniomaxillofacial osteosynthesis implants. *Mater. Sci. Eng. C* **2019**, *103*, 109826. [CrossRef]
31. Hernández-Escobar, D.; Champagne, S.; Yilmazer, H.; Dikici, B.; Boehlert, C.J.; Hermawan, H. Current status and perspectives of zinc-based absorbable alloys for biomedical applications. *Acta Biomater.* **2019**, *97*, 1–22. [CrossRef] [PubMed]
32. Lee, J.-W.; Han, H.-S.; Han, K.-J.; Park, J.; Jeon, H.; Ok, M.-R.; Seok, H.-K.; Ahn, J.-P.; Lee, K.E.; Lee, D.-H.; et al. Long-term clinical study and multiscale analysis of in vivo biodegradation mechanism of Mg alloy. *Proc. Natl. Acad. Sci. USA* **2016**, *113*, 716–721. [CrossRef] [PubMed]
33. Windhagen, H.; Radtke, K.; Weizbauer, A.; Diekmann, J.; Noll, Y.; Kreimeyer, U.; Schavan, R.; Stukenborg-Colsman, C.; Waizy, H. Biodegradable magnesium-based screw clinically equivalent to titanium screw in hallux valgus surgery: Short term results of the first prospective, randomized, controlled clinical pilot study. *Biomed. Eng. Online* **2013**, *12*, 62. [CrossRef] [PubMed]
34. Klauser, H. Internal fixation of three-dimensional distal metatarsal I osteotomies in the treatment of hallux valgus deformities using biodegradable magnesium screws in comparison to titanium screws. *Foot Ankle Surg.* **2018**, *25*, 398–405. [CrossRef]
35. Higgins, J.; Green, S. *Cochrane Handbook for Systematic Reviews of Interventions*; Version 5.1.0; The Cochrane Collaboration: London, UK, 2011; Available online: www.cochrane-handbook.org (accessed on 7 June 2020).

36. Cumpston, M.; Li, T.; Page, M.J.; Chandler, J.; Welch, V.A.; Higgins, J.P.; Thomas, J. Updated guidance for trusted systematic reviews: A new edition of the Cochrane Handbook for Systematic Reviews of Interventions. *Cochrane Database Syst. Rev.* **2019**, *10*, ED000142. [CrossRef]
37. Moher, D.; Shamseer, L.; Clarke, M.; Ghersi, D.; Liberati, A.; Petticrew, M.; Petticrew, M.; Shekell, P.; Stewart, L.A. PRISMA-P Group Preferred reporting items for systematic review and meta-analysis protocols (PRISMA-P) 2015 statement. *Syst. Rev.* **2015**, *4*, 1. [CrossRef]
38. Moher, D.; Liberati, A.; Tetzlaff, J.; Altman, U.G. Preferred Reporting Items for Systematic Reviews and Meta-Analyses: The PRISMA Statement. *PLoS Med.* **2009**, *6*, e1000097. [CrossRef]
39. Sterne, J.A.C.; Hernán, M.A.; Reeves, B.C.; Savović, J.; Berkman, N.D.; Viswanathan, M.; Henry, D.; Altman, D.G.; Ansari, M.T.; Boutron, I.; et al. ROBINS-I: A tool for assessing risk of bias in non-randomised studies of interventions. *BMJ* **2016**, *355*, i4919. [CrossRef]
40. McGuinness, L.A.; Higgins, J.P.T. Risk-of-bias VISualization (robvis): An R package and Shiny web app for visualizing risk-of-bias assessments. *Res. Synth. Methods* **2020**, 1–7. [CrossRef]
41. Sterne, J.A.C.; Savović, J.; Page, M.J.; Elbers, R.G.; Blencowe, N.S.; Boutron, I.; Cates, C.J.; Cheng, H.-Y.; Corbett, M.S.; Eldridge, S.M.; et al. RoB 2: A revised tool for assessing risk of bias in randomised trials. *BMJ* **2019**, *366*, l4898. [CrossRef]
42. Santiago, J.; Batista, V.E.D.S.; Verri, F.R.; Honório, H.; De Mello, C.; Almeida, D.; Pellizzer, E.P. Platform-switching implants and bone preservation: A systematic review and meta-analysis. *Int. J. Oral Maxillofac. Surg.* **2016**, *45*, 332–345. [CrossRef] [PubMed]
43. A Atieh, M.; Ibrahim, H.M.; Atieh, A.H. Platform Switching for Marginal Bone Preservation Around Dental Implants: A Systematic Review and Meta-Analysis. *J. Periodontol.* **2010**, *81*, 1350–1366. [CrossRef]
44. de Carvalho Sales-Peres, S.H.; de Azevedo-Silva, L.J.; Bonato, R.C.S.; de Carvalho Sales-Peres, M.; da Silvia Pinto, A.C.; Junior, J.F.S. Coronavirus (Sars-CoV-2) and the risk of obesity for critically illness and ICU admitted: Meta- análise of epidemiological evidence. *Obesity Res. Clin. Pract.* **2020**. [CrossRef]
45. De Medeiros, F.; Kudo, G.; Leme, B.; Saraiva, P.; Verri, F.R.; Honório, H.; Pellizzer, E.; Junior, J.F.S. Dental implants in patients with osteoporosis: A systematic review with meta-analysis. *Int. J. Oral Maxillofac. Surg.* **2017**, *47*, 480–491. [CrossRef] [PubMed]
46. Borenstein, M.; Hedges, L.V.; Higgins, J.P.T.; Rothstein, H.R. *Introduction to Meta-Analysis*; Wiley: Hoboken, NJ, USA, 2009.
47. Leonhardt, H.; Ziegler, A.; Lauer, G.; Franke, A. Osteosynthesis of the Mandibular Condyle With Magnesium-Based Biodegradable Headless Compression Screws Show Good Clinical Results During a 1-Year Follow-Up Period. *J. Oral Maxillofac. Surg.* **2020**. [CrossRef]
48. Carvalho, M.D.V.; De Moraes, S.L.D.; Lemos, C.A.A.; Junior, J.F.S.; Vasconcelos, B.C.D.E.; Pellizzer, E.P. Surgical versus non-surgical treatment of actinic cheilitis: A systematic review and meta-analysis. *Oral Dis.* **2018**, *25*, 972–981. [CrossRef]
49. Annibali, S.; Bignozzi, I.; Cristalli, M.P.; La Monaca, G.; Graziani, F.; Polimeni, A. Peri-implant marginal bone level: A systematic review and meta-analysis of studies comparing platform switching *versus* conventionally restored implants. *J. Clin. Periodontol.* **2012**, *39*, 1097–1113. [CrossRef]
50. Deeks, J.J.H.J.; Altman, D.G. Chapter 10: Analysing data and undertaking meta-analyses. In *Cochrane Handbook for Systematic Reviews of Interventions*; Version.6.0 (Updated July 2019); Higgins, J.P.T., Thomas, J., Chandler, J., Cumpston, M., Li, T., Page, M.J., Welch, V.A., Eds.; Cochrane: Chichester, UK, 2019; Available online: www.training.cochrane.org/handbook (accessed on 6 June 2020).
51. Kania, A.; Nowosielski, R.; Gawlas-Mucha, A.; Babilas, R. Mechanical and Corrosion Properties of Mg-Based Alloys with Gd Addition. *Materials* **2019**, *12*, 1775. [CrossRef]
52. Kubásek, J.; Dvorsky, D.; Šedý, J.; Msallamova, S.; Levorová, J.; Foltán, R.; Vojtěch, D. The Fundamental Comparison of Zn–2Mg and Mg–4Y–3RE Alloys as a Perspective Biodegradable Materials. *Materials* **2019**, *12*, 3745. [CrossRef]
53. Witte, F.; Kaese, V.; Haferkamp, H.; Switzer, E.; Meyer-Lindenberg, A.; Wirth, C.; Windhagen, H. In vivo corrosion of four magnesium alloys and the associated bone response. *Biomaterials* **2005**, *26*, 3557–3563. [CrossRef]
54. Janning, C.; Willbold, E.; Vogt, C.; Nellesen, J.; Meyer-Lindenberg, A.; Windhagen, H.; Thorey, F.; Witte, F. Magnesium hydroxide temporarily enhancing osteoblast activity and decreasing the osteoclast number in peri-implant bone remodelling. *Acta Biomater.* **2010**, *6*, 1861–1868. [CrossRef] [PubMed]

55. Li, R.W.; Kirkland, N.; Truong, J.; Wang, J.; Smith, P.N.; Birbilis, N.; Nisbet, D.R. The influence of biodegradable magnesium alloys on the osteogenic differentiation of human mesenchymal stem cells. *J. Biomed. Mater. Res. Part A* **2014**, *102*, 4346–4357. [CrossRef] [PubMed]
56. Zhao, D.W.; Witte, F.M.; Lu, F.; Wang, J.; Li, J.; Qin, L. Current status on clinical applications of magnesium-based orthopaedic implants: A review from clinical translational perspective. *Biomaterials* **2017**, *112*, 287–302. [CrossRef] [PubMed]
57. Street, J.; Bao, M.; DeGuzman, L.; Bunting, S.; Peale, F.V.; Ferrara, N.; Steinmetz, H.; Hoeffel, J.; Cleland, J.L.; Daugherty, A.; et al. Vascular endothelial growth factor stimulates bone repair by promoting angiogenesis and bone turnover. *Proc. Natl. Acad. Sci. USA* **2002**, *99*, 9656–9661. [CrossRef] [PubMed]
58. Yoshizawa, S.; Brown, A.; Barchowsky, A.; Sfeir, C. Magnesium ion stimulation of bone marrow stromal cells enhances osteogenic activity, simulating the effect of magnesium alloy degradation. *Acta Biomater.* **2014**, *10*, 2834–2842. [CrossRef] [PubMed]
59. Saris, N.E.; Mervaala, E.; Karppanen, H.; Khawaja, J.A.; Lewenstam, A. Magnesium: An update on physiological, clinical and analytical aspects. *Clin. Chim. Acta.* **2000**, *294*, 1–26. [CrossRef]
60. Schlingmann, K.P.; Waldegger, S.; Konrad, M.; Chubanov, V.; Gudermann, T. TRPM6 and TRPM7—Gatekeepers of human magnesium metabolism. *Biochim. Biophys. Acta (BBA) Mol. Basis Dis.* **2007**, *1772*, 813–821. [CrossRef]
61. Witte, F.; Abeln, I.; Switzer, E.; Kaese, V.; Windhagen, H.; Meyer-Lindenberg, A. Evaluation of the skin sensitizing potential of biodegradable magnesium alloys. *J. Biomed. Mater. Res. Part A* **2008**, *86*, 1041–1047. [CrossRef]
62. Zhang, X.-B.; Yuan, G.; Niu, J.; Fu, P.; Ding, W. Microstructure, mechanical properties, biocorrosion behavior, and cytotoxicity of as-extruded Mg–Nd–Zn–Zr alloy with different extrusion ratios. *J. Mech. Behav. Biomed. Mater.* **2012**, *9*, 153–162. [CrossRef]
63. Staiger, M.P.; Pietak, A.; Huadmai, J.; Dias, G. Magnesium and its alloys as orthopedic biomaterials: A review. *Biomaterials* **2006**, *27*, 1728–1734. [CrossRef]
64. Diekmann, J.; Bauer, S.; Weizbauer, A.; Willbold, E.; Windhagen, H.; Helmecke, P.; Lucas, A.; Reifenrath, J.; Nolte, I.; Ezechieli, M. Examination of a biodegradable magnesium screw for the reconstruction of the anterior cruciate ligament: A pilot in vivo study in rabbits. *Mater. Sci. Eng. C* **2016**, *59*, 1100–1109. [CrossRef] [PubMed]
65. Seitz, J.-M.; Eifler, R.; Bach, F.-W.; Maier, H.J. Magnesium degradation products: Effects on tissue and human metabolism. *J. Biomed. Mater. Res. Part A* **2013**, *102*, 3744–3753. [CrossRef]
66. Willbold, E.; Gu, X.; Albert, D.; Kalla, K.; Bobe, K.; Brauneis, M.; Janning, C.; Nellesen, J.; Czayka, W.; Tillmann, W.; et al. Effect of the addition of low rare earth elements (lanthanum, neodymium, cerium) on the biodegradation and biocompatibility of magnesium. *Acta Biomater.* **2015**, *11*, 554–562. [CrossRef] [PubMed]
67. Scarano, A.; Lorusso, F.; Staiti, G.; Sinjari, B.; Tampieri, A.; Mortellaro, C. Sinus Augmentation with Biomimetic Nanostructured Matrix: Tomographic, Radiological, Histological and Histomorphometrical Results after 6 Months in Humans. *Front. Physiol.* **2017**, *8*, 565. [CrossRef] [PubMed]
68. Viehe, R.; Haupt, D.J.; Heaslet, M.W.; Walston, S. Complications of Screw-Fixated Chevron Osteotomies for the Correction of Hallux Abducto Valgus. *J. Am. Podiatr. Med Assoc.* **2003**, *93*, 499–502. [CrossRef] [PubMed]
69. Hanft, J.R.; Kashuk, K.B.; Bonner, A.C.; Toney, M.; Schabler, J. Rigid internal fixation of the Austin/Chevron osteotomy with Herbert screw fixation: A retrospective study. *J. Foot Surg.* **1992**, *31*, 512–518.
70. Trnka, H.-J.; Zembsch, A.; Easley, M.E.; Salzer, M.; Ritschl, P.; Myerson, M.S. The chevron osteotomy for correction of hallux valgus. Comparison of findings after two and five years of follow-up. *JBJS* **2000**, *82*, 1373–1378. [CrossRef]
71. Stoustrup, P.; Kristensen, K.D.; Küseler, A.; Herlin, T.; Pedersen, T.K. Normative values for mandibular mobility in Scandinavian individuals 4–17 years of age. *J. Oral Rehabil.* **2016**, *43*, 591–597. [CrossRef]
72. Xiang, G.-L.; Long, X.; Deng, M.-H.; Han, Q.-C.; Meng, Q.-G.; Li, B. A retrospective study of temporomandibular joint ankylosis secondary to surgical treatment of mandibular condylar fractures. *Br. J. Oral Maxillofac. Surg.* **2014**, *52*, 270–274. [CrossRef]
73. Rangdal, S.; Singh, D.; Joshi, N.; Soni, A.; Sament, R. Functional outcome of ankle fracture patients treated with biodegradable implants. *Foot Ankle Surg.* **2012**, *18*, 153–156. [CrossRef]
74. An, J.; Jia, P.; Zhang, Y.; Gong, X.; Han, X.; He, Y. Application of biodegradable plates for treating pediatric mandibular fractures. *J. Cranio-Maxillofac. Surg.* **2015**, *43*, 515–520. [CrossRef] [PubMed]

75. Luthringer, B.; Witte, F.M.; Willumeit-Römer, R. Magnesium-based implants: A mini-review. *Magnes. Res.* **2014**, *27*, 142–154. [CrossRef] [PubMed]

Article

Biodegradable Magnesium Bone Implants Coated with a Novel Bioceramic Nanocomposite

Mehdi Razavi [1,2,3,4,*], Mohammadhossein Fathi [3,4], Omid Savabi [5], Lobat Tayebi [6] and Daryoosh Vashaee [7,8,*]

1 Biionix^{TM} (Bionic Materials, Implants & Interfaces) Cluster, Department of Internal Medicine, College of Medicine, University of Central Florida, Orlando, FL 32827, USA
2 Department of Materials Science & Engineering, University of Central Florida, Orlando, FL 32816, USA
3 Biomaterials Research Group, Department of Materials Engineering, Isfahan University of Technology, Isfahan 84156-83111, Iran; fathi@cc.iut.ac.ir
4 Dental Materials Research Center, Isfahan University of Medical Sciences, Isfahan 81746-73461, Iran
5 Torabinejad Dental Research Center, School of Dentistry, Isfahan University of Medical Sciences, Isfahan 81746-73461, Iran; savabi@dnt.mui.ac.ir
6 Marquette University School of Dentistry, Milwaukee, WI 53233, USA; lobat.tayebi@marquette.edu
7 Electrical and Computer Engineering Department, North Carolina State University, Raleigh, NC 27606, USA
8 Materials Science and Engineering Department, North Carolina State University, Raleigh, NC 27606, USA
* Correspondence: Mehdi.Razavi@ucf.edu (M.R.); dvashae@ncsu.edu (D.V.); Tel.: +19-19-515-9599 (D.V.)

Received: 22 January 2020; Accepted: 9 March 2020; Published: 13 March 2020

Abstract: Magnesium (Mg) alloys are being investigated as a biodegradable metallic biomaterial because of their mechanical property profile, which is similar to the human bone. However, implants based on Mg alloys are corroded quickly in the body before the bone fracture is fully healed. Therefore, we aimed to reduce the corrosion rate of Mg using a double protective layer. We used a magnesium-aluminum-zinc alloy (AZ91) and treated its surface with micro-arc oxidation (MAO) technique to first form an intermediate layer. Next, a bioceramic nanocomposite composed of diopside, bredigite, and fluoridated hydroxyapatite (FHA) was coated on the surface of MAO treated AZ91 using the electrophoretic deposition (EPD) technique. Our in vivo results showed a significant enhancement in the bioactivity of the nanocomposite coated AZ91 implant compared to the uncoated control implant. Implantation of the uncoated AZ91 caused a significant release of hydrogen bubbles around the implant, which was reduced when the nanocomposite coated implants were used. Using histology, this reduction in the corrosion rate of the coated implants resulted in an improved new bone formation and reduced inflammation in the interface of the implants and the surrounding tissue. Hence, our strategy using a MAO/EPD of a bioceramic nanocomposite coating (i.e., diopside-bredigite-FHA) can significantly reduce the corrosion rate and improve the bioactivity of the biodegradable AZ91 Mg implant.

Keywords: biodegradable magnesium implants; bioceramics; corrosion; bioactivity; orthopedic implant

1. Introduction

Over the last decade, the development of biodegradable orthopedic implants has significantly advanced [1–3]. Completed and ongoing clinical trials for bone repair and regeneration are mostly focused on biodegradable ceramics include calcium phosphate (ClinicalTrials.gov Identifier: NCT02153372, and NCT02803177 in Germany). Although calcium phosphates are known to be bioactive and support osteoblast adhesion and proliferation [4,5], their major limitation is mechanical properties; namely, they are brittle with a poor fatigue resistance [6–8]. Brittleness so far restricted their application to non-load bearing areas, filler or coating [9], rendering it impossible to use them for the repair or regeneration of load-bearing bone defects [10–12]. Biodegradable polymers such as poly-L-lactic acid, poly-glycolic acid, and copolymers with Food and Drug Administration (FDA) approval for human clinical use have also been extensively used in preclinical studies of bone tissue engineering. However, the application of biopolymer materials suffers from poor processability and weak mechanical properties [13]. Magnesium (Mg) alloys are significantly more flexible than bioceramics, mechanically stronger than biopolymers with the advantage of bioabsorption capabilities over other biometals [14–16]. Mg is an essential mineral crucial to bone health and can even stimulate new bone formation, and its physical and mechanical properties are similar to those of human bones [14,17,18]. Given these characteristics, Mg is considered an attractive element for forming a bone implant. However, a series of clinical trials using Mg-based implants failed prematurely due to the Mg's rapid corrosion and high hydrogen-evolution [14]. A strategy to tackle this issue is to reduce Mg's corrosion rate [19], and the promising future of biodegradable Mg implants is dependent on being able to reduce their corrosion. Several treatments have been proposed to reduce Mg's corrosion rate, including Mg purification [20], fluoride conversion coatings [21], alloying [19], anodizing [22], and compositing [23].

An effective technique to reduce the Mg's corrosion rate is surface coating [24]. For a bone implant, the surface coating can also enhance the implant's surface bioactivity (i.e., osteoproductivity), thereby resulting in improved bone-implant integration [25]. As a coating material, silicate glass-ceramics are a suitable option given their low biodegradation and their ability for new bone formation [26,27]. Biocompatible silicate glass-ceramics are diopside ($CaMgSi_2O_6$) [28], akermanite ($Ca_2MgSi_2O_7$) [29], merwinite ($Ca_3MgSi_2O_8$) [30], and bredigite ($Ca_7MgSi_4O_{16}$) [31]. We have previously synthesized and separately coated the mentioned glass-ceramics on Mg implants [29,32–34]. Our results showed that, among tested glass-ceramics, bredigite had the least biodegradation, and diopside indicated the greatest bioactivity.

Furthermore, compared to calcium phosphates coated on the surface of Mg implants, fluoridated hydroxyapatite (FHA: $Ca_{10}(PO_4)_6(OH)_{2-x}F_x$) has indicated the improved bioactivity and biocompatibility [35]. Accordingly, we synthesized a diopside-bredigite-fluoridated hydroxyapatite nanocomposite and coated on the surface of an Mg implant. We used the electrophoretic deposition (EPD) method to coat our nanocomposite on the surface of the AZ91 Mg alloy. EPD was chosen since it offers many advantages as a coating method, including simplicity, cost-effectiveness, and environmentally friendly processing [36,37]. EPD has also already been utilized for coating the bioceramics on the surface of biometals for orthopedic implants [38–40]. However, before coating with EPD, we first treated our AZ91 substrate with micro-arc oxidation (MAO) technique. On an Mg alloy, a conversion coating such as MAO acts as an intermediate layer to reduce not only the Mg's corrosion rate, but also enhance the adhesion between the Mg and final coating [22].

Hence, the main aim of this work was to reduce the corrosion rate and also enhance the bioactivity of a biodegradable AZ91 Mg implant using a nanocomposite coating composed of diopside, bredigite, and fluoridated hydroxyapatite which was prepared using MAO/EPD technique.

2. Materials and Methods

2.1. Preparation of AZ91 Mg Alloy Substrate

A commercial AZ91 Mg alloy (Al 9%, Zn 1%, Mn 0.2%, Fe $< 0.005\%$, all in wt.%) was machined to obtain the substrates with dimensions of $20 \times 15 \times 5$ mm. Samples were then polished with SiC papers from 80 to 600 grit.

2.2. Surface Coating

2.2.1. Nanocomposite Powder Preparation

The diopside, bredigite, and fluoridated hydroxyapatite (FHA) nanoparticles were first separately synthesized according to our previously published protocols [16,41,42]. They were then blended with the ratio of 1/3:1/3:1/3, respectively, to acquire the nanocomposite powder. The nanocomposite powder was then coated on the surface of AZ91 samples using the combined MAO/EPD method.

2.2.2. Micro Arc Oxidation (MAO)

MAO was performed according to our previously published protocol [22]. In brief, an AZ91 sample was used as the anode and a stainless-steel plate as the cathode electrode, an aqueous solution of NaOH (200 g/L) and Na_2SiO_3 (200 g/L) as the electrolyte along with a power supply. MAO treatment was performed in the applied voltage of 60 V for 30 min.

2.3. Electrophoretic Deposition (EPD)

EPD was performed according to our previously published protocol [42]. In brief, an AZ91 sample was used as the cathode and a graphite rod as the anode electrode. The electrolyte was a suspension of nanocomposite particles at a concentration of 10 g particles/100 mL methanol (99.9%). Before starting the EPD process, the nanoparticles were dispersed into the suspension by placing them in an ultrasonic bath for 60 min and then stirring with a magnetic stirrer for 30 min. EPD was then performed under a constant voltage of 100 V for 3 min at room temperature (Figure 1).

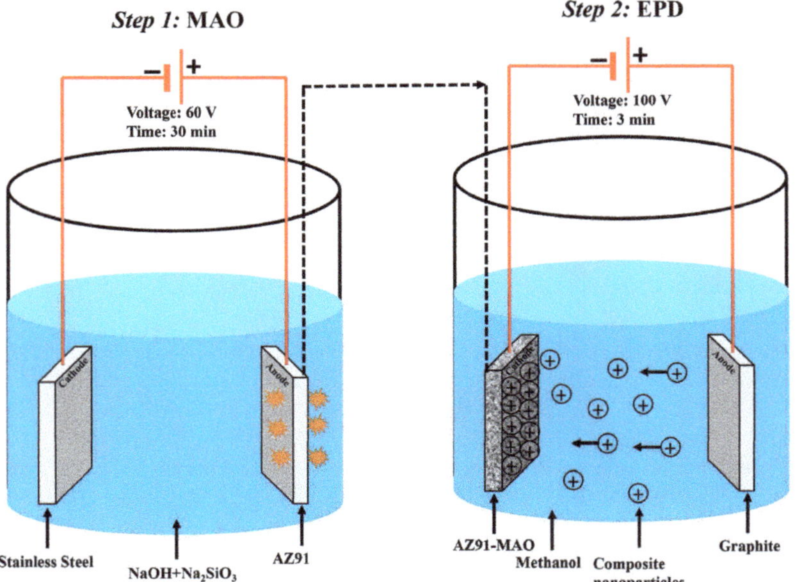

Figure 1. Schematic representation of MAO-EPD coating method: In step 1, i.e., MAO, a power supply was used; an AZ91 sample was used as the anode and a stainless-steel plate as the cathode electrodes; a mixture of NaOH (200 g/L) and Na_2SiO_3 (200 g/L) were also used as the electrolyte solution. MAO was performed in the voltage of 60 V for 30 min. In step 2, i.e., EPD, an AZ91 sample was used as the cathode and a graphite rod as the anode electrode. A suspension of nanocomposite particles at a concentration of 10 g particles/100 mL methanol was also used as the electrolyte. EPD was then performed under a voltage of 100 V for 3 min at room temperature.

2.4. Characterizations

The size and morphology of synthesized nanocomposite particles were measured using a transmission electron microscope (TEM, JEOL JEM-2100). The morphology and chemical composition of surfaces were observed under a scanning electron microscope (SEM) equipped with energy dispersive spectroscopy (EDS) (Philips XL 30: Eindhoven). The topography of surfaces was also observed using a laser scanning microscope (LSM) (Keyence, VK X100/X200). Phase structure analysis was performed using an X-ray diffractometer (XRD, Philips Xpert). The obtained XRD patterns were compared with the standards compiled by the Joint Committee on Diffraction Pattern and Standards (JCDPS). The grain size of synthesized nanocomposite particles was also estimated by broadening XRD peaks using the Williamson–Hall equation (Equation (1)) [43]:

$$\beta \cos\theta = 0.89\ \lambda/D + 2\varepsilon \sin\theta \quad (1)$$

where β is the full width of diffraction peak (rad) in the middle of its height, θ is Bragg's angle (°), and λ is the wavelength of the X-ray (nm) considered after computer fitting of the X-ray data using Gaussian line shape. When $\beta \cos\theta$ is plotted against $\sin\theta$, a straight line is obtained with the slope of 2ε and the intercept as ($0.89\ \lambda/D$) and the grain size, d (nm), can be calculated.

A compression test was performed on our AZ91 Mg alloy according to ASTM E9 standard. The rod samples with a diameter of 3 mm and a length of 6 mm were machined for the experiment. To measure the compressive properties of samples, we used an INSTRON 8562 universal tensile testing machine at a crosshead displacement rate of 0.5 mm/min.

2.5. Corrosion Tests

2.5.1. Electrochemical Test

A PARSTAT 2273 Ametek potentiostat was used for measuring the electrochemical (i.e., polarization and electrochemical impedance spectroscopy (EIS)) properties of samples in the standard simulated body fluid (SBF) solution prepared according to Kokubo's protocol [44]. A three-electrode cell was used include the working electrode (i.e., AZ91 sample), the reference electrode (i.e., calomel), and the counter electrode (i.e., platinum). The experiment started after the sample was incubated in the SBF solution for 60 min to be stabilized, and a scanning rate of 1 $mV.s^{-1}$ was applied for the polarization experiment. The impedance data were recorded with a frequency range of 100 kHz to 10 mHz.

2.5.2. Immersion Test

The immersion test was performed in the SBF, according to ASTM-G31-72 [45], to monitor the corrosion rate of samples. Each sample was individually immersed into a falcon tube containing SBF and incubated at 37 °C for 672 h (28 days). The corrosion rate was determined by measuring the weight loss of each sample after 0, 72 h, 168 h, 336 h, 504 h, and 672 h immersion. The corrosion products formed on the surface of samples during the corrosion were removed using chromic acid (200 g/L CrO_3) [16]. The difference in weight of samples before and after the immersion into chromic acid showed the amount of weight loss, and the corrosion rate of the samples was then calculated using the weight loss as a function of immersion time, according to Equation (2):

$$\text{Corrosion rate} = W/At \quad (2)$$

where W is the weight loss, A is the sample's surface area exposed to the SBF, and t is the immersion time.

2.6. In Vivo Animal Study

The animal experiments in our study were approved by the University Ethics Committee of the Isfahan University of Medical Sciences. Rabbits with average weights of 3 kg were anesthetized by subcutaneous administration of Ketamine (35 mg/kg), Xylazine (5 mg/kg) and Acepromazine (1 mg/kg). After anesthesia, the operation sites were shaved, decortication was carried out, and then the holes with 3 mm diameter were created at the greater trochanter of rabbits using a hand driller. AZ91, MAO, and composite coated rod implants (n = 3) were then implanted into the created bone defects. After the operation, all rabbits received subcutaneous injections of antibiotics. The rabbits were then allowed to move freely in their cages without any external support. The rabbits were sacrificed 2 months post-implantation. Also, a radiography imaging was performed on the implantation site 2 weeks after the surgery and prior to sacrificing the rabbits. For histology, the bone samples were fixed in the formaldehyde solution (4%), dehydrated using ethanol, and then decalcified in a nitric acid solution. The samples were then embedded in paraffin and cut into films. The sectioned samples were stained with Hematoxylin and Eosin (H&E) stain, and the morphological and histological analyses were performed under a microscope to observe the bone regeneration and inflammation around the implant.

2.7. Statistical Analysis

All values were expressed as the mean ± standard deviation (SD). Statistical analysis of all quantitative data was performed using a one-way ANOVA (Analysis of Variance) with post-hoc Tukey test (Astatsa.com; Online Web Statistical Calculators, USA) with any differences considered statistically significant when $p < 0.05$.

3. Results and Discussion

3.1. Characterizations

The microstructure of our AZ91 Mg alloy substrate has been shown in Figure 2a. Using compression test, we determined that this alloy has an elastic modulus of 45 GPa (Figure 2b), which is similar to human cancellous bones (40 GPa) [46]. This lower elastic modulus compared to biometals or bioceramics enhances implant-to-bone stress loading and can minimize bone atrophy due to stress shielding [47].

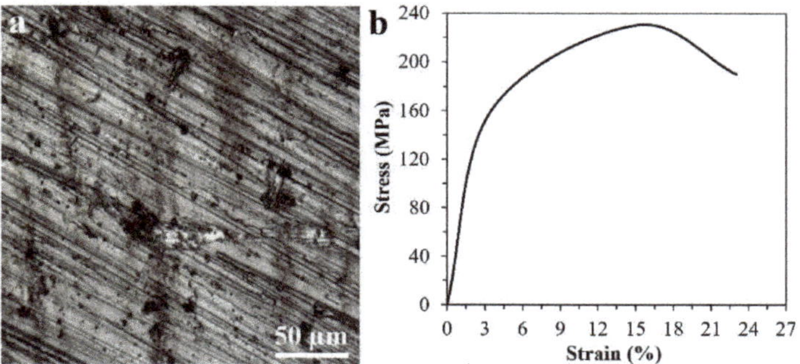

Figure 2. LSM image (**a**) and compressive stress-strain curve (**b**) of our AZ91 Mg alloy.

The results of TEM imaging showed that our composite nanoparticles had a size of 50 ± 20 nm (Figure 3a). Following MAO treatment, a porous and rough surface was formed on AZ91 samples (Figure 3b). The MAO layer has been composed of one external porous layer and one internal compact barrier layer. The external layer is coarse with many microholes which have been formed due to the oxygen bubbles released during the growth process as well as the thermal stress as a result of the rapid solidification of the molten oxide in the relatively cooling electrolyte. However, the internal layer has been attached to the Mg substrate, is compact and uniform, and can act as a barrier to block the exposure of corrosive solutions to the Mg substrate [48]. The compact internal layer can insulate the substrate from the corrosive electrolyte ions while the external layer can absorb more corrosive electrolytes and therefore reduce the corrosion resistance of the Mg alloy substrate. Hence, the external layer of the MAO coating should be sealed by another coating layer. In addition, an MAO layer with a porous and rough surface can offer sites for our composite nanoparticles to be settled in. We then coated our MAO treated AZ91 with composite nanoparticles using the EPD process. The nanocomposite coating could adequately cover the pores of the MAO layer; this can then prevent the MAO layer from being directly exposed to the corrosive solutions. Similar to the MAO layer, the surface of our nanocomposite coating was rough and porous (Figure 3c,d). Using LSM, the surface roughness of AZ91 obtained 5 ± 3 μm; however, following coating with MAO and composite, the surface roughness increased to 12 ± 8, and 150 ± 80 μm, respectively. This can promote bone–implant integration since previous research has suggested that a rough and porous surface can encourage cell attachment and bone in-growth, which can enhance the anchorage of the implant to the bone [36]. Also, using SEM, the thickness of the MAO layer and nanocomposite coat obtained 100 and 250 μm, respectively (Figure 3e,f). The line-scan analysis of the cross-sectional SEM image confirmed that the nanocomposite coat mainly consists of Ca, P, and Mg elements. The intensity of Ca and P gradually decreased from nanocomposite coat to substrate, while Mg had an opposite trend (Figure 3e).

In the XRD pattern of the AZ91 substrate, Mg peaks were detected. When MAO treated AZ91 was tested with XRD, MgO, and Mg_2SiO_4 peaks were detected. MgO is formed by dissolving Mg^{2+} outward from the substrate and the oxidized oxygen O^{2-} inward from electrolyte according to reaction (3):

$$Mg^{2+} + O^{2-} \rightarrow MgO \quad (3)$$

Mg_2SiO_4 peaks indicate the existence of the anion ${SiO_3}^{2-}$. In an aqueous solution, the silicate is transformed into $Si(OH)_4$ by hydroxylation. The water-assisted formation of $Si(OH)_4$, which has 4 silanol groups (Si-OH) forms siloxane groups (i.e., Si–O–Si) and SiO_2 during the strong electrical field and high-temperature anodization based on the reactions shown below (4) and (5) [49]:

$$4H_2O + {SiO_3}^{2-} \rightarrow Si(OH)_4 + 4OH^- \quad (4)$$

$$Si(OH)_4 + Si(OH)_4 + \ldots . \rightarrow XSiO_2 + yH_2O \quad (5)$$

At high temperatures, both SiO_2 and MgO are present in the fused state [49].

However, during the interval stops of anodization sparking and micro-arcing, and by the cooling effect of the electrolyte, the fused state SiO_2 and MgO forms Mg_2SiO_4 according to reaction (6):

$$SiO_2 + 2MgO \rightarrow Mg_2SiO_4 \quad (6)$$

Mg_2SiO_4 is a bioactive ceramic [50], which can also have a protective effect on the AZ91 substrate [22].

The XRD patterns also confirmed the peaks related to diopside, bredigite, and FHA within the composite coat. The grain size of the composite coating was obtained to be approximately 25 nm according to the Williamson–Hall equation, confirming that our composite coating is a nanostructure material (Figure 3g).

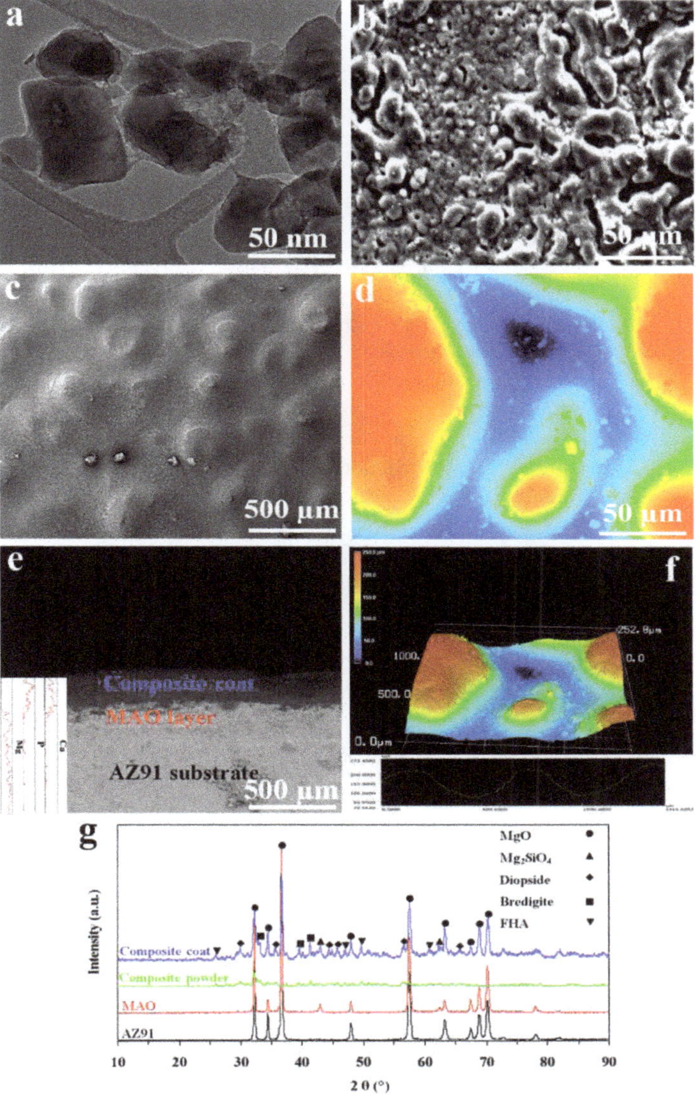

Figure 3. Microstructure characterizations: (**a**) TEM image of composite nanoparticles showing their size which was 50 ± 20 nm; (**b**) SEM image of a MAO treated AZ91 sample showing its rough and porous structure; (**c**–**f**) SEM (**c**,**e**) and LSM (**d**,**f**) images from top (**c**,**d**) and cross-section (**e**,**f**) of nanocomposite coated samples showing that the coating could effectively cover the pores of the MAO layer, the surface of nanocomposite coating was rough and porous, and the thickness of the MAO layer and nanocomposite coating was 100 and 250 µm, respectively. Also, the line-scan analysis of the cross-sectional SEM image confirmed that the nanocomposite coating mainly consists of Ca, P, and Mg elements. The intensity of Ca and P gradually decreased from nanocomposite coat to substrate, while Mg had an opposite trend (**e**); (**g**) XRD patterns of AZ91, MAO, and nanocomposite coated samples showing that AZ91 substrate had peaks related to Mg; MAO treated AZ91 had MgO and Mg_2SiO_4 peaks, and nanocomposite coating had peaks related to diopside, bredigite, and FHA.

3.2. Corrosion Tests

3.2.1. Electrochemical Tests

The values of corrosion current density (i_{corr}), and corrosion potential (E_{corr}) derived from the potentiodynamic polarization curves (Figure 4a) showed that AZ91 sample has a 63,100 nA/cm^2 i_{corr}; This value decreased to 53,700, and 1.99 nA/cm^2 for MAO and nanocomposite coated samples, respectively. The polarization test also recorded an increase in E_{corr} from −1.60 V to −1.56 and −1.45 V for AZ91, MAO, and nanocomposite coated samples, respectively. In general, a decrease in i_{corr} and an increase in E_{corr} is an indication of improvement in corrosion resistance [51]. EIS Nyquist plots showed that the Zim/Zre ratio of AZ91 increased with the MAO and nanocomposite coating, indicating an enhanced capacitive behavior for the solid/liquid interface. The MAO and nanocomposite coated samples showed larger capacitive loops in the EIS spectra than the AZ91 sample. Since a larger diameter loop represents better corrosion resistance [52], the EIS results confirm that the MAO and nanocomposite coating can improve the corrosion resistance of the AZ91 Mg alloy. Also, in the Nyquist plots, two capacitive loops and one inductive loop are seen for samples, similar to previously reported Nyquist plots of pure Mg [53]. The diameter of the loop in the high-frequency range is normally attributed to the charge transfer reaction, which is proportional to the transfer resistance value, i.e., R_t. The larger the R_t, the better is the corrosion resistance of coating [16]. From R_t value, the exchange-current density (j_0) could also be calculated using Equation (7) [54]:

$$J_0 = RT/nFR_t \tag{7}$$

where n is the number of transferred charges, and F is Faraday constant. In Equation (7), j_0 is in opposite proportion to R_t, i.e., the higher the R_t is, the lower would be the corrosion rate [53]. Hence, charge transfer resistance could be utilized to assess the corrosion rate of the samples. This is because an increase in j_0 should correspond to an increase in the corrosion rate. It can be deduced from EIS spectra that R_t of AZ91 samples increased from 137.6 Ω cm^2 to 439.7 Ω cm^2 and 5432.7 Ω cm^2 for MAO and nanocomposite coated samples, suggesting that the nanocomposite coating is more corrosion resistant than AZ91, which is in good agreement with the results of polarization measurements (Figure 4b). Hence, the results of electrochemical tests reveal the increased corrosion resistance afforded by the nanocomposite coating. A delayed corrosion process is critical for a biodegradable implant, as the implant needs to maintain its mechanical functionality for a certain period before the bone defect is fully healed [55]. Therefore, the immersion tests can provide additional information regarding the corrosion rates of the AZ91, MAO, and nanocomposite coated samples for a longer time.

3.2.2. Immersion Tests

The corrosion rate of the AZ91 sample obtained significantly higher than the MAO and nanocomposite coated samples (0.57 ± 0.02 vs. 0.39 ± 0.01 and 0.08 ± 0.01 mg/cm^2/hr, respectively after 72 h immersion in the SBF) showing the effective protection provided by the MAO and nanocomposite coating (Figure 4c). Following the immersion test, local areas of the AZ91 surface were corroded, and many large cracks and pores were detected on the surface due to significant corrosion. Clusters of white particles had also been formed on the AZ91 surface (Figure 4d,g). The surface morphology of MAO treated AZ91 had too been corroded, and some pits and cracks were seen (Figure 4e,h). Comparing the surface morphology of samples following immersion showed that the density of cracks and pits formed on the AZ91 sample due to the corrosion were significantly higher than those formed on MAO and nanocomposite coated samples. It could be clearly seen that the MAO and nanocomposite coated samples had a more uniform and milder corrosion attack when compared to the AZ91 sample. The density of white particles formed on the surface of nanocomposite coated samples was also higher than MAO and AZ91 samples. In fact, the total surface of nanocomposite coated samples had been covered with cauliflower-like white particles (Figure 4f,i). Also, the degree of corrosion attack and formed white particles for the MAO sample was between AZ91 and nanocomposite coated samples (Figure 4e,h).

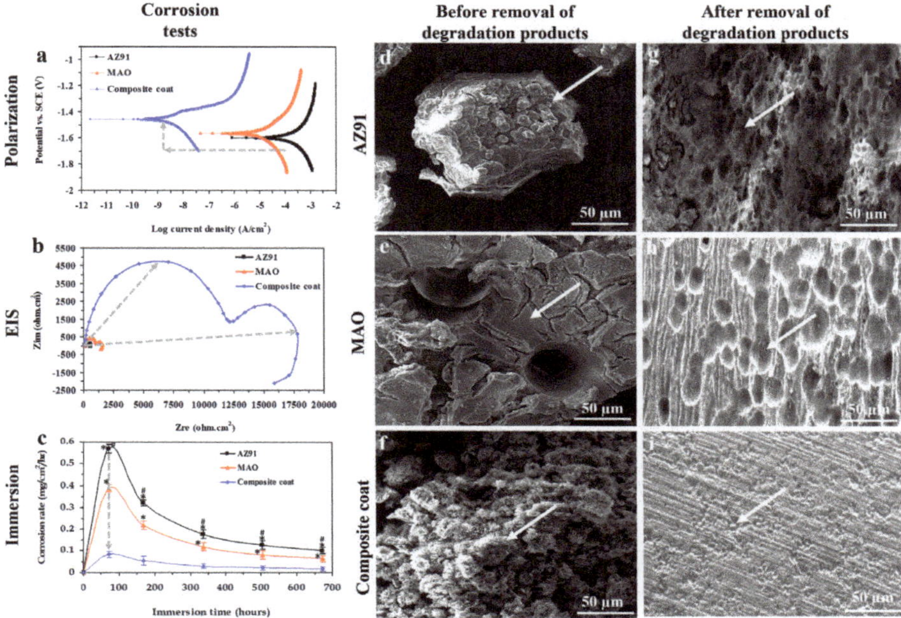

Figure 4. Corrosion tests: (**a**,**b**) Results of electrochemical corrosion tests include potentiodynamic polarization (**a**), EIS (**b**) and (**c**) immersion test showing that the corrosion rate of AZ91 Mg alloy substrate reduced following MAO and nanocomposite coating; (**d–i**) SEM images of AZ91 (**d**,**g**), MAO (**e**,**h**), and nanocomposite coated (**f**,**i**) samples after 672hr in SBF. Images have been taken before (**d–f**) and after (**g–i**) removal of degradation products. SEM images show that the cracks and pits formed on the AZ91 sample due to the corrosion were significantly more than MAO and nanocomposite coated samples. Furthermore, the surface of nanocomposite coated samples had been totally covered with cauliflower-like white particles. Significant differences: * $p < 0.05$: AZ91 vs. MAO or Composite coat, # $p < 0.05$: MAO vs. Composite coat.

Due to the excellent castability, mechanical properties, corrosion resistance, and high maximum solubility of 12.7 wt.% in Mg, aluminum (Al) has been one of the most commonly used alloying elements for Mg-alloy systems in the early development stage of biodegradable orthopedic implants [56]. Mg–Al alloy systems such as AZ alloys, which were already processed for industrial applications, are currently available for further optimization, such as a surface coating as used in our study. In general, an increased Al content in Mg alloys enhances the ultimate tensile strength (UTS) and elongation up to 6 wt.% while reducing the corrosion rate by forming an aluminum oxide film [57]. Although Al is a well-known neurotoxicant linked with Alzheimer's disease and dementia [58], researchers argue that the amount of Al released from such alloy systems with less than 5 wt.% Al is well below the weekly intake limits, and long-term in vivo studies have shown no direct detrimental effect [14,59]. AZ91 Mg alloy used in our study consists of 9 wt.% Al, i.e., higher than the threshold mentioned above (5 wt.%), however, using our nanocomposite coating system, the corrosion rate of AZ91 significantly reduced from 0.57 ± 0.02 to 0.08 ± 0.01 mg/cm^2/hr which will also cause a significant reduction in Al release from our AZ91 substrate. Hence, our composite coated AZ91 offers a reduced corrosion rate as well as high mechanical properties.

The corrosion of Mg alloy proceeds by the following reactions:

$$\text{Anodic reaction: } Mg \rightarrow Mg^{2+} + 2e^-, \quad (8)$$

$$\text{Cathodic reaction: } 2H_2O + 2e^- \rightarrow H_2 + 2OH^-, \quad (9)$$

$$\text{Total reaction: } Mg_{(s)} + 2H_2O_{(aq)} \rightarrow Mg(OH)_{2\,(s)} + H_{2\,(g)}, \quad (10)$$

$$Mg(OH)_{2\,(s)} + 2Cl^-_{(aq)} \rightarrow MgCl_{2\,(aq)} + 2OH^-_{(aq)}. \quad (11)$$

Following the immersion of an Mg alloy in the SBF, the electrolyte penetration followed by chemical dissolution results in the substrate to undergo rapid corrosion, and a magnesium hydroxide ($Mg(OH)_2$) layer is then formed on its surface (reactions (8)–(10)). The deposition of $Mg(OH)_2$ layer on the surface of the Mg alloy substrate can also act as a protective film that can prevent direct exposure of corrosion medium to Mg alloy. $Mg(OH)_2$ would then react with chloride ions in the SBF and form the soluble $MgCl_2$ (reaction (11)) [60]. This is the reason why our AZ91 Mg alloy sample had been subjected to a significant rate in the initial phase of immersion. Next, the formed corrosion product, which mainly contains $Mg(OH)_2$, would thicken with immersion time, and the corrosion rate gradually decreases [61]. Although $Mg(OH)_2$ forms on the surface of Mg alloy, it is too porous to protect the AZ91 substrate from corrosion effectively. Hence, continuous corrosion happens on Mg. This corrosion trend, i.e., fast corrosion initially followed by slow corrosion as time evolves, is also supported by our results (Figure 4c). However, our MAO and nanocomposite coating could act as an effective barrier to protect the AZ91 substrate from corrosion. Ca^{2+} ions and PO_4^{3-} groups from the SBF and Mg^{2+} ions released from the AZ91 substrate also took part in the surface reaction of samples and form a calcium phosphate layer such as $Ca_3Mg_3(PO_4)_4$ on the sample (reaction (12)) which has a cauliflower-like structure [62]. Our SEM results also confirmed the formation of a layer with a cauliflower-like structure on the samples (Figure 4f). The formation of this phosphate coating can further protect the substrate from fast corrosion [63]. Therefore, a plateau in corrosion rates of AZ91, MAO, and nanocomposite coated samples were observed at the last stage of our immersion test, i.e., from 336 h to 672 h (Figure 4c).

$$Mg^{2+} + Ca^{2+} + PO_4^{3-} \rightarrow Ca_3Mg_3(PO_4)_4. \quad (12)$$

3.3. *In Vivo Animal Study*

Following implantation, animals did not exhibit any sign of moribund/lethargic, distress, or local infections. Also, a normal wound healing was reported post-operation. In radiography images, AZ91 implants showed the highest hydrogen bubbles formation at the beginning, followed by a reduction over time. However, hydrogen bubbles formation reduced when MAO implants were used, and almost no hydrogen bubbles were seen around the nanocomposite coated implants (Figure 5d–i). Similarly, previous research has also shown that hydrogen bubbles are found around the Mg implants, which are then disappeared after 2–3 weeks [17,64]. In the first two weeks post-implantation, the rate of hydrogen-evolution from Mg implants is quicker than the hydrogen absorption rate. Over time, the corrosion rate reduces because of the formation of $Mg(OH)_2$ and other corrosion products such as $Ca_3Mg_3(PO_4)_4$ [65]. Hence, the hydrogen bubbles around implants reduced from two weeks to two months. When the volume and weight of explanted implants were measured, the change in volume and weight of nanocomposite coated implants were significantly lower compared to AZ91 and MAO treated implants (volume change: 9.1 ± 0.2% vs. 42.8 ± 3.3% and 32 ± 1.7%; weight change: 4 ± 1 vs. 25 ± 4, and 16 ± 3 mg/cm^2) showing the reduced in vivo corrosion of AZ91 implants when coated with our composite nanoparticles. Using histology, we observed new bone had been formed around the implants for all implants. However, compare to both AZ91 and MAO implants, nanocomposite coated implants showed the highest amount of new bone formation (56 ± 5% vs. 27 ± 1% and 31 ± 2%; Figure 5j–o,r). When inflammatory response in the tissue surrounding implants was compared, AZ91 had the highest inflammatory response; however, it decreased when MAO or composite coated implant was used (15 ± 3% vs. 42 ± 4% and 32 ± 4%; Figure 5j–o,s). This increase in bone formation and reduction in inflammatory response due to the composite coating on implants can be due to several reasons. One reason is a reduced corrosion rate and, therefore hydrogen-evolution since the coating decreases the direct contact of the implant with the body fluid. Furthermore, the production of hydrogen bubbles due to high corrosion of Mg alloys can prevent physiological bone reaction and callus formation [66], thereby resulting in a decrease in new bone formation and higher inflammation around the uncoated implants when compared to the coated ones.

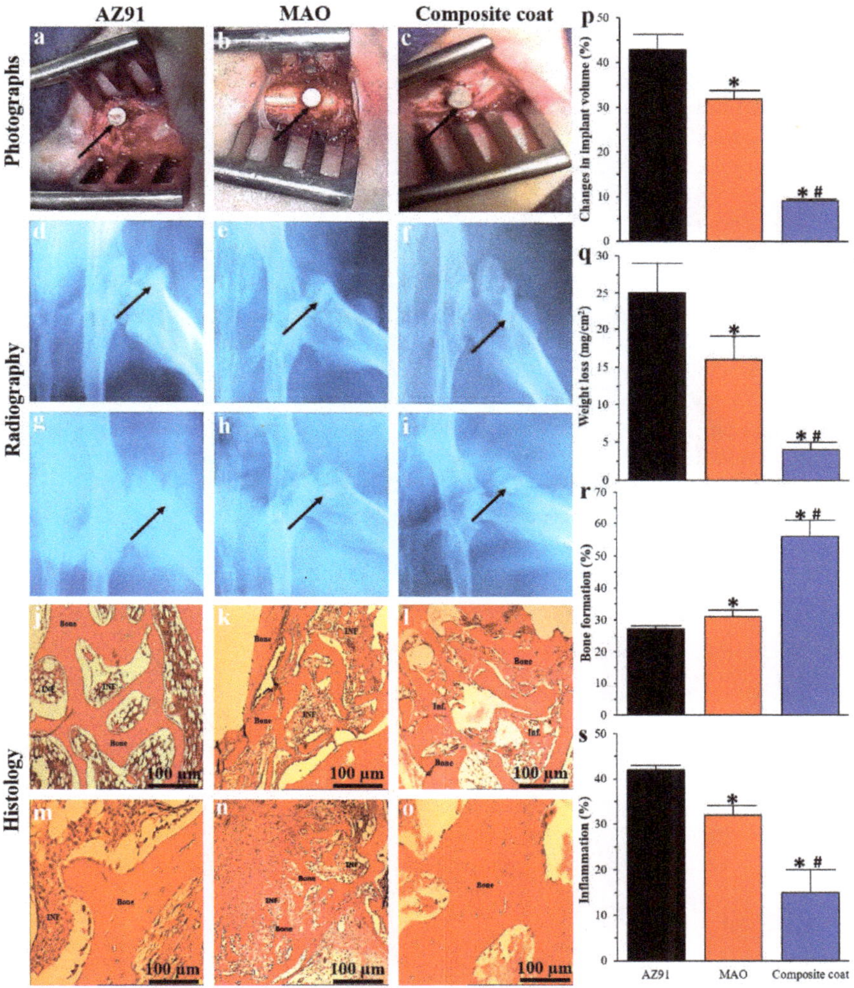

Figure 5. In vivo animal study: Surgical photos (**a–c**), radiography images (**d–i**), and histological images (**i–o**) of AZ91 (**a,d,g,j,m**), MAO (**b,e,h,k,n**), and nanocomposite coated (**c,f,I,l,o**) samples. Radiography images have been taken 2 weeks (**d–f**), and 2 months (**g–i**) after implantation. Black arrows in radiography images show the implants. Histological images have been presented in low (**j–l**) and high (**m–o**) magnification; Change in implant volume (**p**), weight loss (**q**), percentage of bone formation (**r**), and inflammation (**s**) for AZ91, MAO, and nanocomposite samples. Significant differences: * $p < 0.05$: AZ91 vs. MAO or Composite coat, # $p < 0.05$: MAO vs. Composite coat.

Previous studies on biodegradable Mg alloys have shown that AZ91 Mg alloy corrodes at a rate of 1.1 mm/year [67], LAE442 at 2.8 mm/year [68], and WE43 at 3.9 mm/year [69]. Zinc, another biodegradable metallic implant, corrodes at a rate of 0.2 mm/year, which is a critically low corrosion rate for satisfactory biodegradable cardiovascular stents, although zinc corrodes more quickly after 3 months and should be removed away from the artery [70,71]. The corrosion rate for the AZ91 Mg alloy in our study obtained 1.3 mm/year, which is similar to the value reported for AZ91, i.e., 1.1 mm/year [72]. However, following MAO treatment and composite coating, the corrosion rate of AZ91 significantly reduced to 1.2 and 0.00005 mm/year, respectively, which is considerably lower than the corrosion rate of widely studied Mg alloys such as LAE442 or WE43 for orthopedic implant applications. This result demonstrates that our composite coat prepared by the MAO/EPD method is promising for applications with a strict requirement for corrosion rate. In addition to reducing the corrosion rate, our strategy will also give bioactivity, i.e., osteoproductivity to Mg, which makes it a suitable platform for bone implantation and regeneration. Although our nanocomposite coating has been applied on AZ91 Mg alloy, this surface coating material (i.e., diopside-bredigite-FHA) with its coating method (i.e., MAO/EPD) can also be utilized on other biodegradable Mg alloys where a reduced corrosion rate, as well as an enhanced bioactivity, are required.

4. Conclusions

A nanocomposite made of diopside, bredigite, and fluoridated hydroxyapatite bioceramics were successfully coated on a biodegradable AZ91 Mg alloy using micro-arc oxidation followed by an electrophoretic deposition method. Following coating, the corrosion rate of AZ91 reduced from 0.57 ± 0.02 to 0.08 ± 0.01 mg/cm^2/hr, which resulted in a reduced hydrogen-evolution in vivo. Improved bone regeneration ($27 \pm 1\%$ to $56 \pm 5\%$) with a reduced inflammatory response ($42 \pm 4\%$ to $15 \pm 3\%$) was detected in the tissue surrounding the composite coated implant compared to the uncoated ones. Hence, our composite coating strategy can be used on biodegradable Mg bone implants, where a reduced corrosion rate and improved implant osteointegration are required. Our results will help shed not only new light on the possible development of Mg-based orthopedic implants, i.e., bone plates, screws, pins, and nails, but also provide guidelines for the development and surface coating of Mg-based porous scaffolds for bone tissue engineering.

Author Contributions: Mehdi Razavi: Conceptualization, Methodology, Investigation, Data curation, Writing—original draft. Mohammadhossein Fathi, Omid Savabi, Lobat Tayebi and Daryoosh Vashaee: Visualization, Supervision, review & editing. All authors have read and agreed to the published version of the manuscript.

Funding: This research was funded by National Science Foundation (NSF) under grant numbers ECCS-1351533, ECCS-1515005, ECCS-1711253, and Air Force Office of Scientific Research (AFOSR) under contract number FA9550-12-1-0225.

Acknowledgments: The authors are thankful for the contributions of the Isfahan University of Technology and Torabinejad Dental Research Center in this work.

Conflicts of Interest: The authors declare no conflict of interest.

References

1. Razavi, M.; Huang, Y. Effect of hydroxyapatite (HA) nanoparticles shape on biodegradation of Mg/HA nanocomposites processed by high shear solidification / equal channel angular extrusion route. *Mater. Lett.* **2020**, *267*, 127541. [CrossRef]
2. Razavi, M.; Fathi, M.H.; Meratian, M. Microstructure, mechanical properties and bio-corrosion evaluation of biodegradable AZ91-FA nanocomposites for biomedical applications. *Mater. Sci. Eng. A* **2010**, *527*. [CrossRef]
3. Razavi, M.; Fathi, M.; Savabi, O.; Razavi, S.M.; Beni, B.H.; Vashaee, D.; Tayebi, L. Controlling the degradation rate of bioactive magnesium implants by electrophoretic deposition of akermanite coating. *Ceram. Int.* **2013**, *40*, 3865–3872. [CrossRef]
4. Davies, J.E. In Vitro modeling of the bone/implant interface. *Anat. Rec.* **1996**. [CrossRef]
5. Anselme, K. Osteoblast adhesion on biomaterials. *Biomaterials* **2000**. [CrossRef]

6. Hench, L.L. Bioceramics: From concept to clinic. *J. Am. Ceram. Soc.* **1991**, *74*, 1487–1510. [CrossRef]
7. Cao, W.; Hench, L.L. Bioactive materials. *Ceram. Int.* **1996**. [CrossRef]
8. Suchanek, W.; Yoshimura, M. Processing and properties of hydroxyapatite-based biomaterials for use as hard tissue replacement implants. *J. Mater. Res.* **1998**. [CrossRef]
9. Krüger, R.; Groll, J. Fiber reinforced calcium phosphate cements—On the way to degradable load bearing bone substitutes? *Biomaterials* **2012**, *33*, 5887–5900. [CrossRef]
10. Dorozhkin, S.V. Calcium orthophosphates. *J. Mater. Sci.* **2007**, *42*, 1061–1095. [CrossRef]
11. Dorozhkin, S.V. Calcium orthophosphates in nature, biology and medicine. *Materials* **2009**, *2*, 399–498. [CrossRef]
12. Dorozhkin, S.V. Calcium Orthophosphates as Bioceramics: State of the Art. *J. Funct. Biomater.* **2010**, *1*, 22–107. [CrossRef] [PubMed]
13. Meng, L.; Xie, F.; Zhang, B.; Wang, D.K.; Yu, L. Natural Biopolymer Alloys with Superior Mechanical Properties. *ACS Sustain. Chem. Eng.* **2019**, *7*, 2792–2802. [CrossRef]
14. Han, H.S.; Loffredo, S.; Jun, I.; Edwards, J.; Kim, Y.C.; Seok, H.K.; Witte, F.; Mantovani, D.; Glyn-Jones, S. Current status and outlook on the clinical translation of biodegradable metals. *Mater. Today* **2019**, *23*, 57–71. [CrossRef]
15. Witte, F.; Feyerabend, F.; Maier, P.; Fischer, J.; Störmer, M.; Blawert, C.; Dietzel, W.; Hort, N. Biodegradable magnesium–hydroxyapatite metal matrix composites. *Biomaterials* **2007**, *28*, 2163–2174. [CrossRef]
16. Razavi, M.; Fathi, M.; Savabi, O.; Vashaee, D.; Tayebi, L. In Vitro study of nanostructured diopside coating on Mg alloy orthopedic implants. *Mater. Sci. Eng. C* **2014**, *41*, 168–177. [CrossRef]
17. Staiger, M.P.; Pietak, A.M.; Huadmai, J.; Dias, G. Magnesium and its alloys as orthopedic biomaterials: A review. *Biomaterials* **2006**, *27*, 1728–1734. [CrossRef]
18. Razavi, M.; Fathi, M.; Savabi, O.; Vashaee, D.; Tayebi, L. Biodegradable magnesium alloy coated by fluoridated hydroxyapatite using MAO/EPD technique. *Surf. Eng.* **2014**, *30*, 545–551. [CrossRef]
19. Razavi, M.; Huang, Y. A Magnesium-based Nanobiocomposite Processed by a Novel Technique Combining High Shear Solidification and Hot Extrusion. *Recent Pat. Nanotechnol.* **2019**, *13*, 38–48. [CrossRef]
20. Song, G. Control of biodegradation of biocompatable magnesium alloys. *Corros. Sci.* **2007**, *49*, 1696–1701. [CrossRef]
21. Chiu, K.Y.; Wong, M.H.; Cheng, F.T.; Man, H.C. Characterization and corrosion studies of fluoride conversion coating on degradable Mg implants. *Surf. Coat. Technol.* **2007**, *202*, 590–598. [CrossRef]
22. Razavi, M. In Vitro Evaluations of Anodic Spark Deposited AZ91 Alloy as Biodegradable Metallic Orthopedic Implant. *Annu. Res. Rev. Biol.* **2014**, *4*, 3716–3733. [CrossRef]
23. Razavi, M.; Fathi, M.H.; Meratian, M. Fabrication and characterization of magnesium-fluorapatite nanocomposite for biomedical applications. *Mater. Charact.* **2010**, *61*. [CrossRef]
24. Razavi, M.; Fathi, M.; Savabi, O.; Vashaee, D.; Tayebi, L. Micro-arc oxidation and electrophoretic deposition of nano-grain merwinite ($Ca_3MgSi_2O_8$) surface coating on magnesium alloy as biodegradable metallic implant. *Surf. Interface Anal.* **2014**, *46*, 387–392. [CrossRef]
25. Razavi, M.; Fathi, M.; Savabi, O.; Vashaee, D.; Tayebi, L. Improvement of Biodegradability, Bioactivity, Mechanical Integrity and Cytocompatibility Behavior of Biodegradable Mg Based Orthopedic Implants Using Nanostructured Bredigite ($Ca_7MgSi_4O_{16}$) Bioceramic Coated via ASD/EPD Tec. *Ann. Biomed. Eng.* **2014**, *42*. [CrossRef]
26. Wu, C.; Ramaswamy, Y.; Zreiqat, H. Porous diopside ($CaMgSi_2O_6$) scaffold: A promising bioactive material for bone tissue engineering. *Acta Biomater.* **2010**, *6*, 2237–2245. [CrossRef]
27. Razavi, M.; Fathi, M.; Savabi, O.; Tayebi, L.; Vashaee, D. Improvement of in vitro behavior of an Mg alloy using a nanostructured composite bioceramic coating. *J. Mater. Sci. Mater. Med.* **2018**, *29*. [CrossRef]
28. Iwata, N.Y.; Lee, G.H.; Tsunakawa, S.; Tokuoka, Y.; Kawashima, N. Preparation of diopside with apatite-forming ability by sol-gel process using metal alkoxide and metal salts. *Colloids Surf. B Biointerfaces* **2004**, *33*, 1–6. [CrossRef]
29. Wu, C.; Chang, J. Synthesis and apatite-formation ability of akermanite. *Mater. Lett.* **2004**, *58*, 2415–2417. [CrossRef]
30. Hafezi-Ardakani, M.; Moztarzadeh, F.; Rabiee, M.; Talebi, A.R. Synthesis and characterization of nanocrystalline merwinite ($Ca_3Mg(SiO_4)_2$) via sol-gel method. *Ceram. Int.* **2011**, *37*, 175–180. [CrossRef]

31. Wu, C.; Chang, J.; Zhai, W.; Ni, S. A novel bioactive porous bredigite ($Ca_7MgSi_4O_{16}$) scaffold with biomimetic apatite layer for bone tissue engineering. *J. Mater. Sci. Mater. Med.* **2007**, *18*, 857–864. [CrossRef] [PubMed]
32. Razavi, M.; Fathi, M.; Savabi, O.; Razavi, S.M.; Heidari, F.; Manshaei, M.; Vashaee, D.; Tayebi, L. In Vivo study of nanostructured diopside ($CaMgSi_2O_6$) coating on magnesium alloy as biodegradable orthopedic implants. *Appl. Surf. Sci.* **2014**, *313*, 60–66. [CrossRef]
33. Razavi, M.; Fathi, M.; Savabi, O.; Vashaee, D.; Tayebi, L. In Vivo biocompatibility of Mg implants surface modified by nanostructured merwinite/PEO. *J. Mater. Sci. Mater. Med.* **2015**, *26*, 184. [CrossRef] [PubMed]
34. Razavi, M.; Fathi, M.; Savabi, O.; Vashaee, D.; Tayebi, L. Regenerative influence of nanostructured bredigite ($Ca_7MgSi_4O_{16}$)/anodic spark coating on biodegradable AZ91 magnesium alloy implants for bone healing. *Mater. Lett.* **2015**, *155*. [CrossRef]
35. Razavi, M.; Fathi, M.; Savabi, O.; Boroni, M. A review of degradation properties of Mg based biodegradable implants. *Res. Rev. Mater. Sci. Chem.* **2012**, *1*, 15–58.
36. Boccaccini, A.R.; Keim, S.; Ma, R.; Li, Y.; Zhitomirsky, I. Electrophoretic deposition of biomaterials. *J. R. Soc. Interface* **2010**, *7*, S581–S613. [CrossRef]
37. Corni, I.; Ryan, M.P.; Boccaccini, A.R. Electrophoretic deposition: From traditional ceramics to nanotechnology. *J. Eur. Ceram. Soc.* **2008**, *28*, 1353–1367. [CrossRef]
38. Kwok, C.T.; Wong, P.K.; Cheng, F.T.; Man, H.C. Characterization and corrosion behavior of hydroxyapatite coatings on Ti6Al4V fabricated by electrophoretic deposition. *Appl. Surf. Sci.* **2009**, *255*, 6736–6744. [CrossRef]
39. Razavi, M.; Fathi, M.; Savabi, O.; Vashaee, D.; Tayebi, L. In Vitro Analysis of Electrophoretic Deposited Fluoridated Hydroxyapatite Coating on Micro-arc Oxidized AZ91 Magnesium Alloy for Biomaterials Applications. *Metall. Mater. Trans. A Phys. Metall. Mater. Sci.* **2014**, *46*. [CrossRef]
40. Chen, Q.; Cordero-Arias, L.; Roether, J.A.; Cabanas-Polo, S.; Virtanen, S.; Boccaccini, A.R. Alginate/Bioglass® composite coatings on stainless steel deposited by direct current and alternating current electrophoretic deposition. *Surf. Coat. Technol.* **2013**, *233*, 49–56. [CrossRef]
41. Razavi, M.; Fathi, M.; Savabi, O.; Beni, B.H.; Vashaee, D.; Tayebi, L. Surface microstructure and in vitro analysis of nanostructured akermanite ($Ca_2MgSi_2O_7$) coating on biodegradable magnesium alloy for biomedical applications. *Colloids Surf. B Biointerfaces* **2014**, *117*, 432–440. [CrossRef] [PubMed]
42. Razavi, M.; Fathi, M.; Savabi, O.; Vashaee, D.; Tayebi, L. In vivo study of nanostructured akermanite/PEO coating on biodegradable magnesium alloy for biomedical applications. *J. of Biomed. Mater. Res. Part A.* **2015**, *103*, 1798–1808. [CrossRef] [PubMed]
43. Williamson, G.K.; Hall, W.H. X-ray line broadening from filed aluminium and wolfram. *Acta Metall.* **1953**, *1*, 22–31. [CrossRef]
44. Kokubo, T.; Takadama, H. How useful is SBF in predicting In Vivo bone bioactivity. *Biomaterials* **2006**, *27*, 2907–2915. [CrossRef]
45. ASTM. *ASTM G31—Standard Practice for Laboratory Immersion Corrosion Testing of Metals*; American Society For Testing Materials; ASTM: Philadelphia, PA, USA, 1999.
46. Raman, R.K.S.; Jafari, S.; Harandi, S.E. Corrosion fatigue fracture of magnesium alloys in bioimplant applications: A review. *Eng. Fract. Mech.* **2015**, *137*, 97–108. [CrossRef]
47. Ridzwan, M.I.Z.; Shuib, S.; Hassan, A.Y.; Shokri, A.A.; Ibrahim, M.N.M. Problem of stress shielding and improvement to the hip implant designs: A review. *J. Med. Sci.* **2007**, *7*, 460–467. [CrossRef]
48. Razavi, M.; Fathi, M.; Savabi, O.; Vashaee, D.; Tayebi, L. Biodegradation, bioactivity and In Vivo biocompatibility analysis of plasma electrolytic oxidized (PEO) biodegradable Mg implants. *Phys. Sci. Int. J.* **2014**, *4*, 708–722. [CrossRef]
49. Guo, H.F.; An, M.Z.; Huo, H.B.; Xu, S.; Wu, L.J. Microstructure characteristic of ceramic coatings fabricated on magnesium alloys by micro-arc oxidation in alkaline silicate solutions. *Appl. Surf. Sci.* **2006**, *252*, 7911–7916. [CrossRef]
50. Kheirkhah, M.; Fathi, M.; Salimijazi, H.R.; Razavi, M. Surface modification of stainless steel implants using nanostructured forsterite (Mg2SiO4) coating for biomaterial applications. *Surf. Coatings Technol.* **2015**, *276*, 580–586. [CrossRef]
51. Cui, X.; Li, Y.; Li, Q.; Jin, G.; Ding, M.; Wang, F. Influence of phytic acid concentration on performance of phytic acid conversion coatings on the AZ91D magnesium alloy. *Mater. Chem. Phys.* **2008**, *111*, 503–507. [CrossRef]

52. Song, G.; Bowles, A.L.; StJohn, D.H. Corrosion resistance of aged die cast magnesium alloy AZ91D. *Mater. Sci. Eng. A* **2004**, *366*, 74–86. [CrossRef]
53. Song, G.; Atrens, A.; Wu, X.; Zhang, B. Corrosion behaviour of AZ21, AZ501 and AZ91 in sodium chloride. *Corros. Sci.* **1998**, *40*, 1769–1791. [CrossRef]
54. Udhayan, R.; Bhatt, D.P. On the corrosion behaviour of magnesium and its alloys using electrochemical techniques. *J. Power Sources* **1996**, *63*, 103–107. [CrossRef]
55. Witte, F. The history of biodegradable magnesium implants: A review. *Acta Biomateri.* **2010**, *6*, 1680–1692. [CrossRef]
56. Chen, Y.; Xu, Z.; Smith, C.; Sankar, J. Recent advances on the development of magnesium alloys for biodegradable implants. *Acta Biomater.* **2014**, *10*, 4561–4573. [CrossRef]
57. Zheng, Y.F.; Gu, X.N.; Witte, F. Biodegradable metals. *Mater. Sci. Eng. R Rep.* **2014**, *77*, 1–34. [CrossRef]
58. El-Rahman, S.S.A. Neuropathology of aluminum toxicity in rats (glutamate and GABA impairment). *Pharmacol. Res.* **2003**, *47*, 189–194. [CrossRef]
59. Angrisani, N.; Reifenrath, J.; Zimmermann, F.; Eifler, R.; Meyer-Lindenberg, A.; Vano-Herrera, K.; Vogt, C. Biocompatibility and degradation of LAE442-based magnesium alloys after implantation of up to 3.5 years in a rabbit model. *Acta Biomater.* **2016**, *44*, 355–365. [CrossRef]
60. Razavi, M.; Huang, Y. Assessment of magnesium-based biomaterials: From bench to clinic. *Biomater. Sci.* **2019**, *7*, 2241–2263. [CrossRef]
61. Zhang, Y.; Yan, C.; Wang, F.; Li, W. Electrochemical behavior of anodized Mg alloy AZ91D in chloride containing aqueous solution. *Corros. Sci.* **2005**, *47*, 2816–2831. [CrossRef]
62. Razavi, M.; Fathi, M.; Savabi, O.; Hashemi Beni, B.; Razavi, S.M.; Vashaee, D.; Tayebi, L. Coating of biodegradable magnesium alloy bone implants using nanostructured diopside (CaMgSi2O6). *Appl. Surf. Sci.* **2014**, *288*, 130–137. [CrossRef]
63. Fathi, M.; Meratian, M.; Razavi, M. Novel magnesium-nanofluorapatite metal matrix nanocomposite with improved biodegradation behavior. *J. Biomed. Nanotech.* **2011**, *7*, 441–445. [CrossRef] [PubMed]
64. Razavi, M.; Fathi, M.; Meratian, M. Bio-corrosion behavior of magnesium-fluorapatite nanocomposite for biomedical applications. *Mater. Lett.* **2010**, *64*, 2487–2490. [CrossRef]
65. Witte, F.; Fischer, J.; Nellesen, J.; Crostack, H.-A.; Kaese, V.; Pisch, A.; Beckmann, F.; Windhagen, H. In Vitro and In Vivo corrosion measurements of magnesium alloys. *Biomaterials* **2006**, *27*, 1013–1018. [CrossRef]
66. Witte, F.; Kaese, V.; Haferkamp, H.; Switzer, E.; Meyer-Lindenberg, A.; Wirth, C.J.; Windhagen, H. In Vivo corrosion of four magnesium alloys and the associated bone response. *Biomaterials* **2005**, *26*, 3557–3563. [CrossRef]
67. Serre, C.M.; Papillard, M.; Chavassieux, P.; Voegel, J.C.; Boivin, G. Influence of magnesium substitution on a collagen–apatite biomaterial on the production of a calcifying matrix by human osteoblasts. *J. Biomed. Mater. Res.* **1998**, *42*, 626–633. [CrossRef]
68. Chiu, C.; Lu, C.T.; Chen, S.H.; Ou, K.L. Effect of hydroxyapatite on the mechanical properties and corrosion behavior of Mg-Zn-Y alloy. *Materials* **2017**, *10*. [CrossRef]
69. Laws, P. Biodegradable Magnesium Alloys and Uses Thereof. U.S. Patent 20090081313A1, 26 March 2009.
70. Jiang, L.; Xu, F.; Xu, Z.; Chen, Y.; Zhou, X.; Wei, G.; Ge, H. Biodegradation of AZ31 and WE43 magnesium alloys in simulated body fluid. *Int. J. Electrochem. Sci.* **2015**, *10*, 10422–10432.
71. Bowen, P.K.; Drelich, J.; Goldman, J. Zinc exhibits ideal physiological corrosion behavior for bioabsorbable stents. *Adv. Mater.* **2013**, *25*, 2577–2582. [CrossRef]
72. Li, H.; Zheng, Y.; Qin, L. Progress of biodegradable metals. *Prog. Nat. Sci. Mater. Int.* **2014**, *24*, 414–422. [CrossRef]

Article

In Vitro Corrosion Behavior of Biodegradable Iron Foams with Polymeric Coating

Radka Gorejová [1], Renáta Oriňaková [1,*], Zuzana Orságová Králová [1], Matej Baláž [2], Miriam Kupková [3], Monika Hrubovčáková [3], Lucia Haverová [1], Miroslav Džupon [3], Andrej Oriňak [1], František Kaľavský [1] and Karol Kovaľ [3]

1 Department of Physical Chemistry, Faculty of Science, Pavol Jozef Šafárik University in Košice, Moyzesova 11, 041 54 Košice, Slovakia; radka.gorejova@student.upjs.sk (R.G.); zuzana.orsagova.kralova@upjs.sk (Z.O.K.); markusovabuckova@gmail.com (L.H.); andrej.orinak@upjs.sk (A.O.); frantisek.kalavsky@upjs.sk (F.K.)

2 Institute of Geotechnics, Slovak Academy of Sciences, Watsonova 45, 040 01 Košice, Slovakia; balazm@saske.sk

3 Institute of Materials Research, Slovak Academy of Sciences, Watsonova 47, 040 01 Košice, Slovakia; mkupkova@saske.sk (M.K.); mhrubovcakova@saske.sk (M.H.); mdzupon@saske.sk (M.D.); kkoval@saske.sk (K.K.)

* Correspondence: renata.orinakova@upjs.sk; Tel.: +421-55-234-2324

Received: 17 December 2019; Accepted: 28 December 2019; Published: 2 January 2020

Abstract: Research in the field of biodegradable metallic scaffolds has advanced during the last decades. Resorbable implants based on iron have become an attractive alternative to the temporary devices made of inert metals. Overcoming an insufficient corrosion rate of pure iron, though, still remains a problem. In our work, we have prepared iron foams and coated them with three different concentrations of polyethyleneimine (PEI) to increase their corrosion rates. Scanning electron microscopy (SEM) coupled with energy dispersive X-ray analysis (EDX), Fourier-transform infrared spectroscopy (FT-IR), and Raman spectroscopy were used for characterization of the polymer coating. The corrosion behavior of the powder-metallurgically prepared samples was evaluated electrochemically using an anodic polarization method. A 12 weeks long in vitro degradation study in Hanks' solution at 37 °C was also performed. Surface morphology, corrosion behavior, and degradation rates of the open-cell foams were studied and discussed. The use of PEI coating led to an increase in the corrosion rates of the cellular material. The sample with the highest concentration of PEI film showed the most rapid corrosion in the environment of simulated body fluids.

Keywords: iron foam; polyethyleneimine (PEI); biodegradation; powder metallurgy; coating

---

1. Introduction

In recent years, the development of biodegradable orthopedical scaffolds has advanced significantly [1–7]. Resorbable materials are intended to serve as temporary support for damaged tissue. Compared to the standard medical devices typically made of stainless steel, cobalt-chromium, or titanium alloys [8], this new group of materials possess a particular advantage in the form of in vivo self-adsorbing capacity. Corrosion is therefore no longer seen as a problem, and appropriate biodegradable devices can be made by targeted designing and influencing of their degradation rates.

Iron-based biodegradable materials (Fe-BM) are considered a suitable alternative to permanent metallic implants [9–15]. They showed satisfactory cytocompatibility in previous studies and their mechanical properties could match those of natural bone [9,16,17]. Hydrogen evolution, too rapid degradation, or suppressed antibacterial performance, problems associated with the other most-studied biodegradable metal—magnesium, are not present in the case of iron [18,19]. However, the disadvantage of very slow degradation in physiological pH has to be overcome. There have been several reports studying the corrosion behavior of Fe-BM using different approaches to solve this issue. One of the most used methods to fasten the degradation is alloying with another element(s). Manganese, platinum, sulfur, carbon, palladium, etc. were tested in different ratios to the iron [10,15,20–22]. Even though these additions managed to accelerate iron degradation, mechanical properties or overall biocompatibility are often impaired. Degradation of BM depends on various factors and it is known that besides the composition of the specimen, the preparation method and its geometrical form plays also an important role [23]. The porous structure is beneficial for healthy vascularization and tissue ingrowth and is typically used in the field of orthopedic implants [24–26].

Another way to enhance corrosion, but also improve the biological performance of prepared material, is the usage of different coatings. Three groups of coating materials are usually used. The first group consists of inorganic ceramic coatings where hydroxyapatite (HAp) and other calcium phosphates (tricalcium phosphates (TCP), biphasic calcium phosphates (BCP)) have a leading position due to their similarity to the inorganic component of natural bone, osteoconductivity, and osseointegration properties [23,27,28]. Representatives of the second group are the metal–ceramic composites (calcium silicate-iron e.g., [29].)

Polymers are the third group of the coating materials for bioabsorbable metals. Poly-lactic-acid (PLA), poly-lactic-co-glycolic acid (PLGA), or polyethyleneglycol (PEG) are used to the highest extent [25,30,31]. It is known that the passivation layer of corrosion products can be formed on the surface of the specimen which retards further corrosion. Yusop et al. [31] found that pH in the proximity to the metal surface can be lowered by the polymer degradation and therefore the solubility of this passive layer is enhanced as long as the solubility of these corrosion products (mostly calcium or magnesium phosphates, iron hydroxides, etc.) is higher in the lower pH. This can lead to higher corrosion rates of the studied implants. Polymeric coating, though, can not only enhance the corrosion rate but also improve the biological performance of the scaffold. When PEG was used, the positive effect on the material biocompatibility was observed [32].

Polyethyleneimine (PEI) is an organic polymer soluble in water and ethanol with a high density of amino groups which can be protonated [33,34]. This polycation exists in linear or branched form and its properties depend on molecular weight and structure [35]. PEI has been studied for several decades [36] and found its place in various biological applications. It can be utilized as a drug carrier [35], in tumor imaging [37], or in gene transduction into mesenchymal stem cells (MSCs) [38]. The cytotoxic effect of the PEI relies upon the size, structure, and its ratio. Xia et al. [35] found, that by a careful selection of PEI size, it is possible to achieve minimal or no cytotoxicity. Yao et al. [38] confirmed that not only the size, but also the concentration of the coated layer, has an influence on the resulting cytotoxicity, which can be adjusted by careful choosing. Moreover, the polycationic character of PEI due to the amino groups' protonation can interact with negatively charged bacteria [34]. Several studies confirmed improvement of the biocompatibility of PEI-coated materials [36,38]. In addition to this, the PEI structure provides possibilities to modify it with various polymers, create layers (e.g., PEG, chitosan), and furthermore, to load it with drugs that could possibly take a place in the bone healing process [34,38–40].

In our work, we have prepared foam-like scaffolds from the carbonyl iron powder (CIP) via the powder metallurgy process. Inspired by our previous work on Fe-PEG [30] material, we used polyethyleneimine as a coating material, which was deposited on the surface of the Fe sample using a cost-effective dip-coating method. The morphology of the sample's surface was studied prior to coating and after depositing the PEI layer in three different concentrations. To determine the corrosion rate in a physiological environment, electrochemical potentiodynamic tests and in vitro immersion tests were carried out using Hanks' solution to mimic body fluids. The composition and appearance of the corrosion products created after 4, 8, and 12 weeks of immersion were examined and the influence of the polymeric layer on corrosion of the iron scaffold was discussed.

2. Materials and Methods

2.1. Material Preparation

Porous iron samples were prepared from carbonyl iron powder (CIP) by BASF (type CC d50, 3.8–5.3 µm; 99.5% Fe, 0.05% C, 0.01% N, 0.18% O) by the impregnation of the polyurethane (PUR) foam (Filtren, TM 25133). The impregnating suspension consisted of 7 g of CIP iron powder, 6 mL of distilled water, and 0.2 g of gelatine (Sigma-Aldrich) dissolved at 60 °C for better adhesion of iron slurry to the PUR foam. Cylindrical (Ø 5 mm, h 15 mm) foams were impregnated for 24 h and thermally treated in a tube furnace (ANETA 1) at 450 °C for 1 h in N_2 atmosphere (for PUR matrix elimination) and sintered at 1120 °C for 1 h in a reduction atmosphere (10% H_2, 90% N_2) to obtain the final structure. CIP pellets (Ø 10 mm, h 2 mm) used for Raman spectroscopy experiments were prepared by cold pressing iron powder at 600 MPa, subsequent sintering at 1120 °C for 1 h in a reduction atmosphere (10% H_2, 90% N_2), and coated as described below.

2.2. PEI Coating Preparation

Polyethyleneimine (Sigma-Aldrich; 50% (*w*/*v*) in H_2O) film was achieved by a dip-coating process. Samples were ultrasonically cleaned in acetone and ethanol, in each for 10 min, and dipped into three different PEI solutions (5, 10, and 15 wt % corresponding to PEI1, PEI2, and PEI3, respectively) for 90 min and then dried at 37 °C for 12 h.

2.3. Microstructure and Surface Characterization

The microstructure of porous iron foams before and after 4, 8, and 12 weeks of corrosion was observed using an optical microscope (Olympus GX71, OLYMPUS Europa Holding GmbH, Hamburg, Germany). Samples were molded into the methyl-methacrylate resin (Dentacryl), hardened, and grinded.

Scanning electron microscopy (SEM, Jeol Ltd., Tokyo, Japan) and energy dispersive X-ray analysis (EDX) (JOEL JSM-7001F with INCA EDX analyzer, Oxford Instruments, Abingdon, Oxfordshire, UK) were used for surface morphology characterization.

The specific surface area (S_{BET}) was determined by the low-temperature nitrogen adsorption method using a NOVA 1200e Surface Area and Pore Size Analyzer (Quantachrome Instruments, Hartley Wintney, UK). The values were calculated using Brunauer–Emmett–Teller (BET) theory.

The Fourier-transform infrared spectroscopy (FT-IR) spectra were recorded on the Tensor 29 infrared spectrometer (Bruker, Karlsruhe, Germany) using the attenuated total reflection (ATR) method.

The Raman spectra were recorded using a Renishaw inVia spectrophotometer (Renishaw UK Sales Ltd., Wotton-under-Edge, UK). All spectra were recorded through 4x-objective using a 532 nm laser from 100 to 4000 cm^{-1} at a 50% laser power. The samples were exposed to the laser for 10 s with 3 accumulations.

2.4. Electrochemical Corrosion Testing

The electrochemical measurements were conducted in Hanks' solution (8 NaCl, 0.4 KCl, 0.14 $CaCl_2$, 0.06 $MgSO_4.7H_2O$, 0.06 $NaH_2PO_4.2H_2O$, 0.35 $NaHCO_3$, 1.00 glucose, 0.60 KH_2PO_4, and 0.10 $MgCl_2.6H_2O$ in g/L) with pH 7.4 ± 0.2 at 37 ± 1 °C using a potentiostat (Autolab PGSTAT 302N). A three-electrode system with Ag/AgCl/KCl (3 mol/L) as a reference electrode, platinum counter electrode, and iron sample as the working electrode were used. The potentiodynamic polarization tests were carried out from −1000 to −300 mV (vs. Ag/AgCl/KCl (3 mol/L)) at a scanning rate of 0.1 mV/s. The corrosion rate was determined using the Tafel extrapolation method and calculated from Equation (1), where *CR* is corrosion rate, j_{corr} is corrosion current density (µA/cm^2), *K* is a constant (3.27×10^{-3})

determining output units of *CR*, *EW* is equivalent weight (27.92 g/eq for Fe), and *d* is the iron foam density (0.024 g/cm^3 [41]).

$$CR = \frac{j_{corr} K\, EW}{d} \tag{1}$$

2.5. Immersion Test

Before the static immersion test, all uncoated samples were ultrasonically cleaned in acetone and ethanol for 10 min, air-dried, and weighed. Static immersion tests were conducted for 12 weeks at 37 °C. Corrosion rates were calculated from Equation (2), where m_f is sample weight after degradation (g), m_i is sample weight at the beginning of the experiment (g), *K* is the constant (8.76 × 10^4), *A* is the sample area (cm^2), *t* is the exposure time (h), and *d* is the material density (g/cm^3). Samples were immersed in 120 mL of Hanks' solution and the uniform access of the corrosion medium to the whole sample surface was ensured.

$$CR = \frac{\left(m_i - m_f\right) 8.76 \times 10^{-4}}{A\, t\, d} \tag{2}$$

pH of Hanks' solution was measured, and total iron content was determined using atomic absorption spectroscopy on AAnalyst 100 after 4, 8, and 12 weeks of corrosion.

3. Results and Discussion

3.1. Material Characterization

3.1.1. Morphology of the Sintered Iron Foam

Cellular iron-based samples intended to serve as a potential orthopedic implant were prepared via the powder-metallurgical route. A little shrinkage of the specimens occurred after sintering when compared to the size of green compacts. Open porosity was well-preserved, which indicates a good material capacity for further tissue growth through the implant. Pores in the micrometer range (600 to 2000 μm) were present alongside smaller pores in the range of 0.5 to 6 μm, as shown in Figure 1. The surface of the sintered foams was humpy, as shown in Figure 1b, which can be attributed to the spherical character of the iron powder particles serving as raw material. Metallographic cross-sections of the pure iron foams, as shown in Figure 1c,d, confirmed these observations and showed a highly micro-porous structure.

Thinning of the cell walls occurred at their centers and the widest wall size was observed at the cell joints. Evaporation of gases after PUR foam elimination led to the creation of a third type of porosity, which was localized randomly, only in some regions of the samples, as shown in Figure 1c. This uncertain porosity should impair the mechanical properties of the specimen and should be considered and eliminated in the future fabrication process.

3.1.2. Characterization of the Polymer Coating

Sintered iron foams were ultrasonically cleaned and dip-coated with three different concentrations of PEI. The thin polymeric coating was observed after solvent evaporation. Ethanol (96 vol %) was selected as a solvent to achieve fast evaporation and to minimize the risk of material corrosion during the manufacturing processes. The presence of the polymeric layer was confirmed by the EDX method, where nitrogen, carbon, and oxygen were spotted for the coated samples while only iron was detected for the uncoated specimen, as shown in Figure 2. The average (from 10 measurements) weight of the resultant coating for different PEI concentrations and corresponding weight percentage is summarized in Table 1. While the PEI1 coating forms almost 2.0 wt % of the sample, it is 5.0 wt % for the PEI2 and 6.6 wt % for the PEI3 sample. The small difference between the weight of the PEI2 and PEI3 coating should be attributed to the higher saturation on the sample surface and depletion of the free space available for deposition. Similar space occupation by the polymeric layer for PEI2 and PEI3 can be seen in Figure 3i,l, while the uncoated areas are present when the sample is coated with only PEI1 (5 wt % of PEI) solution.

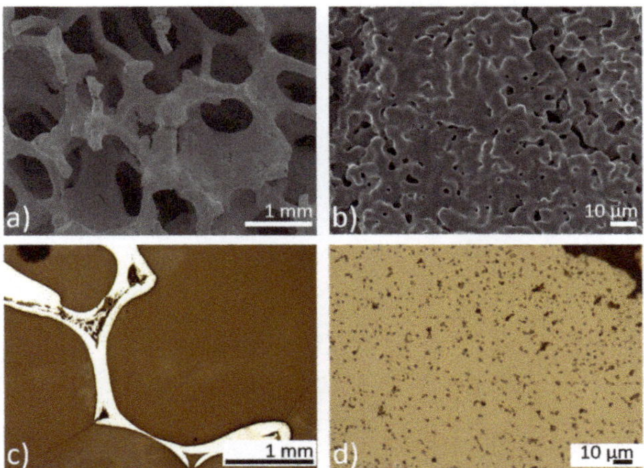

Figure 1. (**a**,**b**) The scanning electron microscopy (SEM) micrographs of the sintered iron scaffold; (**c**,**d**) metallographic cross-sections of the sintered iron scaffold. Comparison of the different porosities present in the material.

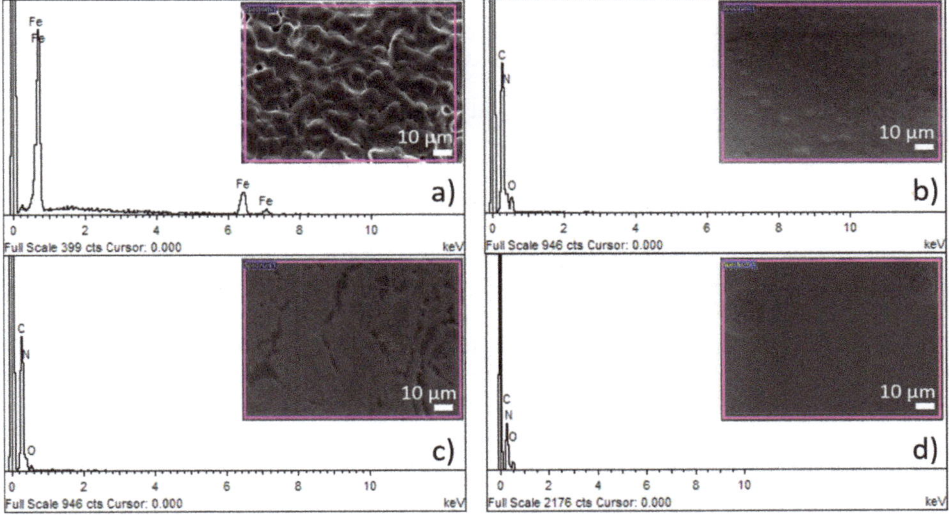

Figure 2. Chemical composition of the surface of (**a**) pure Fe; (**b**) Fe-PEI1; (**c**) Fe-PEI2; and (**d**) Fe-PEI3 samples studied by the energy dispersive X-ray analysis (EDX) method.

Table 1. Average weight (mg) and content (wt %) of polyethyleneimine (PEI) coating deposited on the surface of the iron foams.

	Fe-PEI1	Fe-PEI2	Fe-PEI3
Average PEI weight (mg)	15.9	41.4	53.6
Average PEI content (wt %)	1.9	5.0	6.6

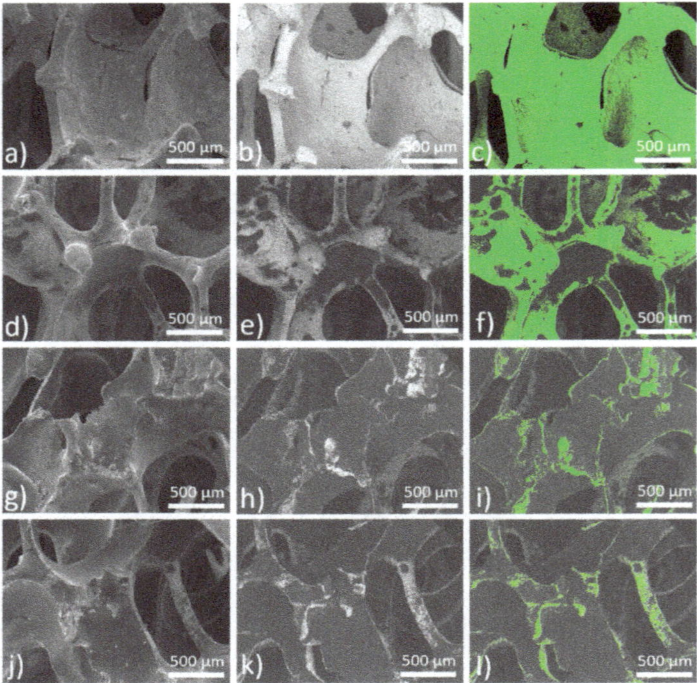

Figure 3. SEM micrographs of the prepared (**a–c**) Fe; (**d–f**) Fe-PEI1; (**g–i**) Fe-PEI2; and (**j–l**) Fe-PEI3 materials. Iron matrix is highlighted in green for better contrast in (**c**,**f**,**i**,**l**).

Polymer distribution on the surface of iron matrix for the samples with different content of the polymer is depicted in Figure 3. SEM micrographs were taken in two different scanning modes (Secondary Electron Imaging (SEI) and Composition (COMPO)) for better evaluation of the polymer surface distribution. Difference between the heavy elements (e.g., Fe) and the light elements (e.g., C, O, N) is displayed as a color difference—the heavier is the element, the lighter is the color in which it is displayed. In Figure 3c,f,i,l, pure iron is graphically highlighted in green for better contrast. It can be seen that the coating on the Fe-PEI1 sample does not cover the entire surface and the polymer is mostly localized in the cell valleys, while the coverage of the wall edges is incomplete, as shown in Figure 3c. With the higher polymer concentration, coverage of the material increases, however, the edges of the walls still remain uncoated. Surface smoothing with increasing coverage of the iron substrate by the polymer layer led to the creation of the homogeneous surface. Polymer addition led to the micropores filling with coating material, as shown in Figure 4. The specific surface area (S_{BET}) significantly decreased with the increasing polymer concentration, as shown in Table 2. This finding is similar to the results observed for the Fe-PEG material [30], where surface area increased after the first addition of polymer but decreased continuously with increasing polymer concentration. It can be seen that coating with PEI does not lead to the creation of islands of polymer and its use resulted in the smoothing of the coated surface. In the case of very porous substances with large specific surface areas, like activated carbon or fibrous silica, the addition of PEI also decreased the S_{BET} value [42,43]. The surface area plays an important role in the evaluation of the corrosion measurements and therefore should not be neglected in further analysis.

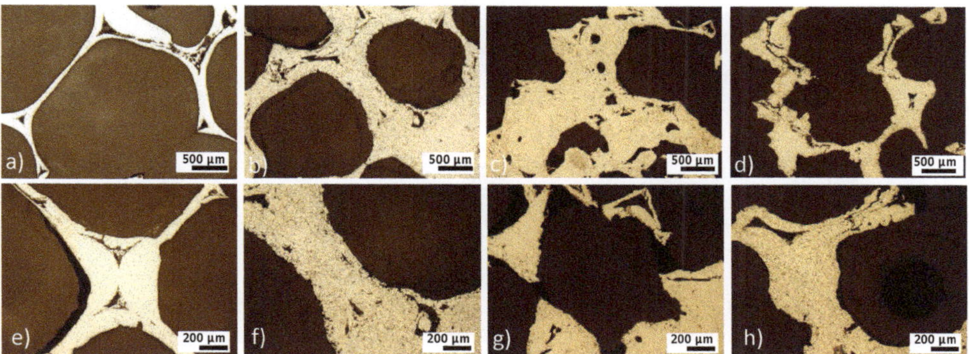

Figure 4. Metallographic cross-sections of the (**a**,**e**) Fe; (**b**,**f**) Fe-PEI1; (**c**,**g**) Fe-PEI2; and (**d**,**h**) Fe-PEI3 before corrosion.

Table 2. Specific surface area (S_{BET}) of the PEI coated (Fe-PEI) and the uncoated (pure Fe) foams.

S_{BET} (g m $^{-2}$)			
Fe	**Fe-PEI1**	**Fe-PEI2**	**Fe-PEI3**
1.19	0.92	0.61	0.04

The polymeric layer was confirmed and analyzed by different methods. FT-IR and Raman spectra were recorded before the corrosion of material to study the PEI layer and its interaction with the Fe matrix, as shown in Figure 5.

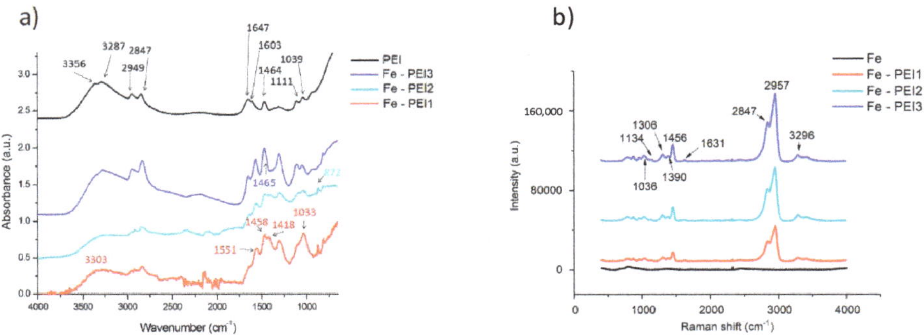

Figure 5. The spectra of pure iron and PEI coated scaffolds recorded using (**a**) FT-IR and (**b**) Raman spectroscopy.

The vibrations of the functional groups of PEI can be found at the following wavenumber in the spectrum of pure PEI: the stretching vibrations of the $-NH_2$ group at 3356 and 3287 cm^{-1}, the asymmetric and symmetric stretching vibrations of $-CH_2$ group at 2949 and 2847 cm^{-1}, the bending vibration of the $-NH_2$ group at 1603 cm^{-1}, the in-plane bending vibration of the $-CH_2$ group at 1464 cm^{-1}, and the stretching vibration of the C–N group at 1111 cm^{-1}. These positions are in accordance with recent literature [44–46]. After the interaction of PEI with Fe, the spectrum has significantly changed, as the majority of the peaks were shifted and also their intensity changed, as shown in Figure 5a. Moreover, there are differences among the Fe-PEI samples. Basically, for the samples with the low PEI content, the intensity of the peaks corresponding to the vibrations of the $-NH_2$ group has significantly decreased in the case of these samples, meaning that this group can be mainly responsible for the interaction with iron. Furthermore, the new peak located at 872 cm^{-1} was evidenced. It is possible that the most effective interaction between Fe and the polymer is achieved when a smaller amount of PEI is used.

Figure 5b depicts the Raman spectrum of the PEI in different concentrations deposited on the surface of the CIP pellets. The bands at 1456 cm^{-1} and bands at 1306, 1134, and 1036 cm^{-1} correspond to the methylene $-CH_2$ group (wagging and twisting motions) and could be also found in the spectra of the pure polyethylene and ethylenediamine [47]. The 1456 and 1306 cm^{-1} bands are also present in the spectrum of the pure 50 wt % PEI solution, which was observed in [47]. Bands at the 1631 cm^{-1} correspond to the amino group ($-NH_2$). Intensive bands corresponding to the C–H bond are present at 2700 to 3100 cm^{-1}, which is in accordance with the literature [48]. Different conformational changes could appear during the polymer adsorption to the surface. Symmetric and asymmetric valence vibrations are slightly shifted in the spectrum due to the amino groups in the PEI structure. In the pure 50 wt % PEI spectrum are these bands at 2956 and 2866 cm^{-1}, but due to the adsorption processes to the metallic surface, these could be shifted (from 2866 cm^{-1} to 2847 cm^{-1} in this case). Chaufer [49] studied PEI adsorption onto Zr and assumed that the Lewis acid–base bonding occurs between the amino groups and the metal, which was also reported in the case of silver [47]. However, the phenomenon was observed when the $-NH_2$ group band (1600 cm^{-1}) was shifted to the lower values. We have observed shift to the higher values (1631 cm^{-1}), therefore this type of interaction probably could not be applied to our system. Despite this fact, the analysis of the FT-IR spectra of coated foams discussed earlier in this work confirmed the interaction between the amino group and the metal surface.

3.2. Degradation Study

3.2.1. Potentiodynamic Polarization Tests

For the evaluation of the degradation rates of coated and uncoated samples, a potentiodynamic polarization test was performed. Corrosion current density (j_{corr}), corrosion potential (E_{corr}), and the polarization resistance are summarized in Table 3. Potentiodynamic curves obtained during the measurement in the Hanks' solution at 37 ± 1 °C from −1000 to −300 mV are shown in Figure 6.

Table 3. Electrochemical parameters of the Fe and Fe-PEI samples obtained from the Tafel analysis of polarization curves measured at 37 ± 1 °C in the Hanks' solution. Corrosion current (i_{corr}), corrosion current density (j_{corr}), corrosion potential (E_{corr}), polarization resistance (PR), corrosion rate (CR).

Sample	E_{corr} (V)	i_{corr} (A)	j_{corr} (μA cm^{-2})	PR (Ω cm^{-2})	CR (mm y^{-1})
Fe	−0.627	11.91×10^{-5}	1.18×10^{-2}	0.017	0.045
Fe-PEI1	−0.722	37.11×10^{-5}	4.52×10^{-2}	0.041	0.172
Fe-PEI2	−0.687	102.95×10^{-5}	3.10×10^{-2}	0.039	0.118
Fe-PEI3	−0.658	5.39×10^{-5}	15.49×10^{-2}	4.460	0.590

The coating of the iron foams with the PEI has resulted in the shift of the corrosion potential to the more negative values obtained for all three different concentrations of polymer. The lowest value of E_{corr} was observed for the Fe-PEI1, followed by the Fe-PEI2, Fe-PEI3, and the pure iron, which exhibited the most positive value (−627.0 mV). Reported corrosion potentials of the pure iron observed by the anodic polarization method were −860.7 mV [41], −484.0 mV [50] or −510.0 mV [51]. For example, in [52], the authors studied electrochemical degradation of the pure iron bars and phosphated iron bars in Hanks' solution and observed the corrosion potential of pure iron to be −670 mV, which is similar to that observed in this study. It can be seen, from the different results for the same composition of the sample (Fe), that the corrosion rate is dependent also on the preparation method and sample geometry. The foam-like structure of the samples prepared by the powder metallurgy method resulted in the shift of the electrochemical corrosion potential to more negative values when compared to that of standard pure iron.

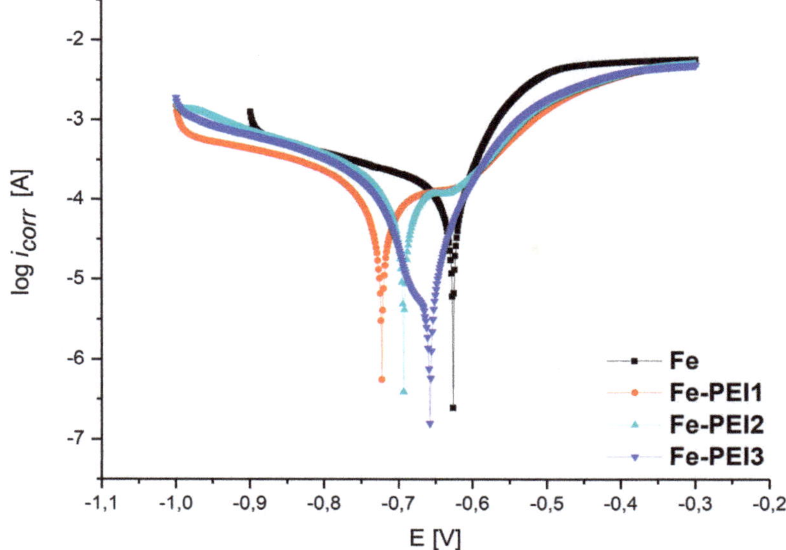

Figure 6. Polarization curves of the iron-based foams with or without PEI coating obtained in the Hanks' solution at 37 ± 1 °C.

The highest corrosion current density was observed for the Fe-PEI3 sample, indicating the highest ability to corrode. Corrosion rates (*CR*) calculated from Equation (1) confirmed that the highest corrosion rate from all the samples was the Fe-PEI3 sample, even though its polarization resistance (*PR*) was the highest when compared to other samples. The determining parameter for such a behavior is the material surface area, as shown in Table 2. The specific surface area of the pure iron is, due to its inhomogeneity, almost 30 times higher when compared to the Fe-PEI3. This fact emphasizes the need to know the real surface areas during the evaluation of degradation behavior of polymer-coated samples with the tendency to lower S_{BET} values. Porous iron coated with PLGA [31] reached the corrosion rate of 0.420 mm y^{-1}, which is similar to the results of Fe-PEI3 material (0.590 mm y^{-1}). An important difference can be seen in the corrosion rates of pure iron, which is 5 times lower than that reported in [53]. This fact can be also attributed to the different surface areas of the pure iron sample affected by the preparation method and to the highly-porous structure. Similar values of *CR* (0.04 ± 0.01) were observed in [16] for 3D-printed Fe-Mn samples, while in [51], the authors reported the i_{corr} of pure Fe to be 1.68×10^{-5} A, which emphasizes even more the influence of the preparation method on the degradation behavior of biodegradable materials.

3.2.2. Static Degradation Tests

Static immersion tests provide complex information about the degradation processes of metallic samples and also about the changes in the corrosion medium. The most important advantage of this type of corrosion testing is its ability to simulate real-body conditions in a more authentic way than dynamic electrochemical tests. Pictures of iron and PEI-coated iron foams after degradation tests are shown in Figure 7. After 8 weeks of corrosion, all samples were completely covered with corrosion products in brown, red, and orange color forming the rust. With the prolonged time of immersion, surface roughness and inhomogeneity increased for all samples, as shown in Figures 8–10. Uniform corrosion, as shown in Figure 8a,b, was observed in the initial stage of degradation whereas pitting corrosion, as shown in Figure 9f,g, typical for the environment with high concentration of chloride ions [54], occurred with prolonged time of immersion. In Figure 10 are presented cross-sections of the

Fe and Fe-PEI samples after 12 weeks of corrosion. The main difference, compared to the un-corroded material shown in Figure 4, could be observed in the thinning of the cell walls accompanied by cracking at the narrowest points, as shown in Figure 10c,d, which can lead to the worsening of the mechanical properties of such material.

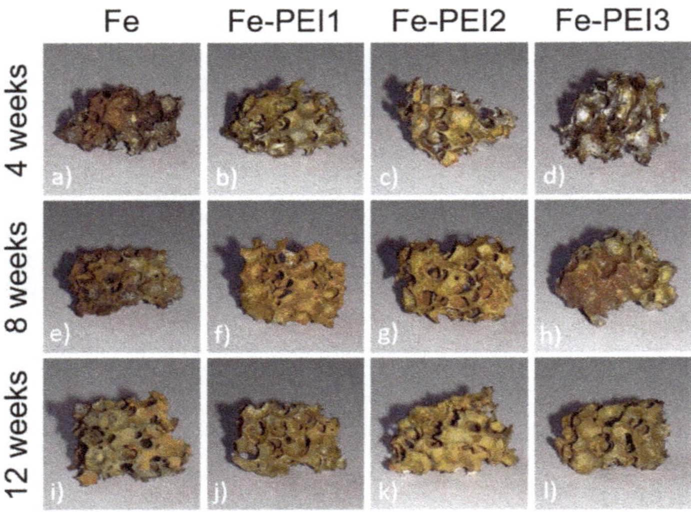

Figure 7. Cellular scaffold made of (**a**,**e**,**i**) pure Fe; (**b**,**f**,**j**) Fe-PEI1; (**c**,**g**,**k**) Fe-PEI2; and (**d**,**h**,**l**) Fe-PEI3 after a static immersion test in Hanks' solution at 37 °C.

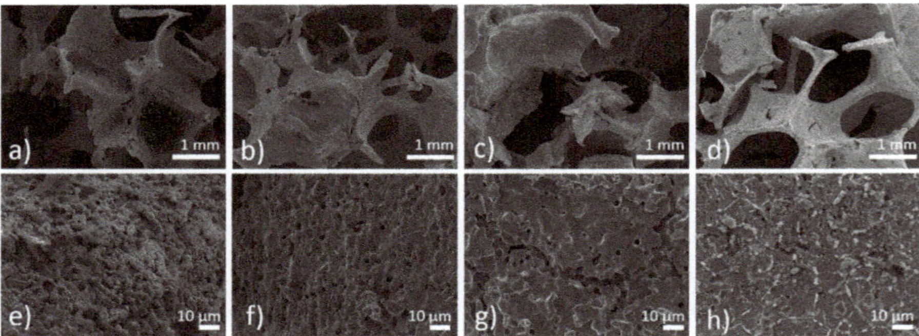

Figure 8. The SEM micrographs of the (**a**,**e**) Fe; (**b**,**f**) Fe-PEI1; (**c**,**g**) Fe-PEI2; and (**d**,**h**) Fe-PEI3 samples after 4 weeks of corrosion in the Hanks´ solution at 37 °C.

Corrosion rates of the coated and uncoated iron-based foams were calculated from the weight loss values recorded after 4, 8, and 12 weeks. The slowest corrosion in the Hanks' solution at 37 °C was observed for the pure iron, followed by the Fe-PEI1, Fe-PEI2, and Fe-PEI3, which is in accordance with the results obtained for Fe and Fe-PEI3 during the potentiodynamic tests. *CR* at different stages of the immersion experiment is listed in Table 4. Yusop et al. [31] have studied PLGA-coated iron and found the corrosion rate to be 0.76 mm y^{-1}, which is comparable to that of Fe-PEI3 determined in this study, as shown in Table 4. The degradation rate of pure iron did not change significantly during the testing period, whereas the *CRs* of Fe-PEI1 and Fe-PEI2 decreased with prolonged time. The Fe-PEI3 sample corroded fastest for the second month and its degradation slowed down during the third month due to the formation of a passivation layer of corrosion products. The thickest polymer layer deposited on the Fe-PEI3 sample could, therefore, serve as a corrosion barrier in the initial stage of the degradation process.

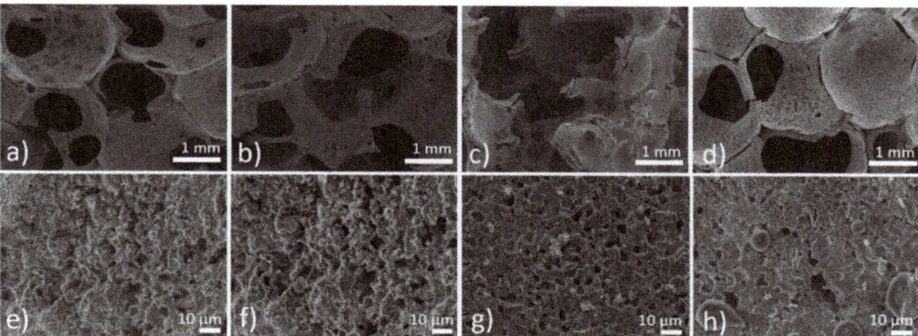

Figure 9. The SEM micrographs of the (**a**,**e**) Fe; (**b**,**f**) Fe-PEI1; (**c**,**g**) Fe-PEI2; and (**d**,**h**) Fe-PEI3 samples after 12 weeks of corrosion in the Hanks′ solution at 37 °C.

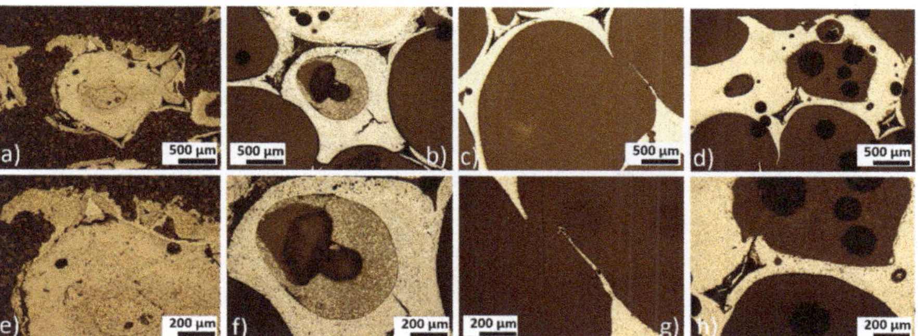

Figure 10. Metallographic cross-sections of the (**a**,**e**) pure iron; (**b**,**f**) Fe-PEI1; (**c**,**g**) Fe-PEI2; and (**d**,**h**) Fe-PEI3 after 12 weeks of corrosion.

Table 4. Corrosion rates of Fe, Fe-PEI1, Fe-PEI2, and Fe-PEI3 calculated from the weight-loss experiments in Hanks' solution at 37 °C for 12 weeks.

	CR [mm y $^{-1}$]		
Week of Immersion	**4**	**8**	**12**
Fe	0.004 ± 0.0015	0.005 ± 0.0030	0.005 ± 0.0034
Fe-PEI1	0.024 ± 0.0052	0.006 ± 0.0008	0.015 ± 0.0047
Fe-PEI2	0.148 ± 0.0420	0.037 ± 0.0113	0.021 ± 0.0209
Fe-PEI3	0.697 ± 0.0398	1.547 ± 0.0793	0.199 ± 0.0109

The composition of the iron-based sample surface after 12 weeks of degradation is shown in Figure 11. All the samples were completely covered with the corrosion products comprising mostly of iron hydroxides. No evidence of nitrogen was observed, assuming total degradation of the PEI coating after three months, which was also confirmed by the FT-IR analysis, as shown in Figure 12.

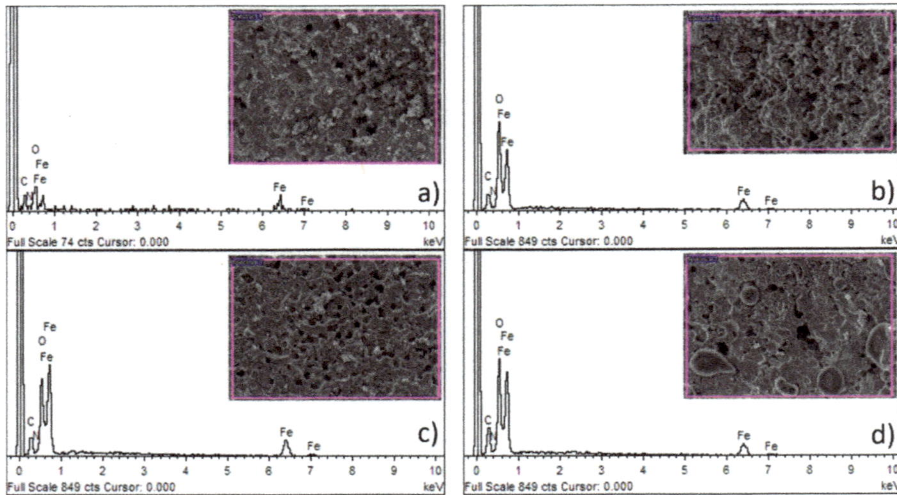

Figure 11. The chemical composition of the surface of (**a**) pure Fe; (**b**) Fe-PEI1; (**c**) Fe-PEI2; and (**d**) Fe-PEI3 after 12 weeks of corrosion in simulated body fluids studied by the EDX method.

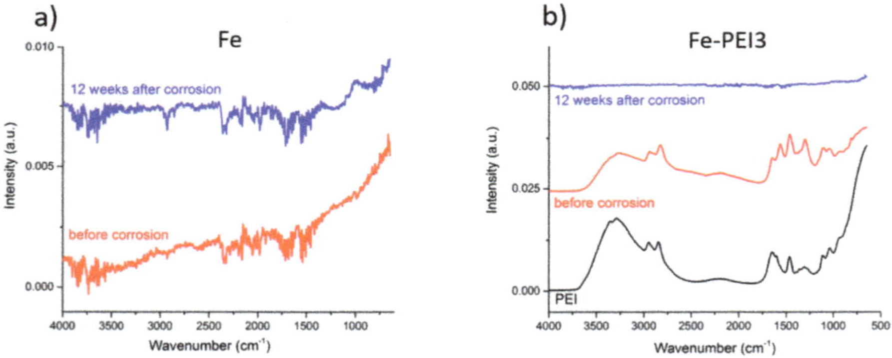

Figure 12. FT-IR spectra of (**a**) pure Fe; (**b**) Fe-PEI3 after 12 weeks of corrosion in Hanks' solution. Spectrum of pure PEI is added for better comparison.

The influence of the corrosion environment on the Fe and Fe-PEI was investigated by FT-IR, as shown in Figure 13a,b, respectively. In Figure 13b, the spectrum of PEI is also included for comparison. The individual intensities of the spectra were adjusted in order to create a comparable figure, namely, the intensities of the polymer and Fe-PEI samples were decreased 10 and 5 times, respectively. The spectra of iron, as shown in Figure 13a, before and after corrosion are almost identical; the differences stay within the measurement error. Neither of the samples exhibit a significant band in FT-IR. The difference between the spectrum of the pure PEI and Fe-PEI specimen was already described earlier. After 12 weeks of corrosion, no bands can be observed, thus the organic layer was completely decomposed. Reactions ongoing during iron degradation in Hanks' solution were already described [23].

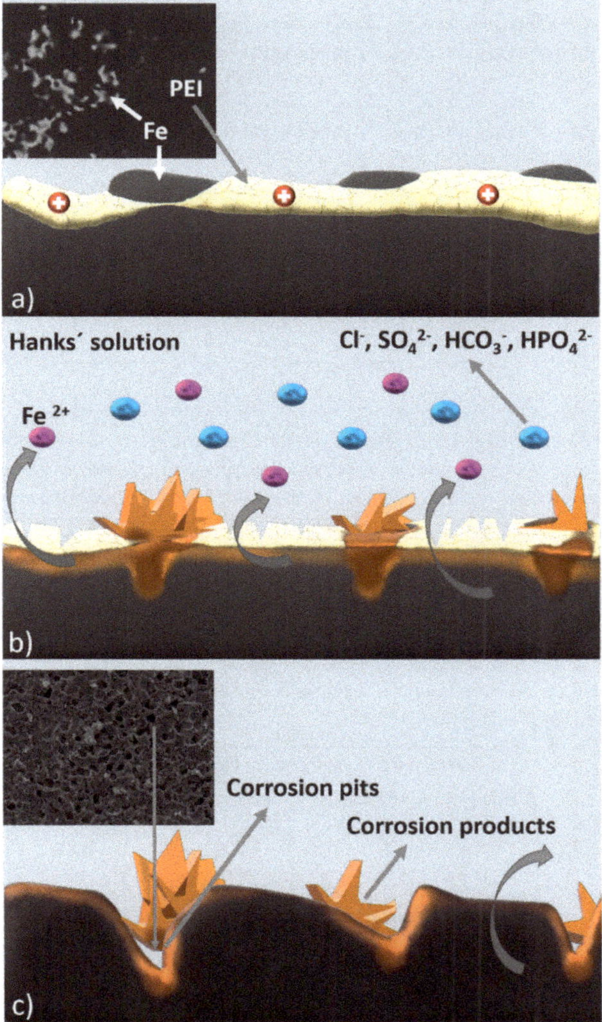

Figure 13. Schematic representation of corrosion processes ongoing on the surface of the Fe-PEI material. (**a**) Polymer-coated sample on-air; (**b**) polymer-coated sample after immersion into Hanks' solution—PEI layer disruption; (**c**) formation of corrosion pits after 12 weeks of biodegradation.

Yusop et al. [31] have suggested accelerated corrosion as a consequence of lowering of pH induced by the polymer hydrolysis. The pH of the Hanks' solution was measured during the immersion study and results are listed in Table 5. After 4 weeks, pH of the corrosion medium was slightly shifted to the higher values. In the case of pure iron and the Fe-PEI3 sample, the constant rise of pH was observed with resultant values of 8.06 and 8.73, respectively. pH values of all of the tested samples were higher at the end of the immersion study and exceeded the value of pH = 8. More basic pH can be attributed to the creation of the corrosion products (e.g., $Fe(OH)_2$, $Fe(OH)_3$, $FeCl_2OH$). Fe-PEI1 and Fe-PEI2 degradation resulted in the lowering of the pH after 8 weeks of the immersion. It is possible that in the case of higher PEI concentration (PEI3), the degradation of the polymeric layer is more rapid in the initial stage (before the fourth week). It increases solubility of the layer of corrosion products, which results in enhanced corrosion, and therefore in the higher content of corrosion products in the Hanks' solution responsible for basic pH. As long as PEI can act as a buffering agent, residual acidity created at the beginning of the degradation process can be compensated by the protonation of PEI functional amino groups.

Table 5. pH of Hanks' solution after 4 to 12 weeks long immersion of iron-based samples coated with PEI and pure iron foams.

pH ± 0.2				
Week of Immersion	**0**	**4**	**8**	**12**
Fe	7.40	7.46	7.75	8.06
Fe-PEI1	7.40	7.43	7.06	8.08
Fe-PEI2	7.40	7.69	7.53	8.23
Fe-PEI3	7.40	7.48	8.25	8.73

Degradation of the polymer caused creation of the localized cracks and pits which served as an entrance for corrosive medium into the depth of the sample, as shown in Figure 13. This can explain the corrosion rate enhancement of coated samples and presence of the pitting corrosion. Moreover, chelation processes between polymer and metal ions [55] could contribute to the higher content of the iron in the solution. Atomic absorption spectroscopy was used to determine the total concentration of the iron (mg mL^{-1}) in the medium, as shown in Table 6. All the observed values increased with prolonged time of immersion, with the highest amount of 0.0337 mg mL^{-1} for Fe-PEI3. Information about the iron concentration released into solution is important for further determination of material cytotoxicity.

Table 6. Total iron concentration observed in the Hanks' solution after 4, 8, and 12 weeks of immersion of cellular scaffolds with polymer (PEI) coating.

Total Iron Concentration (mg mL^{-1})			
Week of Immersion	**4**	**8**	**12**
Fe	0.0243	0.0265	0.0316
Fe-PEI1	0.0151	0.0278	0.0284
Fe-PEI2	0.0184	0.0224	0.0251
Fe-PEI3	0.0139	0.0283	0.0337

Zhang [51] reported that iron concentration lower than 0.075 mg mL^{-1} is safe for cells, therefore it can be assumed that biodegradable foams studied in this paper should not possess cytotoxicity based on the amount of released metal ions into the body. Cytotoxicity and hemocompatibility studies of the material are therefore necessary for its further evaluation.

4. Conclusions

Iron-based foams with open-cell porosity were prepared and analyzed. The deposition of the polymeric (PEI) layer on the surface of the samples led to the changes in their morphology. A significant decrease in the surface area was observed after the application of coating (from 1.19 mg m^{-2} for pure iron to 0.04 mg m^{-2} for Fe-PEI3). Moreover, desirable corrosion rate enhancement mediated through the polymer cracking and corrosion medium penetration enabling took place in the case of polymer-coated samples.

Based on the results reported in this study, coating with polymers can lead to changes in the corrosion rates of metallic samples. Variations in their concentrations seem to be an appropriate way to design devices with desired degradation behavior. PEI is a flexible polymer that can be further functionalized, which makes a room for its future modification. Loading PEI with drugs that could help in bone treatment processes should be an interesting challenge for future research.

Author Contributions: Conceptualization, R.O. and A.O.; Investigation, R.G., Z.O.K., M.B., M.K., M.H., L.H., M.D., F.K., and K.K.; Supervision, R.O.; Writing—original draft, R.G.; Writing—review & editing, R.O. All authors have read and agreed to the published version of the manuscript.

Funding: This research received no external funding.

Acknowledgments: This work was supported by the projects APVV-16-0029 and APVV-18-0357 of the Slovak Research and Development Agency, VEGA 1/0074/17 and VEGA 2/0044/18 of the Slovak Scientific Grant Agency, and VVGS-PF-2019-1049 of the Internal Research Grant System of Faculty of Science of P. J. Šafárik University. The authors would like to thank Zuzana Jurašeková, from the Department of Biophysics and Center for Interdisciplinary Biosciences, Faculty of Science, P.J. Šafárik University, for facilitating the measurement of the Raman spectra.

Conflicts of Interest: The authors declare no conflict of interest.

References

1. Wu, G.; Ibrahim, J.M. Surface design of biodegradable magnesium alloys—A review. *Surf. Coat. Technol.* **2013**, *233*, 2–12. [CrossRef]
2. Yun, Y.; Dong, Z.; Lee, N.; Liu, Y.; Xue, D.; Guo, X.; Kuhlmann, J.; Doepke, A.; Halsall, H.B.; Heineman, W.; et al. Revolutionizing biodegradable metals. *Mater. Today* **2009**, *12*, 22–32. [CrossRef]
3. Han, H.S.; Loffredo, S.; Jun, I.; Edwards, J.; Kim, Y.C.; Seok, H.K.; Witte, F.; Mantovani, D.; Glyn-Jones, S. Current status and outlook on the clinical translation of biodegradable metals. *Mater. Today* **2019**, *23*, 57–71. [CrossRef]
4. Li, H.; Zheng, Y.; Qin, L. Progress of biodegradable metals. *Prog. Nat. Sci. Mater. Int.* **2014**, *24*, 414–422. [CrossRef]
5. Tan, L.; Yu, X.; Wan, P.; Yang, K. Biodegradable materials for bone repairs: a review. *J. Mater. Sci. Technol.* **2013**, *6*, 503–513. [CrossRef]
6. Yusop, A.H.; Bakir, A.A.; Shaharom, N.A.; Abdul Kadir, M.R.; Hermawan, H. Porous Biodegradable Metals for Hard Tissue Scaffolds: A Review. *Int. J. Biomater.* **2012**, *2012*. [CrossRef] [PubMed]
7. Seitz, J.M.; Durisin, M.; Goldman, J.; Drelich, J.W. Recent Advances in Biodegradable Metals for Medical Sutures: A Critical Review. *Adv. Healthc. Mater.* **2015**, *4*, 1915–1936. [CrossRef]
8. Mahyudin, F.; Widhiyanto, L.; Hermawan, H. Biomaterials in orthopaedics. In *Biomaterials and Medical Devices*; Springer: Cham, Switzerland, 2016; pp. 161–181.
9. Kraus, T.; Moszner, F.; Fischerauer, S.; Fiedler, M.; Martinelli, E.; Eichler, J.; Witte, F.; Willbold, E.; Schinhammer, M.; Meischel, M.; et al. Biodegradable Fe-based alloys for use in osteosynthesis: Outcome of an in vivo study after 52 weeks. *Acta Biomater.* **2014**, *10*, 3346–3353. [CrossRef]
10. Schinhammer, M.; Steiger, P.; Moszner, F.; Löffler, J.F.; Uggowitzer, P.J. Degradation performance of biodegradable FeMnC(Pd) alloys. *Mater. Sci. Eng. C* **2013**, *33*, 1882–1893. [CrossRef]
11. Huang, T.; Cheng, J.; Zheng, Y.F. In vitro degradation and biocompatibility of Fe–Pd and Fe–Pt composites fabricated by spark plasma sintering. *Mater. Sci. Eng. C* **2014**, *35*, 43–53. [CrossRef]
12. Schinhammer, M.; Hänzi, A.C.; Löffler, J.F.; Uggowitzer, P.J. Design strategy for biodegradable Fe-based alloys for medical applications. *Acta Biomater.* **2010**, *6*, 1705–1713. [CrossRef] [PubMed]
13. Chou, D.T.; Wells, D.; Hong, D.; Lee, B.; Kuhn, H.; Kumta, P.N. Novel processing of iron-manganese alloy-based biomaterials by inkjet 3-D printing. *Acta Biomater.* **2013**, *9*, 8593–8603. [CrossRef] [PubMed]
14. Oriňaková, R.; Oriňak, A.; Markušová, L.; Labbanczová, E.; Kupková, M.; Hrubovčáková, M.; Fedorková, A. Biodegradable Open Cell Iron Foams for Potential Skeletal Application. *Powder Metall. Prog.* **2012**, *12*, 219–223.
15. Hermawan, H.; Alamdari, H.; Mantovani, D.; Dubé, D. Iron–manganese: New class of metallic degradable biomaterials prepared by powder metallurgy. *Powder Metall.* **2008**, *51*, 38–45. [CrossRef]
16. Hong, D.; Chou, D.T.; Velikokhatnyi, O.I.; Roy, A.; Lee, B.; Swink, I.; Issaev, I.; Kuhn, H.A.; Kumta, P.N. Binder-jetting 3D printing and alloy development of new biodegradable Fe-Mn-Ca/Mg alloys. *Acta Biomater.* **2016**, *45*, 375–386. [CrossRef]
17. Cheng, J.; Huang, T.; Zheng, Y.F. Microstructure, mechanical property, biodegradation behavior, and biocompatibility of biodegradable Fe-Fe2O3 composites. *J. Biomed. Mater. Res. Part A* **2014**, *102*, 2277–2287. [CrossRef]

18. Bakhsheshi-Rad, H.R.; Ismail, A.F.; Aziz, M.; Hadisi, Z.; Omidi, M.; Chen, X. Antibacterial activity and corrosion resistance of Ta_2O_5 thin film and electrospun PCL/MgO-Ag nanofiber coatings on biodegradable Mg alloy implants. *Ceram. Int.* **2019**, *45*, 11883–11892. [CrossRef]
19. Bakhsheshi-Rad, H.R.; Akbari, M.; Ismail, A.F.; Aziz, M.; Hadisi, Z.; Pagan, E.; Daroonparvar, M.; Chen, X. Coating biodegradable magnesium alloys with electrospun poly-L-lactic acid-åkermanite-doxycycline nanofibers for enhanced biocompatibility, antibacterial activity, and corrosion resistance. *Surf. Coat. Technol.* **2019**, *377*, 124898. [CrossRef]
20. Hufenbach, J.; Wendrock, H.; Kochta, F.; Kühn, U.; Gebert, A. Novel biodegradable Fe-Mn-C-S alloy with superior mechanical and corrosion properties. *Mater. Lett.* **2017**, *186*, 330–333. [CrossRef]
21. Cheng, J.; Zheng, Y.F. In vitro study on newly designed biodegradable Fe-X composites (X = W, CNT) prepared by spark plasma sintering. *J. Biomed. Mater. Res. Part B Appl. Biomater.* **2013**, *4*, 485–497. [CrossRef]
22. Zhang, Q.; Cao, P. Degradable porous Fe-35wt.%Mn produced via powder sintering from NH_4HCO_3 porogen. *Mater. Chem. Phys.* **2015**, *163*, 394–401. [CrossRef]
23. Gorejová, R.; Haverová, L.; Oriňaková, R.; Oriňak, A.; Oriňak, M. Recent advancements in Fe-based biodegradable materials for bone repair. *J. Mater. Sci.* **2019**, *54*, 1913–1947. [CrossRef]
24. Čapek, J.; Vojtěch, D.; Oborná, A. Microstructural and mechanical properties of biodegradable iron foam prepared by powder metallurgy. *Mater. Des.* **2015**, *83*, 468–482. [CrossRef]
25. Li, Y.; Jahr, H.; Lietaert, K.; Pavanram, P.; Yilmaz, A.; Fockaert, L.; Leeflang, M.A.; Pouran, B.; Gonzalez-Garcia, Y.; Weinans, H.; et al. Additively manufactured biodegradable porous iron. *Acta Biomater.* **2018**, *77*, 380–393. [CrossRef]
26. Posada, V.M.; Orozco, C.; Ramirez Patino, J.F.; Fernandez-Morales, P. Human Bone Inspired Design of an Mg Alloy-Based Foam. *Mater. Sci. Forum* **2018**, *933*, 291–296. [CrossRef]
27. Daud, N.M.; Sing, N.B.; Yusop, A.H.; Majid, F.A.A.; Hermawan, H. Degradation and in vitro cell-material interaction studies on hydroxyapatite-coated biodegradable porous iron for hard tissue scaffolds. *J. Orthop. Transl.* **2014**, *2*, 177–184.
28. Oriňaková, R.; Oriňak, A.; Kupková, M.; Hrubovčáková, M.; Markušová-Bučková, L.; Giretová, M.; Medvecký, L.; Dobročka, E.; Petruš, O.; Kaľavský, F. In vitro degradation and cytotoxicity evaluation of iron biomaterials with hydroxyapatite film. *Int. J. Electrochem. Sci.* **2015**, *10*, 8158–8174.
29. Wang, S.; Xu, Y.; Zhou, J.; Li, H.; Chang, J.; Huan, Z. In vitro degradation and surface bioactivity of iron-matrix composites containing silicate-based bioceramic. *Bioact. Mater.* **2017**, *2*, 10–18. [CrossRef]
30. Haverová, L.; Oriňaková, R.; Oriňak, A.; Gorejova, R.; Baláž, M.; Vanýsek, P.; Kupkova, M.; Hrubovčáková, M.; Mudroň, P.; Radoňák, J.; et al. An In Vitro Corrosion Study of Open Cell Iron Structures with PEG Coating for Bone Replacement Applications. *Metals* **2018**, *8*, 499. [CrossRef]
31. Yusop, A.H.M.; Daud, N.M.; Nur, H.; Kadir, M.R.A.; Hermawan, H. Controlling the degradation kinetics of porous iron by poly(lactic-co-glycolic acid) infiltration for use as temporary medical implants. *Sci. Rep.* **2015**, *5*, 11194. [CrossRef]
32. Oriňaková, R.; Gorejová, R.; Macko, J.; Oriňak, A.; Kupková, M.; Hrubovčáková, M.; Ševc, J.; Smith, R.M. Evaluation of in vitro biocompatibility of open cell iron structures with PEG coating. *Appl. Surf. Sci.* **2019**, *475*, 515–518. [CrossRef]
33. Vancha, A.R.; Govindaraju, S.; Parsa, K.V.L.; Jasti, M.; González-García, M.; Ballestero, R.P. Use of polyethyleneimine polymer in cell culture as attachment factor and lipofection enhancer. *BMC Biotechnol.* **2004**, *4*, 1–12. [CrossRef] [PubMed]
34. Yang, F.; Hu, S.; Lu, Y.; Yang, H.; Zhao, Y.; Li, L. Effects of coatings of polyethyleneimine and thyme essential oil combined with chitosan on sliced fresh channa argus during refrigerated storage. *J. Food Process Eng.* **2015**, *38*, 225–233. [CrossRef]
35. Xia, T.; Kovochich, M.; Liong, M.; Meng, H.; Kabehie, S.; George, S.; Zink, J.I.; Nel, A.E. Polyethyleneimine coating enhances the cellular uptake of mesoporous silica nanoparticles and allows safe delivery of siRNA and DNA constructs. *ACS Nano* **2009**, *3*, 3273–3286. [CrossRef] [PubMed]
36. Islam, M.A.; Park, T.E.; Singh, B.; Maharjan, S.; Firdous, J.; Cho, M.H.; Kang, S.K.; Yun, C.H.; Choi, Y.J.; Cho, C.S. Major degradable polycations as carriers for DNA and siRNA. *J. Control. Release* **2014**, *193*, 74–89. [CrossRef] [PubMed]

37. Li, J.; Zheng, L.; Cai, H.; Sun, W.; Shen, M.; Zhang, G.; Shi, X. Polyethyleneimine-mediated synthesis of folic acid-targeted iron oxide nanoparticles for invivo tumor MR imaging. *Biomaterials* **2013**, *34*, 8382–8392. [CrossRef]
38. Yao, X.; Zhou, N.; Wan, L.; Su, X.; Sun, Z.; Mizuguchi, H.; Yoshioka, Y.; Nakagawa, S.; Zhao, R.C.; Gao, J.Q. Polyethyleneimine-coating enhances adenoviral transduction of mesenchymal stem cells. *Biochem. Biophys. Res. Commun.* **2014**, *447*, 383–387. [CrossRef]
39. Dong, P.; Hao, W.; Wang, X.; Wang, T. Fabrication and biocompatibility of polyethyleneimine/heparin self-assembly coating on NiTi alloy. *Thin Solid Films* **2008**, *516*, 5168–5171. [CrossRef]
40. Bergstrand, A.; Rahmani-Monfared, G.; Östlund, Å.; Nydén, M.; Holmberg, K. Comparison of PEI-PEG and PLL-PEG copolymer coatings on the prevention of protein fouling. *J. Biomed. Mater. Res. Part A* **2009**, *88*, 608–615. [CrossRef]
41. Oriňaková, R.; Oriňak, A.; Bučková, L.M.; Giretová, M.; Medvecký, L.; Labbanczová, E.; Kupková, M.; Hrubovčáková, M.; Kovaľ, K. Iron based degradable foam structures for potential orthopedic applications. *Int. J. Electrochem. Sci.* **2013**, *8*, 12451–12465.
42. Yin, C.Y.; Aroua, M.K.; Daud, W.M.A.W. Metal-polyethyleneimine-activated carbon interaction parameter at equilibrium adsorption capacity. *J. Appl. Sci.* **2010**, *10*, 1192–1195. [CrossRef]
43. Dhiman, M.; Chalke, B.; Polshettiwar, V. Efficient Synthesis of Monodisperse Metal (Rh, Ru, Pd) Nanoparticles Supported on Fibrous Nanosilica (KCC-1) for Catalysis. *ACS Sustain. Chem. Eng.* **2015**, *3*, 3224–3230. [CrossRef]
44. Li, J.; Tang, W.; Yang, H.; Dong, Z.; Huang, J.; Li, S.; Wang, J.; Jin, J.; Ma, J. Enhanced-electrocatalytic activity of Ni 1−x Fe x alloy supported on polyethyleneimine functionalized MoS 2 nanosheets for hydrazine oxidation. *RSC Adv.* **2014**, *4*, 1988–1995. [CrossRef]
45. Singh, S.; Thomas, V.; Martyshkin, D.; Kozlovskaya, V.; Kharlampieva, E.; Catledge, S.A. Spatially controlled fabrication of a bright fluorescent nanodiamond-array with enhanced far-red Si-V luminescence. *Nanotechnology* **2014**, *25*, 045302. [CrossRef] [PubMed]
46. Liu, Y.; Liu, J.; Yao, W.; Cen, W.; Wang, H.; Weng, X.; Wu, Z. The effects of surface acidity on CO_2 adsorption over amine functionalized protonated titanate nanotubes. *RSC Adv.* **2013**, *3*, 18803–18810. [CrossRef]
47. Sanchez-Cortes, S.; Berenguel, R.M.; Madejón, A.; Pérez-Méndez, M. Adsorption of polyethyleneimine on silver nanoparticles and its interaction with a plasmid DNA: A surface-enhanced Raman scattering study. *Biomacromolecules* **2002**, *3*, 655–660. [CrossRef] [PubMed]
48. Rezaei, F.; Jones, C.W. Stability of Supported Amine Adsorbents to SO_2 and NOx in Postcombustion CO_2 Capture. 1. Single-Component Adsorption. *Ind. Eng. Chem. Res.* **2013**, *52*, 12192–12201. [CrossRef]
49. Chaufer, B.; Rabiller-Baudry, M.; Bouguen, A.; Labbé, J.P.; Quémerais, A. Spectroscopic characterization of zirconia coated by polymers with amine groups. *Langmuir* **2000**, *16*, 1852–1860. [CrossRef]
50. Kupková, M.; Hrubovčáková, M.; Kupka, M.; Oriňáková, R.; Turoňová, A.M. Corrosion behaviour of powder metallurgy biomaterials from phosphated carbonyl-iron powders. *Int. J. Electrochem. Sci.* **2015**, *10*, 671–681.
51. Zhang, E.; Chen, H.; Shen, F. Biocorrosion properties and blood and cell compatibility of pure iron as a biodegradable biomaterial. *J. Mater. Sci. Mater. Med.* **2010**, *21*, 2151–2163. [CrossRef]
52. Chen, H.; Zhang, E.; Yang, K. Microstructure, corrosion properties and bio-compatibility of calcium zinc phosphate coating on pure iron for biomedical application. *Mater. Sci. Eng. C* **2014**, *34*, 201–206. [CrossRef] [PubMed]
53. Kupková, M.; Hrubovčáková, M.; Kupka, M.; Oriňaková, R.; Turoňová, A.M. Sintering behaviour, graded microstructure and corrosion performance of sintered Fe-Mn biomaterials. *Int. J. Electrochem. Sci.* **2015**, *10*, 9256–9268.
54. Wei, J.; Zhou, B.; Wan, T.; Liu, K.; Gong, S.; Wu, J.; Xu, S. Effect of Sulfate and Chloride ions on pitting corrosion behavior of 2Cr12MoV Steel at pH 6 and 90 °C. *Int. J. Electrochem. Sci.* **2018**, *13*, 11596–11606. [CrossRef]
55. Kislenko, V.N.; Oliynyk, L.P. Complex formation of polyethyleneimine with copper(II), nickel(II), and cobalt(II) ions. *J. Polym. Sci. Part A Polym. Chem.* **2002**, *40*, 914–922. [CrossRef]

Article

Advantage of Alveolar Ridge Augmentation with Bioactive/Bioresorbable Screws Made of Composites of Unsintered Hydroxyapatite and Poly-L-lactide

Shintaro Sukegawa [1,2,*], Hotaka Kawai [2], Keisuke Nakano [2], Kiyofumi Takabatake [2], Takahiro Kanno [3], Hitoshi Nagatsuka [2] and Yoshihiko Furuki [1]

1. Department of Oral and Maxillofacial Surgery, Kagawa Prefectural Central Hospital, Takamatsu 765-8557, Japan; furukiy@ma.pikara.ne.jp
2. Department of Oral Pathology and Medicine, Graduate School of Medicine, Dentistry and Pharmaceutical Sciences, Okayama University, Okayama 700-8525, Japan; de18018@s.okayama-u.ac.jp (H.K.); pir19btp@okayama-u.ac.jp (K.N.); gmd422094@s.okayama-u.ac.jp (K.T.); jin@okayama-u.ac.jp (H.N.)
3. Department of Oral and Maxillofacial Surgery, Shimane University Faculty of Medicine, Shimane 693-8501, Japan; tkanno@med.shimane-u.ac.jp

* Correspondence: gouwan19@gmail.com or s-sukegawa@chp-kagawa.jp; Tel.: +81-87-811-3333

Received: 29 September 2019; Accepted: 5 November 2019; Published: 8 November 2019

Abstract: We studied human bone healing characteristics and the histological osteogenic environment by using devices made of a composite of uncalcined and unsintered hydroxyapatite (u-HA) and poly-L-lactide (PLLA). In eight cases of fixation, we used u-HA/PLLA screws for maxillary alveolar ridge augmentation, for which mandibular cortical bone block was used in preimplantation surgery. Five appropriate samples with screws were evaluated histologically and immunohistochemically for runt-related transcription factor 2 (RUNX2), transcription factor Sp7 (Osterix), and leptin receptor (LepR). In all cases, histological evaluation revealed that bone components had completely surrounded the u-HA/PLLA screws, and the bone was connected directly to the biomaterial. Inflammatory cells did not invade the space between the bone and the u-HA/PLLA screw. Immunohistochemical evaluation revealed that many cells were positive for RUNX2 or Osterix, which are markers for osteoblast and osteoprogenitor cells, in the tissues surrounding u-HA/PLLA. In addition, many bone marrow–derived mesenchymal stem cells were notably positive for both LepR and RUNX2. The u-HA/PLLA material showed excellent bioactive osteoconductivity and a highly biocompatibility with bone directly attached. In addition, our findings suggest that many bone marrow–derived mesenchymal stem cells and mature osteoblast are present in the osteogenic environment created with u-HA/PLLA screws and that this environment is suitable for osteogenesis.

Keywords: poly-L-lactide; uncalcined and unsintered hydroxyapatite; biocompatibility; osteoconductivity; mesenchymal stem cell

---

1. Introduction

Titanium fixation devices have been used widely as a standard for maxillofacial surgery because they are easy to operate and relatively inexpensive; however, plate removal may be necessary, and various complications can be caused by the metal [1]. Therefore, bioresorbable fixation devices made of synthetic polymers are currently used widely as an alternative material for internal fixation. An ideal bioresorbable osteosynthesis device should have the proper modulus and high strength, retain that strength as long as bone healing requires support, and be safely absorbed and disassembled without a foreign body reaction that delays the bone-healing process.

Bioabsorbable fixation devices made of high-strength uncalcined and unsintered hydroxyapatite (u-HA) and poly-L-lactide (PLLA) composites have been developed to solve the mechanical and biological problems of life-long implants [2]. Currently, Super FIXSORB MX® (Teijin Medical Technologies Co., Ltd. Osaka, Japan), also known as OSTEOTRANS MX, can be used as a commercially available u-HA/PLLA osteosynthesis bioresorbable device. This bioresorbable device, which consists of u-HA and PLLA, is manufactured by a compression molding reinforcement process and a forging process incorporating machining. Because of its composition and the special manufacturing process, this device has higher mechanical strength and bioactivity [2–5]. The bioactivity of bioresorbable plates is a major advantage, and their bone conduction and bone-binding ability [6,7], complete long-term replacement of the human bone [8], and biocompatibility [6–8] have been reported. In addition, we have previously reported the presence of osteoblast differentiation markers in the environment surrounding u-HA/PLLA materials [7], which has already shown that u-HA/PLLA materials are bioactive materials with excellent bone regeneration ability.

However, the bone-healing properties of this device and the histological environment for bone healing remain unclear. In this study, we investigated bone-healing characteristics and the histological environment for u-HA/PLLA composite devices to understand the in vivo environment when this device is used in maxillofacial clinical treatment.

2. Materials and Methods

2.1. Preparation of Uncalcined and Unsintered Hydroxyapatite/Poly-L-lactide Composite Screws

In this study, we used the Super FIXSORB MX®screw (Teijin Medical Technologies Co., Ltd. Osaka, Japan), comprising a forged composite of u-HA/PLLA (containing 30 weight fractions of raw uncalcined, unsintered HA particles in composites). The screws have a diameter of 2.0 mm and a length of 8–12 mm; u-HA particle size ranges from 0.2 to 20 μm (average size, 3–5 μm); the ratio of HA weight to PLLA weight is 30/70; the ratio of calcium to phosphorus is 1.69 (moles); and CO_3^{2-} level is 3.8 (percentage of moles). The composite material used in this study was the same as that reported in the past [2].

2.2. Subjects

This study included eight consecutive patients (two men and six women; age range, 33–59 years) who needed maxillary alveolar ridge augmentation as preimplantation surgery because their residual bone width was <4 mm; informed consent to participate in the study was obtained from all the patients. All operations were performed by a single oral and maxillofacial surgeon (Shintaro Sukegawa) from April 2018 to March 2019 in the Department of Oral and Maxillofacial Surgery at Kagawa Prefectural Central Hospital, Takamatsu, Kagawa, Japan. The cases of this study are shown in Table 1.

Table 1. Details of patients and u-HA/PLLA screws used.

Patient Number	Sex (Male/Female)	Age (Years)	Screw Length (mm)	Number of Screws	Presence or Absence of u-HA/PLLA Screws in the Specimen	Period from the Screw Placement to Evaluation (Day)
1	Female	44	8.0	2	Presence	246
2	Female	57	8.0	2	Presence	209
3	Female	59	8.0	2	Absence	219
4	Female	55	12.0	3	Presence	203
5	Male	55	12.0	1	Absence	223
6	Female	33	12.0	1	Presence	209
7	Male	50	8.0	2	Presence	226
8	Female	58	8.0	2	Absence	212

u-HA/PLLA, uncalcined and unsintered hydroxyapatite/poly-L-lactide.

2.3. Surgical Bone Augmentation Procedure

In the surgical operation, the amount of material necessary for bone augmentation was collected from the buccal cortical bone block of the mandibular ramus. The cortical bone block was fixed to the recipient site by using u-HA/PLLA screws. The screw fixing method consisted of the following steps: (1) drilling to form a bone hole, (2) forming a tap with a screw tap, and (3) insertion of u-HA/PLLA screws into the holes formed by self-tapping. The screw insertion torque was 5 N. The number of screws used was selected to obtain stable fixation of the bone block (Figure 1a).

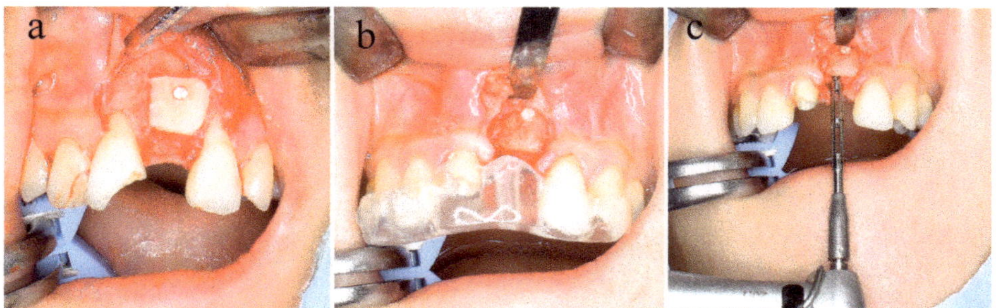

Figure 1. (**a**) In the surgical operation, the amount necessary for bone augmentation was collected from the buccal cortical bone block of the mandibular ramus, and the buccal cortical bone block was fixed to the recipient site with uncalcined and unsintered hydroxyapatite/poly-L-lactide (u-HA/PLLA) screws. (**b**,**c**) Six months later, dental implantation surgery was planned. The position of the u-HA/PLLA screw was confirmed. At the time of implant placement, specimens were collected using a 2.0 mm diameter trephine bar.

2.4. Sample Collection

After approximately 6 months, to allow for bone healing, dental implant placement was planned with the use of computed tomography. (Figure 1b). At the time of implant placement, specimens were collected with a 2.0 mm diameter trephine bar (ACE Surgical Supply Company, Inc., Brockton, MA, USA) (Figure 1c). All procedures were performed by the same expert surgeon (Shintaro Sukegawa) at the same institution. This study was approved by the Ethics Committee of the Kagawa Prefectural Central Hospital (Approval No. 879).

2.5. Preparation for Histological Evaluation

All samples were immediately fixed in 4% paraformaldehyde for 12 h and then decalcified in 10% ethylenediaminetetraacetic acid at 4 °C for 14 days. Samples were dehydrated with a graded series, soaked in xylol several times, and embedded in paraffin. Thin serial sections were made from samples embedded in paraffin. The sections were used for hematoxylin–eosin staining and immunohistochemical study.

2.6. Immunohistochemistry

The expressions of runt-related transcription factor 2 (RUNX2), transcription factor Sp7 (Osterix), and leptin receptor (LepR) were evaluated in an immunohistochemical study. The prepared sample paraffin-embedded block was sectioned in thicknesses of 3 μm. These sections were deparaffinized in xylene for 15 min and rehydrated in graded ethanol solution. To prevent endogenous peroxidase activity, the sections were incubated in 0.3% H_2O_2 and methanol for 30 min. Antigen retrieval was achieved by heat treatment with 10-mM citrate buffer solution at a pH of 9.0. After treatment with normal serum, the sections were incubated with the primary antibodies for RUNX2 (Abcam plc., Cambridgesshire, ab23981, UK, dilution of 1:500), Osterix (Abcam plc., Cambridgesshire, ab22552, UK, dilution of 1:100), and LepR (Proteintech. 20966-1, USA, dilution of 1:50) at 4 °C overnight. To tag the primary antibody, EnVision peroxidase detecting reagent (Dako, Carpinteria, CA, USA) was applied. We identified the immunoreactive site by using the avidin–biotin complex method (Vector Laboratories, Burlingame, CA, USA). Detection was performed with 3,3′-Diaminobenzidine (DAB), and the staining results were observed with an optical microscope.

For double-fluorescent Immunohistochemistry (IHC), the abovementioned LepR and RUNX2 antibodies were used as primary antibodies. Antibodies were diluted with Can Get Signal (Toyobo, Osaka, Japan). Anti-mouse IgG Alexa Fluor 488 (Life technologies, Waltham, MA, USA) and anti-rabbit IgG Alexa Fluor 568 (Carlsbad, CA, USA) were used as secondary antibodies at a dilution of 1:200. After the reactions, the specimens were stained with 1 mg/mL of DAPI (Dojindo Laboratories, Kumamoto, Japan). The staining results were observed with a fluorescence microscope.

3. Results

3.1. Clinical Evaluation

Six months after anterior maxillary alveolar bone augmentation of preimplantation surgery, bone width in all patients was sufficient for placing the dental implant. The transplanted cortical bone block was fully engrafted in all patients. No complications were observed in any of the patients after dental implant placement, and all results with the final prosthesis set were satisfactory. We used the trephine bar to obtain eight specimens from the bone-constructed area with u-HA/PLLA screws at the time of implant placement. Of these specimens, five in which the implant was placed in the same location as the u-HA/PLLA screw were examined histologically.

3.2. Histopathological Evaluations

Eight specimens were examined for resected material by using hematoxylin–eosin staining. In five specimens, we were able to observe both screws and the surrounding bone. In 2 cases (Cases 1 and 2), bone components had completely surrounded the u-HA/PLLA screws (Figure 2a,c). The high magnification field revealed that the bone was directly connected to the biomaterial and that inflammatory cells did not invade the space between the bone and the u-HA/PLLA screw. There was no cellular infiltration between the bone and the u-HA/PLLA screw, and the materials were completely continuous. No foreign body reaction was induced around the u-HA/PLLA screws (Figure 2a,b,d,e). These findings indicate that u-HA/PLLA screw has high bone compatibility.

In other cases (Cases 3–5) fibrous tissue was observed surrounding the u-HA/PLLA. This histological character of the fibrous tissue was uniform, and there was no inflammation and bleeding. Foreign body giant cells were not present in stromal tissue (Figure 3b,c,e,f,h,i). Furthermore, this fibrous tissue contained bone tissue and was continuous (Figure 3c,f,i). These findings are different from those reported in other two cases (Cases 1 and 2), but findings of both these cases suggested that u-HA/PLLA screws are highly biocompatible. In addition, the results of all the three cases (Cases 3–5) indicate the potential for bone making ability around the u-HA/PLLA.

3.3. Immunohistochemical Evaluations

To investigate the characteristics of fibrous tissue, we performed immunostaining. Because of absence of inflammation and existing new bone, we selected the marker for preosteoblast (RUNX2, Osterix) and mesenchymal stem cell marker (LepR). In the fibrous tissue, RUNX2- or Osterix-positive cells were observed; these cells were spindle shape and scattered (Figure 4a,b). LepR-positive cells were also observed in fibrous tissue and their shape was similar to RUNX2- or Osterix-positive cells (Figure 4c). It was confirmed that a number of RUNX2-positive cells expressed LepR (Figure 4d–f).

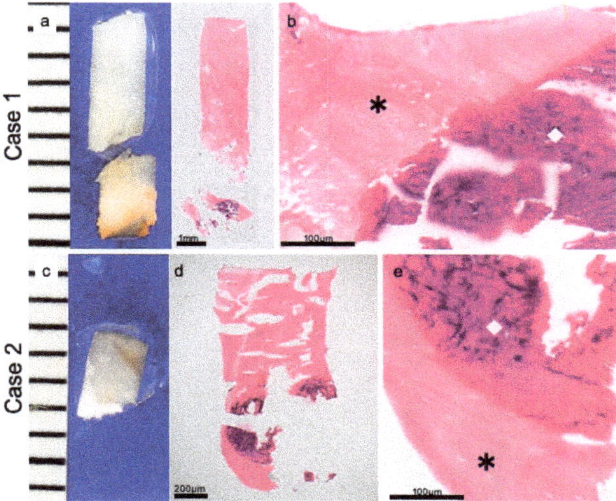

Figure 2. Direct contact of material and bone. Case 1: (**a**) macro findings and hematoxylin-eosin staining. (**b**) Hematoxylin–eosin staining. Case 2: (**c**) macro findings. (**d**,**e**) Hematoxylin–eosin staining. In both cases, the uncalcined and unsintered hydroxyapatite/poly-L-lactide (u-HA/PLLA) screw (◇) and bone (*) are in direct contact, with no connective tissue interposed between them.

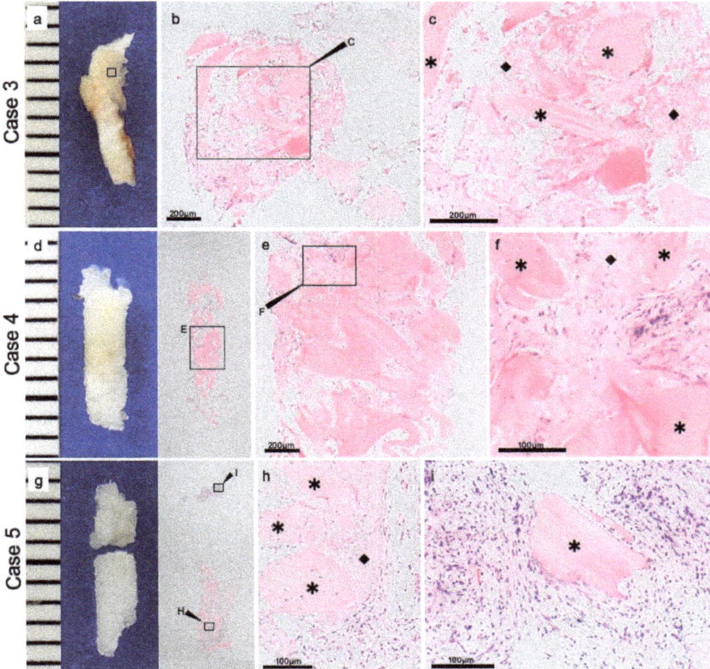

Figure 3. The fibrous tissue around the screw. Case 3: (**a**) macro findings. (**b**,**c**) Hematoxylin–eosin staining. Case 4: (**d**) macro findings. (**e**,**f**) Hematoxylin–eosin staining. Case 5: (**g**) macro findings. (**h**,**i**) Hematoxylin–eosin staining. In the tissue surrounding, the diamond (◆) indicate the u-HA/PLLA screw, and the star (*) indicate the bone.

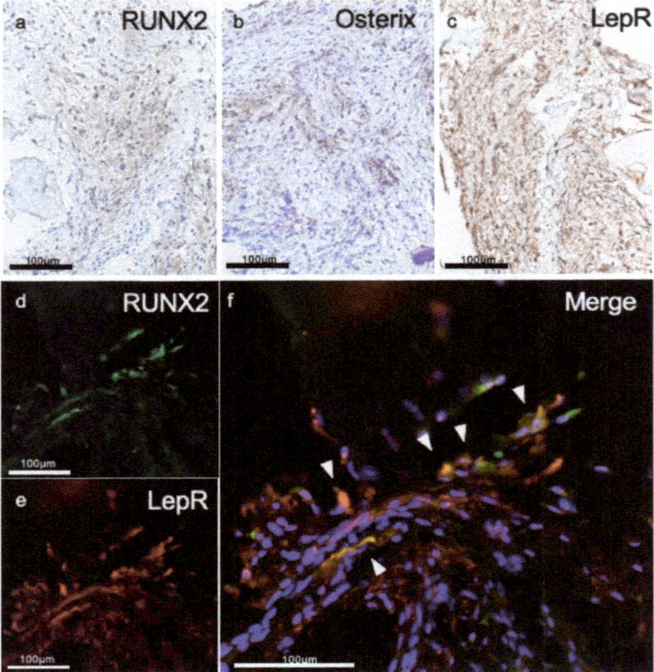

Figure 4. The characteristics of fibrous tissue around the screw. Immunohistochemistry for (**a**) RUNX2, (**b**) Osterix, and (**c**) leptin receptor (LepR). Spindle cells are positive for each marker. Double florescent IHC for fibrous tissue surrounding screw. Several cells are positive for (**d**) RUNX2 or (**e**) LepR, and (**f**) some cells were positive for both markers (arrow head).

4. Discussion

In this study, the use of u-HA/PLLA screws in maxillofacial bone augmentation with bone block as preimplantation surgery yielded reliable results. The u-HA/PLLA screws were highly biocompatible with the bone. Additionally, in the tissues surrounding the u-HA/PLLA screw, many cells were positive for RUNX2 and Osterix, which are markers for osteoprogenitor cells. Furthermore, several cells positive for both LepR and RUNX2.

Bioresorbable osteosynthesis devices made of a variety of synthetic polymers such as polyglycolide, PLLA, polydioxanone, or glycolide–lactide copolymers have been developed [9]. However, clinical studies of conventional resorbable devices have demonstrated various complications, including mechanical weakness, osteolytic changes around the device [10,11], and degradation of the tissue [12]. To overcome these limitations, composite materials comprising bioactive ceramics as fillers and PLLA as matrices have been developed. These composites are intended to provide both biocompatibility with the bone and desirable mechanical properties, including polymer ductility and stiffness equal to or better than that of cortical bone. The composite material consisting of u-HA as a bioactive ceramic and PLLA as a matrix has great advantages. The u-HA/PLLA composite material has an initial bending strength of 280 MPa, higher than that of human cortical bone (120–210 MPa), and an elastic modulus of 12 GPa, which is mechanically stronger than any other bioactive ceramic/polymer composites available to date [13]. The u-HA/PLLA composite material has not only high strength but also bioactivity advantages [14].

Surprisingly, the u-HA/PLLA screw was bound directly to the human bone in all specimens in this study in which the screw was detected. Hydroxyapatite was reported to have formed directly on the surface of the composite material after immersion in a simulated body fluid in in vitro research [2]. When cells touch the surface of material, they usually attach, adhere, and spread. The first stage of this interaction between cell and material depends on the characteristics of the material surface to determine the behavior of the cell upon contact with the material. Osteoblasts have been found to preferentially attach to HA particles via filopodia, demonstrating that HA provides a favorable anchoring site for human osteoblast adhesion [15]. Hydroxyapatite on the surface of u-HA/PLLA screw may play an important role in direct bonding with human bone. This osteoconductive feature is a major advantage of this material.

Of histological importance is the lack of foreign body reaction around the u-HA/PLLA material. In this study, new bone is formed through direct contact with material and human bone. For regenerative bone formation, the implant material must elicit minimal inflammatory reaction. In the past, we quantitatively evaluated the presence of CD68, considered a marker for macrophages, around u-HA/PLLA screws. CD68 was rarely observed around the u-HA/PLLA material [7]. In this study, inflammatory cells and giant cells against foreign bodies were not observed in any of the specimens examined, indicating high biocompatibility of the u-HA/PLLA material.

There are two type of pathways in the bone-healing process: intramembranous ossification and endochondral ossification [16]. Intramembrane ossification is a process formed by mesenchymal cells that condense to become functional osteoblasts without cartilage formation. Endochondral ossification, in contrast, is a process of replacing a cartilage template composed of chondrocytes differentiated from mesenchymal stem cells with bone containing osteoblasts and osteoclasts. These cells are functionally responsible for bone formation that directs the deposition and calcification of bone matrix. Osteoblasts are induced through the expression of osteoblast-specific transcription factor RUNX2 and Osterix from immature mesenchymal stromal cells [17,18]. In our immunostaining evaluation, RUNX2- or Osterix-positive cells were localized in the tissue surrounding u-HA/PLLA material. These cells were present in connective tissue and did not secrete bone matrix. Numerous preosteoblasts were present in the surrounding stroma of u-HA/PLLA, which suggests that this microenvironment might form bone tissue.

In recent studies, bone marrow derived mesenchymal stem cells have been identified as LepR-positive cells in cell lineage analysis [19]. Yang et al. reported that approximately 60% of LepR-positive cells expressed RUNX2 and that among the LepR-positive cells, the RUNX2-positive subpopulation had higher stem cell capacity than did the RUNX2-negative subpopulation [19]. Furthermore, cells that were both LepR and RUNX2 positive showed pluripotency in an in vitro culture system [20]. The findings by Yang et al. suggest that LepR- and RUNX2-positive cells are located upstream of the differentiation tree of bone marrow mesenchymal cells. Our results also showed that LepR- and RUNX2- double positive cells were expressed in the environment surrounding u-HA/PLLA screws. It has already been shown that u-HA/PLLA materials are bioactive materials with excellent bone regeneration ability. However, the bone formation environment has not been elucidated. This study is the first report demonstrating the surrounding environment with LepR- and RUNX2-positive cells following the placement of u-HA/PLLA screws. The results of this study suggested that the bone formation environment performed with u-HA/PLLA screws may enable the expression of bone marrow–derived mesenchymal cells, which is a good environment for bone formation. Moreover, these findings indicate that the screw induce recruitment of bone marrow derived mesenchymal stem cells (RUNX2/LeptinR double positive).

A limitation of this study was that it was not possible to evaluate the tissue changes over long term for materials implanted in the human body. We evaluated the timing of dental implant placement after bone formation and samples collected at the same time as the dental implant placement. Because resorbable osteosynthesis cannot be removed without complications, investigation through animal experiments may enable evaluation at other times. On the contrary, to the best of our knowledge, the results of this study will help elucidate the bone-healing properties of, and the histological environment created by, u-HA/PLLA bioresorbable material in human maxillofacial bone. This is interesting and important with regard to the in vivo response to u-HA/PLLA bioresorbable materials.

5. Conclusions

The u-HA/PLLA screws demonstrated excellent bioactive osteoconductivity and high biocompatibility with maxillofacial bone in this study. In the tissues surrounding the u-HA/PLLA material, many cells were positive for RUNX2 and Osterix, the markers for osteoprogenitor cells. In addition, several bone marrow–derived mesenchymal stem cells were positive for both LepR and RUNX2. These results suggest that the bone formation stimulated by u-HA/PLLA screws may provide a good environment for bone regenerative formation in maxillofacial surgery.

Author Contributions: S.S. participated in the design of the study, acquisition of data, and drafted the manuscript. H.K. performed and coordinated the study and contributed to the drafting of the manuscript. K.N., K.T., H.N., and Y.F. collected the data and compiled all medical records. T.K. and Y.F. conceived the study and participated in its design. All authors approved the final version of the manuscript prior to submission.

Funding: This work was supported by JSPS KAKENHI Grant Numbers JP19K19158, JP19K19159.

Conflicts of Interest: The authors declare that they have no conflict of interest.

References

1. Schmidt, B.L.; Perrott, D.H.; Mahan, D.; Kearns, G. The removal of plates and screws after le Fort I osteotomy. *J. Oral Maxillofac. Surg.* **1998**, *56*, 184–188. [CrossRef]
2. Shikinami, Y.; Okuno, M. Bioresorbable devices made of forged composites of hydroxyapatite (HA) particles and poly-L-lactide (PLLA): Part I. Basic characteristics. *Biomaterials* **1999**, *20*, 859–877. [CrossRef]
3. Hasegawa, S.; Ishii, S.; Tamura, J.; Furukawa, T.; Neo, M.; Matsusue, Y.; Shikinami, Y.; Okuno, M.; Nakamura, T. A 5–7 year in vivo study of high-strength hydroxyapatite/poly(L-lactide) composite rods for the internal fixation of bone fractures. *Biomaterials* **2006**, *27*, 1327–1332. [CrossRef] [PubMed]
4. Shikinami, Y.; Okuno, M. Bioresorbable devices made of forged composites of hydroxyapatite (HA) particles and poly-L-lactide (PLLA). Part II: Practical properties of miniscrews and miniplates. *Biomaterials* **2001**, *22*, 3197–3211. [CrossRef]
5. Sukegawa, S.; Kanno, T.; Manabe, Y.; Matsumoto, K.; Sukegawa-Takahashi, Y.; Masui, M.; Furuki, Y. Biomechanical Loading Evaluation of Unsintered hydroxyapatite/poly-L-lactide Plate System in bilateral Sagittal Split ramus Osteotomy. *Materials* **2017**, *10*, 764. [CrossRef] [PubMed]
6. Sukegawa, S.; Kanno, T.; Katase, N.; Shibata, A.; Takahashi, Y.; Furuki, Y. Clinical evaluation of an unsintered hydroxyapatite/poly-L-lactide osteoconductive composite device for the internal fixation of maxillofacial fractures. *J. Craniofac. Surg.* **2016**, *27*, 1391–1397. [CrossRef] [PubMed]
7. Sukegawa, S.; Kawai, H.; Nakano, K.; Kanno, T.; Takabatake, K.; Nagatsuka, H.; Furuki, Y. Feasible advantage of bioactive/bioresorbable devices made of forged composites of hydroxyapatite particles and poly-L-lactide in alveolar bone augmentation: A preliminary study. *Int. J. Med. Sci.* **2019**, *16*, 311–317. [CrossRef] [PubMed]
8. Sukegawa, S.; Kanno, T.; Kawai, H.; Shibata, A.; Takahashi, Y.; Nagatsuka, H.; Furuki, Y. Long-term bioresorption of bone fixation devices made from composites of unsintered hydroxyapatite particles and poly-L-lactide. *J. Hard Tissue Biol.* **2015**, *24*, 219–224. [CrossRef]
9. Middleton, J.C.; Tipton, A.J. Synthetic biodegradable polymers as orthopedic devices. *Biomaterials* **2000**, *21*, 2335–2346. [CrossRef]
10. Matsusue, Y.; Nakamura, T.; Iida, H.; Shimizu, K. A long-term clinical study on drawn poly-L-lactide implants in orthopaedic surgery. *J. Long-Term Eff. Med. Implant.* **1997**, *7*, 119–137.
11. Kallela, P.L.; Suuronen, R.; Ranta, P.; Iizuka, T.; Lindqvist, C.; Lindqvist, C. Osteotomy site healing following mandibular sagittal split osteotomy and rigid fixation with polylactide biodegradable screws. *Int. J. Oral Maxillofac. Surg.* **1999**, *28*, 166–170. [CrossRef] [PubMed]
12. Bergsma, J.E.; de Bruijn, W.C.; Rozema, F.R.; Bos, R.R.; Boering, G. Late degradation tissue response to poly(L-lactide) bone plates and screws. *Biomaterials* **1995**, *16*, 25–31. [CrossRef]

13. Furukawa, T.; Matsusue, Y.; Yasunaga, T.; Nakagawa, Y.; Okada, Y.; Shikinami, Y.; Okuno, M.; Nakamura, T. Histomorphometric study on high-strength hydroxyapatite/poly(L-lactide) composite rods for internal fixation of bone fractures. *J. Biomed. Mater. Res.* **2000**, *50*, 410–419. [CrossRef]
14. Dong, Q.N.; Kanno, T.; Bai, Y.; Sha, J.; Hideshima, K. Bone Regeneration Potential of Uncalcined and Unsintered Hydroxyapatite/Poly L-lactide Bioactive/Osteoconductive Sheet Used for Maxillofacial Reconstructive Surgery: An In Vivo Study. *Materials* **2019**, *11*, 12. [CrossRef] [PubMed]
15. Zhang, Y.; Tanner, K.E.; Gurav, N.; Di Silvio, L. In Vitro osteoblastic response to 30 vol% hydroxyapatite-polyethylene composite. *J. Biomed. Mater. Res. A* **2007**, *81*, 409–417. [CrossRef] [PubMed]
16. Karsenty, G. The complexities of skeletal biology. *Nature* **2003**, *423*, 316–318. [CrossRef] [PubMed]
17. Komori, T. Regulation of bone development and extracellular matrix protein genes by RUNX2. *Cell Tissue Res.* **2010**, *339*, 189–195. [CrossRef] [PubMed]
18. Nakashima, K.; Zhou, X.; Kunkel, G.; Zhang, Z.; Deng, J.M.; Behringer, R.R.; de Crombrugghe, B. The novel zinc finger-containing transcription factor osterix is required for osteoblast differentiation and bone formation. *Cell* **2002**, *108*, 17–29. [CrossRef]
19. Mizoguchi, T.; Pinho, S.; Ahmed, J.; Kunisaki, Y.; Hanoun, M.; Mendelson, A.; Ono, N.; Kronenberg, H.M.; Frenette, P.S. Osterix marks distinct waves of primitive and definitive stromal progenitors during bone marrow development. *Dev. Cell* **2014**, *29*, 340–349. [CrossRef] [PubMed]
20. Yang, M.; Arai, A.; Udagawa, N.; Hiraga, T.; Lijuan, Z.; Ito, S.; Komori, T.; Moriishi, T.; Matsuo, K.; Shimoda, K.; et al. Osteogenic factor Runx2 marks a subset of leptin receptor-positive cells that sit Atop the bone marrow stromal cell hierarchy. *Sci. Rep.* **2017**, *7*, 4928. [CrossRef] [PubMed]

Article

Biological Assessment of Zn–Based Absorbable Metals for Ureteral Stent Applications

Devi Paramitha [1,2], Stéphane Chabaud [2], Stéphane Bolduc [2,*] and Hendra Hermawan [1,*]

[1] Department of Mining, Metallurgical and Materials Engineering & CHU de Québec Research Center, Laval University, Quebec City, QC G1V 0A6, Canada; devi.paramitha.1@ulaval.ca

[2] Centre de Recherche en Organogénèse Expérimentale/LOEX, Division of Regenerative Medicine, CHU de Québec Research Center, Laval University, Quebec City, QC G1J 1Z4, Canada; stephane.chabaud@crchudequebec.ulaval.ca

* Correspondence: stephane.bolduc@fmed.ulaval.ca (S.B.); hendra.hermawan@gmn.ulaval.ca (H.H.)

Received: 12 September 2019; Accepted: 9 October 2019; Published: 12 October 2019

Abstract: The use of ureteral stents to relieve urinary tract obstruction is still challenged by the problems of infection, encrustation, and compression, leading to the need for early removal procedures. Biodegradable ureteral stents, commonly made of polymers, have been proposed to overcome these problems. Recently, absorbable metals have been considered as potential materials offering both biodegradation and strength. This work proposed zinc-based absorbable metals by firstly evaluating their cytocompatibility toward normal primary human urothelial cells using 2D and 3D assays. In the 2D assay, the cells were exposed to different concentrations of metal extracts (i.e., 10 mg/mL of Zn–1Mg and 8.75 mg/mL of Zn–0.5Al) for up to 3 days and found that their cytoskeletal networks were affected but were recovered at day 3, as observed by immunofluorescence. In the 3D ureteral wall tissue construct, the cells formed a multilayered urothelium, as found in native tissue, with the presence of tight junctions at the superficial layer and laminin at the basal layer, indicating a healthy tissue condition even with the presence of the metal samples for up to 7 days of exposure. The basal cells attached to the metal surface as seen in a natural spreading state with pseudopodia and fusiform morphologies, indicating that the metals were non-toxic.

Keywords: absorbable metal; cytotoxicity; stent; ureteral; urothelial cells; zinc alloy

---

1. Introduction

The urinary tract is part of the renal system, and helps in maintenance of a homeostatic condition by draining the urine from the kidneys to the bladder and out of the body [1]. Obstruction to this tract can cause a build-up of urine leading to hydronephrosis or the swelling of a kidney, which later leads to the failure of other organ systems in the body [2]. Ureteral stents are widely used in patients with urological disorders when a relief of ureteral obstruction is needed, or when the maintenance of ureteral patency is required for healing purposes after ureteral and upper tract reconstruction, endoscopy, or trauma [3,4]. An ideal ureteral stent maintains excellent urine flow to optimize upper tract drainage, is resistant to encrustation, avoids infections, and is biodegradable [5]. The majority of urologists consider a routine placement of stents to be 1 to 6 weeks [6,7] following ureteral dilatation. This temporary need has led to the concept of biodegradable ureteral stents. Introduced in the late 1980s, biodegradable ureteral stent indication is dedicated for temporary treatment such as patients in line for surgery and the early-phase resolution of bladder outlet obstruction in patients waiting for the effect of a medical therapy [8]. Poly(glycolic acid), poly(lactic acid), and poly(lactic-co-glycolic acid) are among the biodegradable polymeric materials used for ureteral stents [5,9]. However, polymeric stents possess limited ability to resist external compression forces, such as those created by a malignant extrinsic ureteral obstruction, due to their intrinsic low radial strength, which is often reinforced by a metal skeleton made of non-degradable alloys [10,11]. At this point, metals that can degrade seem to constitute the ideal material for biodegradable ureteral stents.

Nowadays, absorbable metals or metals that degrade gradually in vivo with an appropriate host response then dissolve completely upon assisting tissue healing [12] are available. Iron, magnesium, zinc, and their alloys are among the absorbable metals that have been studied, mostly for cardiovascular and orthopaedic applications. Coronary stents made of iron and its alloys have shown their safety and efficacy when tested in animals [13], while those made of magnesium alloys have been clinically tested for critical limb ischemia in human with encouraging results [14]. Meanwhile, zinc alloys have been more recently proposed as alternatives to iron and magnesium in view of their moderate corrosion rate, and their cytocompatibility has been proved toward various cell types [15–17]. In relation to ureteral stents, a preliminary study was done on the corrosion and antibacterial properties of Mg–Y alloy in artificial urine, suggesting the potential of absorbable magnesium in urological applications [18]. Since that preliminary description, studies on the potential use of absorbable magnesium for this application were reported. First, an in vitro and in vivo corrosion and compatibility study of Mg–6Zn was done in simulated body fluids (SBF) and a rat bladder model [19]. Another study introduced the use of human urothelial cells using three different culture methods with the presence of Mg–Y. The culture methods used were direct culture, where cells were seeded directly on the material; direct exposure culture, which was done by seeding the cells first and then putting the material on top of the cell layer; and exposure culture method, where cells and material were in the same environment and had interactions but were not in direct contact [20]. One study reported the biodegradation behavior of ZK60 in artificial urine and rat models. The results of the studies showed that absorbable magnesium has the potential to be used in urinary applications [21].

Zinc has been studied as a potential absorbable metal for vascular stent and bone implant applications. Zinc shows a moderate in vitro corrosion rate compared to iron and magnesium. The addition of magnesium has a strengthening effect on zinc; for instance, as-cast Zn–1–2 wt %) Mg possesses a tensile strength of 180 MPa, as high as that of most magnesium alloys, and with the absence of potential mutagenicity and genotoxicity [16]. In addition, an extruded Zn–1.2Mg showed a higher tensile strength of 360 MPa while maintaining an excellent ductility of 20%, and exhibited no potential toxicity [22]. Further alloying with aluminum also increases the mechanical properties of zinc; it has been shown that Zn–0.5Al and Zn–1Al possess tensile strengths of 203 MPa and 223 MPa, respectively, while maintaining high ductility, i.e., 33% and 25%, respectively [23]. Even though aluminum is considered to be potentially toxic, new, absorbable Zn–Al alloys are of interest for their use in the urinary system, where toxicity is less of a concern as it is implanted after the body's natural filters, the liver and the kidney. Therefore, owing to their inherent strength and biodegradability, we propose zinc-based absorbable metals as potential materials for ureteral stents. This work aimed to assess the cytocompatibility of Zn–Mg and Zn–Al alloys towards human urothelial cells in 2D and 3D models.

2. Materials and Methods

2.1. Materials and Specimen Preparation

This study began with a preliminary screening step involving a cell viability test and an electrochemical corrosion test on the three classes of absorbable metals: pure iron, magnesium, and zinc, before then focusing on selected zinc alloys. The first screening tests included a water soluble tetrazolium (WST) mitochondrial assay to assess the cell viability of urothelial cells treated for 3 days with the metal extracts, and a potentiodynamic polarization test to determine the corrosion rate in artificial urine solution, as detailed in another work [24]. The second screening test involved a further cell viability test with the addition of two zinc alloys (Zn–1Mg and Zn–0.5Al) and one magnesium alloy (Mg–2Zn–1Mn, or ZM21), which were prepared via casting and extrusion, as detailed in Mostaed et al. [23]. The alloys were turned into powders (particle size of 2–200 μm) by mechanical filing to be used for indirect 2D cytotoxicity, while small disks (surface area ~200 mm^2) were cut to be used in the direct 3D cytotoxicity assay. Extracts of Zn–1Mg (10 mg/mL) and Zn–0.5Al (8.75 mg/mL) were prepared from the powders based on the 50% inhibition concentration (IC_{50}) determined in the second cell viability test. The powders were sterilized under UV for at least 15 min and mixed with urothelial basal media: Dulbecco–Vogt modification of Eagle's Medium (DMEM; Invitrogen, Burlington, Canada) and Ham's F12 (Flow Laboratory, Mississauga, Canada) in a 3:1 proportion, supplemented with 10% fetal bovine serum (Hyclone, Logan, UT, USA), 5 μg/mL insulin (Sigma, St. Louis, MO, USA), 10 μg/mL epidermal growth factor (Austral Biologicals, San Ramon, CA, USA), 10^{-10} M cholera toxin (ICN, St. Laurent, Canada), 100 U/mL penicillin, and 0.4 μg/mL hydrocortisone (Calbiochem, San Diego, CA, USA). The mixture was then incubated in 8% CO_2 at 37 °C for 72 h. After incubation, the solutions were filtered using 0.22 μm Durapore® PVDF membrane filter (Millex® GV, Merck Millipore, Ltd., Darmstadt, Germany) and stored at 4 °C prior to use.

2.2. *pH and Ion Measurement*

The pH value of the metal extracts was measured using a pH meter (Beckman Coulter PHI 350; Beckman Coulter Life Sciences, Mississauga, Canada). In brief, the pH meter was calibrated using three pH standard solutions (pH 4, 7, and 10) at room temperature. First, the sample solutions were heated to 37 °C. They were then poured into a small beaker and measured repeatedly. The level of ions released into the extraction media was measured by inductive coupled plasma atomic emission spectrometer (ICP/OES, 5110 SVDV, Agilent Technology, Santa Clara, CA, USA) [25–27] for the three metal ions, i.e., zinc, magnesium, and aluminum. In brief, the extracts were digested for 3 days before the ion measurement. At day 1, 3 mL of extract was added into the vial, and a mark was drawn to show the meniscus level of 3 mL. In a fume hood, the crystallizer was filled with paraffin oil prior to vial holder mounting to avoid oil splash. The vial holder was then mounted, not touching the crystallizer bottom surface, and with the vial half immersed in the oil. The oil was heated to 90 °C and 3 mL HNO_3 was added to the vial, upon which the temperature was raised to 115 °C. When the solution inside the vial reduced to 3 mL by observation of the mark, another 3 mL HNO_3 was added. After the solution evaporated, leaving 3 mL inside the vial, the oil bath was lifted and air-dried. It was then placed on a paper towel to absorb the excess oil. The treatment at day 2 started with the addition of 600 μL nanopure water and 900 μL of 30% H_2O_2 into each vial. The oil bath bottle was then heated at 115 °C until the excitement phase appeared. Attention was paid to make sure that no drop of oil was lost by excessive excitement. The solutions were allowed to heat until reduced to approximately 3 mL. The oil bath was then lifted and air-dried and placed on a paper towel to absorb the excess oil. At day 3, the solutions were transferred into a 5 mL volumetric flask while the vial was rinsed with 3 mL nanopure water; the water was then poured into the volumetric flask to reach 5 mL. The solutions were then transferred into centrifuge tubes and ready to be analyzed. The solutions were then analyzed with the ICP/OES instrument.

2.3. *Cell Extraction and Culture*

All procedures involving patients were conducted according to the Helsinki Declaration and were approved by the Research Ethical Committee of CHU de Québec-Université Laval (Project 1012–1341). Donors' consent for tissue harvesting was obtained for each specimen and experimental procedures were performed in compliance with the CHU de Québec guidelines. Dermal biopsies were collected from the skin of healthy donors undergoing plastic surgeries. After extensive washes in phosphate-buffered solution (PBS) with 100 U/mL penicillin, 25 mg/mL gentamicin, and 0.5 mg/mL fungizone (Bristol-Myers Squibb, Montreal, Canada), the skin biopsy was cut into small pieces and incubated in Hepes buffer (pH 7.4) (MP Biomedicals, Montreal, Canada) containing 500 mg/mL thermolysin and 1 mM $CaCl_2$ (Sigma, St. Louis, MO, USA) overnight at 4 °C. Following incubation, the epithelial layer was mechanically removed from the connective tissue containing fibroblasts (Fb). The dermis was transferred into a trypsination unit containing collagenase H solution (0.125 U/mL;

Roche Diagnostics, Mississauga, Canada) in DMEM (Invitrogen, Burlington, Canada), 10% foetal bovine serum (Hyclone, Logan, UT, USA), 100 U/mL penicillin, 25 mg/mL gentamicin, and sodium bicarbonate (Fb media). After 4 h of incubation at 37 °C, cells were harvested by centrifugation at 300 *g* for 10 min. Fb were then seeded into culture flasks containing Fb medium and incubated at 37 °C in a humidified 8% CO_2 atmosphere. Medium was changed three times a week. Normal human urothelial cells (NHUCs) were extracted from a healthy human urological tissue biopsy and were cultured as previously described [28,29]. NHUCs were seeded at a density of 5×10^5 cells in 75 cm^2 culture flasks with 1.5×10^5 irradiated murine NIH/3T3 (ATCC® CRL-1658™) as a feeder layer in urothelial basal media. Medium was changed three times a week. For 2D experiments, each experiment was done in five replicates.

2.4. Cell Viability Measurement in NHUC Monolayer Culture

Cells were seeded in 96 well plate at 10% confluence and were cultured in urothelial basal medium at 37 °C in a humidified 8% CO_2 atmosphere. The next day (Time 0), WST-1 (Roche, Laval, Canada) was used following the manufacturer instructions. The stable tetrazolium salt in WST-1 is transformed into formazan by viable cells. Thus, the viability of cells correlates directly with the amount of formazan. Cell density was measured at the indicated time. Cell viability of metal-extract-treated cultures was determined in comparison to the cell viability in the untreated condition (100%).

2.5. Immunofluorescence for Cytoskeletal Observation

NHUCs at a density of 1×10^5 cells by cm^2 were seeded in 24 well plates containing coverslips and incubated with 0.5 mL of metal extracts for 1 and 3 days at 37 °C with 8% CO_2. The cells were fixed on chamber slides using cold methanol for 10 min at −20 °C, and then washed thoroughly with PBS. Detection of Keratin 8/18 allowed the visualization of cytoskeleton intermediate filaments whereas Hoechst staining allowed detection of nuclei. Next, 50 µL of guinea-pig anti-human keratin 8/18 antibody (ARP, Belmont, MA, USA) diluted 1:50 in PBS–BSA 1% was added and incubated for 45 min. After discarding the first antibody, the cells were rinsed using PBS. Next, 50 µL of Alexa Fluor®594 coupled with secondary antibody (A21203, Thermo Fisher Scientific, Waltham, MA, USA) diluted in PBS–BSA 1% was added, incubated for 30 min in the dark, and washed with PBS three times for 2 min each, and washed with distilled water twice. Next, 50 µL 1:100 dilution of Hoechst 33342 (Thermo Fisher Scientific, Waltham, MA, USA) diluted in PBS–BSA 1% was then added for 10 min of incubation in the dark and washed with distilled water three times. A drop of mounting medium, PBS–glycerol–gelatin (pH 7.6), was put on the slide, and the inverted coverslip was placed on top of the drop. The slides with coverslips were left at 4 °C overnight to make sure the mounting medium was solid. The slides were then viewed under a fluorescence microscope with ApoTome attachment (Zeiss-Axio Imager Z1, Toronto, Canada).

2.6. Flat 3D Ureteral Wall Model Preparation

A flat, 3D construct tissue model (a patch-like tissue) was prepared by adapting our established engineered bladder and ureteral tissue model using the self-assembly method [30,31]. The procedure was adapted from the bladder model as described by Chabaud et al. [32], and characterized by Bureau et al. [33], and Goulet et al. [34]. Normal primary human skin Fb cells were seeded at 3×10^4 cells/cm^2 in 6 well plates. An anchorage paper device had been previously placed in each well as a stroma support system. The cells were cultured in DMEM supplemented with 10% newborn calf serum (NBCS), 100 U/mL penicillin, 25 µg/mL gentamycin (Fb medium), and fresh 50 µg/mL ascorbic acid, and were incubated in 8% CO_2 at 37 °C. The media was changed every two days. After 14 days, a second seeding of Fb took place and the culture was pursued for 14 additional days until stroma sheets were formed. Further steps were done to build a flat 3D ureteral wall (UW) tissue model (hence named "UW model"). NHUCs were seeded at 2×10^5 cells/cm^2 on each stroma sheet in NHUC basal media, supplemented with fresh 50 µg/mL ascorbic acid and incubated in 8% CO_2 at 37 °C for 7 days under submerged conditions. The stroma sheets were then moved into petri dishes to provide elevation at the air–liquid interface for 21 days. The provided interface aimed for the differentiation and maturation of NHUCs into the urothelium layer and formed the UW model.

A direct toxicity test was done by putting metal disks on the top of mature urothelium of the UW model (Figure 1). This simple method mimics the in vivo conditions, and was the first tissue-engineered model used to assess absorbable metal toxicity. The test assessed the cell function in the presence of the metal disk and its corrosion products, and also allowed observation of the effect of compression from the weight of the disk on the tissue, simulating the compression of a stent in the ureter. Disks of ZM21 (355 ± 0.02 mg), Zn–1Mg (801 ± 0.03 mg), and stainless steel 316L (437 ± 0.02 mg control) were used. The alloys were expected to show different toxicities in the 3D culture setting compared to the 2D.

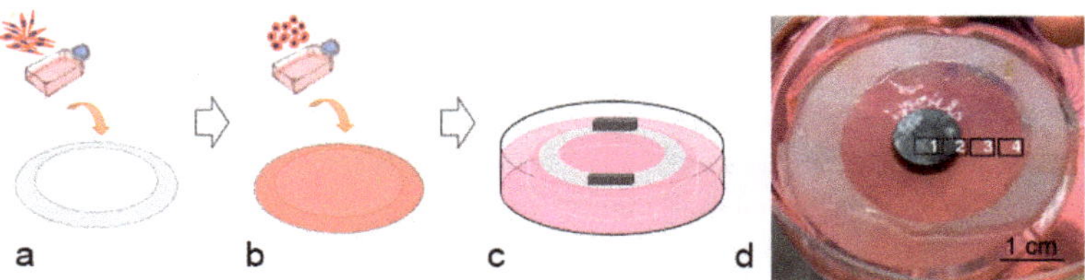

Figure 1. The ureteral wall (UW) model preparation: (**a**) production of stroma, 28 days culture, (**b**) production of tissue equivalent, 7 days culture in submerged condition, (**c**) displacement of tissue equivalent into petri dish, 21 days culture, (**d**) macroscopic view of the UW model with metal in place for direct toxicity test, the areas marked 1–4 are the area used for histological analysis.

2.7. 3D Cell Function Evaluation

Direct toxicity tests were conducted to assess the urothelial cell function. Metal disks were sterilized with UV light exposure for a minimum of 15 min on each side, then dipped in ethanol three times for 5 s each and washed with sterile PBS under a sterilized atmosphere. The disks were put on top of the UW model and incubated for 1 week. The UW tissues were then sacrificed by cutting them into two halves. The first half was fixed with Histochoice tissue fixative (Amresco, Solon, OH, USA) and then embedded in paraffin. Histological sections were made by cutting paraffinized tissue into 5 μm size and staining with Masson's Trichrome. The second half was embedded in frozen tissue medium (optimal cut temperature (OCT) compound; Tissue-Tek, Bayer, Etobicoke, Canada) and then cut into 5 μm for further immunofluorescence (IF) evaluation. The slides from both paraffinized (IF against uroplakin) and OCT (other IFs) were fixed in cold 100% methanol, blocked with PBS–BSA 1%, and incubated with primary antibodies, as described above in the procedure for cytoskeletal observation. Primary antibodies used were ZO-1 for tight junction (1:50; 40–2200, Thermo Fisher Scientific, Waltham, MA, USA), Ki-67 for cell proliferation (1:400; ab15580, Abcam, Cambridge, MA, USA), and laminin-5 for basement membrane (1:400; ab14509, Abcam, Cambridge, MA, USA). Secondary antibodies were coupled with Alexa Fluor® dyes (Thermo Fisher Scientific, Waltham, MA, USA): Alexa Fluor®488 (1:500; A21202) and 594 (1:200; A21468). The slides were then viewed under the fluorescence microscope [30].

2.8. Scanning Electron Microscopy

Metal disk remnants were preserved in 2.5% glutaraldehyde overnight in 4 °C and processed for further observation under electron microscopy, as described by Heckman et al. [35]. After overnight incubation, remnants were washed three times with PBS for 5 min each and then dehydrated by using serial ascending concentrations ethanol solutions from 30%, 50%, 70%, to 90% for 5 min each, and then 3 × 5 min in 100% ethanol. The remnants were dried and then gold sputtered. Observation was done using a scanning electron microscope (JEOL 7500-F, JEOL Ltd., Tokyo, Japan).

2.9. Statistical Analysis

All results are expressed as mean ± standard deviation of each independent experiment. The statistically significant differences between the mean were calculated by ANOVA and post hoc Tukey's test for multiple comparisons, with the level of significance selected at $p < 0.05$ using SPSS 25 (IBM Canada Ltd., Markham, Canada).

3. Results and Discussion

3.1. Screening Tests

The cell viability test revealed that the viability of the urothelial cells exposed to pure iron was lower than in those exposed to pure magnesium and pure zinc (Figure 2a). The viability of the urothelial cells exposed to pure magnesium increased with the metal's incubation time. The viability of urothelial cells exposed to pure zinc remained high for all incubation times (Figure 2a). In artificial urine solution, magnesium corroded faster than zinc and iron (Figure 2b), confirming the previously published results [24]. The constant trend of the cell viability exposed to zinc could be related to the constant release of Zn^{2+} ions into the solution throughout the metal incubation period. Meanwhile, the low cell viability results of pure iron could be related to the formation of iron free radicals, causing oxidative stress. Free radicals react with organic molecules in the cell membrane, initiate a lipid-peroxidation process and eventually lead to cell death [36]. The increasing viability trend of cells exposed to pure magnesium could be related to the formation of magnesium salts, such as $Mg(OH)_2$, $Mg_3(PO4)_2$, $MgCO_3$, $Ca(OH)_2$, $Ca_3(PO_4)_2$, and $CaCO_3$, that in turn reduced the availability of Mg^{2+} ions to interact with the cells [37]. The environment was considered unfavorable for the cells when the corrosion products, i.e., Mg^{2+} and OH^-, reached a certain concentration [20]. From these results, iron was eliminated from further tests because of its low corrosion rate and its high toxicity toward urothelial cells.

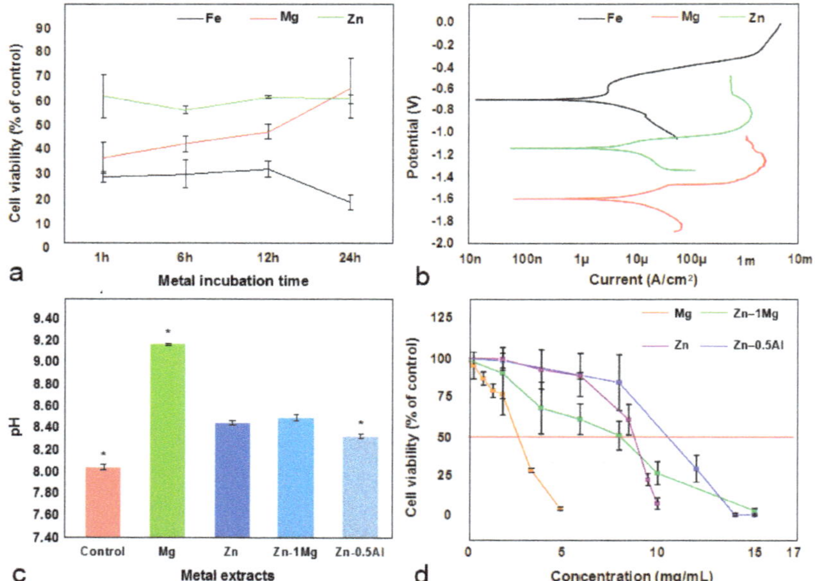

Figure 2. Screening test results: (**a**) viability of urothelial cells as a function of the metal incubation time in artificial urine solution, (**b**) potentiodynamic polarization curves of pure iron, magnesium, and zinc (potential in Volt (V) vs. saturated calomel electrode (SCE)), (**c**) pH value of metal extracts in urothelial basal medium, * $p < 0.01$, (**d**) viability of urothelial cells treated with different concentrations of metal extracts. Red line marks 50% of cell viability.

In the second part of screening test, pure magnesium, pure zinc, Zn–1Mg, and Zn–0.5Al were used. All metal extracts induced an alkalinization of the urothelium basal medium, increasing its pH value from 8.05 to 8.34 for Zn–0.5Al, 8.45 for pure zinc, 8.50 for Zn–1Mg, and the highest value, 9.18, for pure magnesium (Figure 2c). The pH increase was mainly caused by the production of OH^- ions during the corrosion of those metals [38]. The faster the corrosion, the more OH^- was released and the more rapid the alkalinization, until OH^- reached saturation and the pH value became stable [19]. A pH increase influences cellular function and compatibility [37]. Moreover, the viability of urothelial cells decreased as the metal extract concentration increased (Figure 2d), a common condition for many type of cells [15,16,22,39]. Tian et al. [20] reported that starting from pH 8.3, urothelial cells were seen to be unhealthy, with round morphology, and when pH values reached 8.6 the reduction in the density of urothelial cells was significant. Another study on fibroblasts and keratinocytes found that these cells proliferated and migrated better and were more viable in pH 8–8.7 compared to acidic or more alkaline environments [40]. Gu et al. [41] reported that alkaline stress causes severe cytotoxicity in human osteosarcoma cells MG63. Similar results have also been shown for osteoblasts, but the cells have better performance in terms of gene expression and mineralization [42]. Urine has a wide pH range, from 4.5 to 8.0 depending on the body's acid–base equilibrium state. On a regular, average diet, the pH range decreases to 5.0 to 6.5. Consistently acidic urine, i.e., pH $\leq$ 6.0, has been associated with an increased risk of bladder cancer [43]. Urine pH mainly depends on the bicarbonate (HCO_3^-) concentration in blood, where the higher the concentration, the higher the pH is [44]. Other substances in urine include inorganic cations (sodium, potassium, ammonium, calcium, and magnesium) and anions (chloride, sulphate, and phosphate), and organic components such as urate, creatinine, oxalate, and citrate [45].

By plotting the cell viability versus metal extract concentration, the IC_{50} values were determined as the following: pure magnesium (2.5 mg/mL), Zn–1Mg (10 mg/mL), pure zinc (8 mg/mL), and Zn–0.5Al (8.75 mg/mL). Based on this result, Zn alloys were chosen for further biological assessment. The concentrations of metal ions in 10 mg/mL of Zn–1Mg and 8.75 mg/mL of Zn–0.5Al after 72 h of incubation in urothelial basal medium are presented in Table 1. The detected magnesium ion in the Zn–0.5Al extract ca,e from the magnesium content in urothelial basal medium.

Table 1. Ion concentration of metal alloy extracts at IC_{50}.

Metal	Ion Concentration (ppm)		
	Mg	Zn	Al
Zn–1Mg	21.85–30.10	18.26–25.96	0
Zn–0.5Al	21.47–28.59	17.65–20.02	0.030–0.090

Murni et al. [15] reported that 0.49 ppm of zinc ions in a Zn–3Mg alloy extract killed 50% of a normal human osteoblast cell population. Kubasek et al. [16] found that the highest safe concentration of zinc ions for U-2 osteosarcoma and L929 fibroblast cell lines was 7.85 ppm and 5.23 ppm, respectively. In the present study, 25.96 ppm of zinc ions killed half of the urothelial cell population. Different cell line behave differently; one type of cell can be more sensitive to zinc ions than others. Urothelial cells, on the other hand, only present a phenotype similar to the basal cells when cultivated in monolayer culture, which is found in the deepest layer of urothelium in the native tissue. This caused the sensitivity of urothelial cells toward the metal extracts to be higher in the present study. In addition, the urothelial cells used in this study were in passage-2 and were harvested from a primary culture. Generally, primary cultures are more sensitive toward toxic substances than immortal cell lines, because the cell culture induces an adaptation of the cells to the culture conditions [46]. The use of serum (FBS) in the metal extracts or salt solutions gave a protective effect to the cells. It is well known that the use of serum in culture media at either 5% or 10% induces the growth of cells, and studies have reported that cell viability in cytotoxicity tests is higher in tests done with the presence of serum [16,47,48].

3.2. Cytoskeletal Observation

Some changes occurred in the cell size and morphology of NHUCs in the presence of metal ions (Figure 3). These cells still have the capacity to form colonies even when they are experiencing changes, but cells in the Zn–1Mg group shrank and appeared rounded, as in apoptotic stage, compared to Zn–0.5Al group at day 1 (Figure 3a–c). The intensity of cytokeratin (CK) was less in the metal alloy groups compared to controls. Observation at day 3 showed that cells in the Zn–1Mg group survived and those in the Zn–0.5Al group had recovered their normal state (Figure 3d–f). The cells in the Zn–1Mg group did not fully recover, as seen from their size and morphology compared to controls, and the intensity of their CK was also less than the other groups. Rounded morphology found in Zn–1Mg group at day 1 could also have been a result of the alkaline pH (8.50) of the culture media.

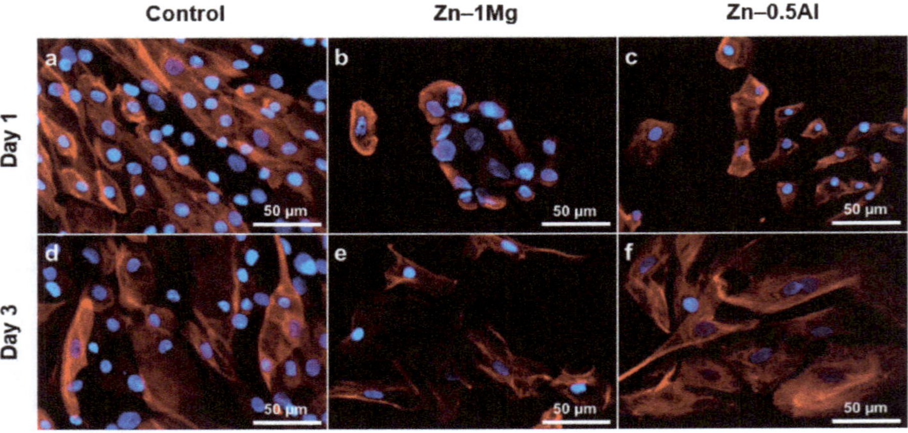

Figure 3. Images of cytoskeletal observation at days 1 and 3 in controls, Zn–1Mg, and Zn–0.5Al groups. Note: anti-cytokeratin 8/18 (red) staining was used to examine the changes of keratin, and Hoechst (blue) was used to stain nucleus of the normal human urothelial cells (NHUCs).

Cytoskeletal changes in the presence of Zn ions were reported by Murni et al. It was observed that the cells underwent changes in size, contour, and skeletal intensity, which corresponded to cell stress and inhibition of cell proliferation. The adaptation of epithelium to changing conditions is usually accompanied by transition in the cytoskeleton of epithelial cells. Cell stress leads to apoptosis events. CK, as a major family of structural protein in epithelial cells, plays a role in mechanical and non-mechanical functions, including protection from cell stress and apoptosis. Meanwhile, epithelial polarity helps in cell–cell adhesion and also in attachment of the epithelial cells to the underlining connective tissue. Besides this structural function, CK also play a role in dynamic processes such as mitosis, mobility, and differentiation [49,50]. CK 8/18 pairs are obligate partners and establish the primary CK pair in many epithelia, including urothelium [51]. CK's roles in apoptosis have been reported by many studies. The apoptosis process, with chromatin condensation as its hallmark, is first preceded by the breakdown of CK8/18. The breakdown results in the collapse of the cytoplasmic and nuclear cytoskeleton [49].

3.3. Cell Function Evaluation

The UW model offered several very interesting characteristics, such as the absence of exogenous material use during the reconstruction, which allowed differentiation of the urothelium close to what happens in native tissue. This adequate urothelium maturation was confirmed by the water-tightness of the 3D construct, similar to that of the native porcine bladder [33,34]. Since urothelium is a highly specialized epithelium, the use of UW wall with an organization close to the native tissue, in combination with direct toxicity, was considered to be a simple method with which to represent in vivo conditions. Table 2 presents the changes observed macroscopically during the incubation of the UW models with the metal disks. The ZM21 group induced significant changes compared to the control and Zn–1Mg groups, as marked by changes in all three parameters.

Table 2. Macroscopic changes on the UW model observed after 1 week.

Parameters	Control	ZM21	Zn–1Mg
Tissue appearance	No change	Thick white structure surrounding the peri-metal site	No change
Medium culture (phenol red as pH indicator)	Yellow after 2 days of culture (acidic)	No color change (alkaline)	Yellow after 2 days of culture (acidic)
Metal corrosion product	Not found	Bubbles surrounding the metal	Not found

Histological sections stained by Masson's Trichrome provided information about the morphology of urothelium in the presence of the metal disks and their corrosion products. The tissue-engineered ureter used in this study consisted of layers of urothelium and stroma (connective tissue) beneath it, which closely mimicked native ureteral tissue. The ability of the urothelial cells to migrate, proliferate, and differentiate to form urothelium is important for maintenance of urothelial integrity and retaining the permeability barrier function [52]. Figure 1d shows the division of the tissue into four areas, from the most proximal area to the metal disk to the most distal one. In all groups, including the control group, which was a disk of the surgical grade stainless steel, in the section stained from area 1 (in direct contact with or under metal disks) no cells or dead cells could be seen (Figure 4a,e,i). In area 2, the layers of urothelium seemed damaged in the control and Zn–1Mg groups, (respectively Figure 4b,j), or consisted of a thin epithelium in the case of the ZM21 group (Figure 4f). The transition from the absence of urothelium to a thin damaged epithelium was gradual, as well depicted in Figure 4j. In area 3, the urothelium seemed unaffected, and the superficial urothelial cells were distinct in the control (Figure 4c) and the Zn–1Mg group (Figure 4k), whereas they formed a wave-like structure in the ZM21 group (Figure 4g). This area is known to be the site of bubbles and thick white structures; it seems obvious that bubbles pushed the urothelium into this shape. The bubbles were made of hydrogen as a result of magnesium corrosion, which can be written as an overall reaction of: $Mg + 2H_2O \rightarrow Mg(OH)_2 + H_2$, where the $Mg(OH)_2$ further reacts with Cl^- ions to form soluble $MgCl_2$ [38]. Finally, in area 4, the urothelium remained unaffected by the metal disks, whatever the condition. Based on the Masson's Trichrome staining results, the metal alloy groups showed similar morphology compared to the controls.

Differentiated urothelial cells, also known as umbrella cells, assume their role in maintaining the barrier function by producing specific proteins such as zonula occludens (ZO)-1. Other proteins, such as Ki67, allow evaluation of the proliferative state of the tissue, which could be overexpressed in the case of a tissue physiological or pathological response (e.g., injury or cancer). Laminin-5 is also an important molecule because its presence in the basal lamina is an absolute requirement for adequate urothelial cell differentiation, as evidenced by Rousseau et al. [53]. Figure 5 presents the ZO-1, Ki67, and laminin-5 immunofluorescence staining of the histological sections of the UW model. ZO-1 staining was done to evaluate the presence of tight junctions between urothelial cells. Figure 5a–c shows tight junctions separating the superficial urothelial cells in each group. These tight junctions, together with uroplakin, play an important role in urothelium, acting as a barrier from urine leakage, ions, or any other compound in the urine. Cell proliferation processes were not observed for the UW model, as shown by the negative result of the Ki67 staining (Figure 5d–f). Laminin-5 staining confirmed that delamination or detachment of basal layer did not occur in either the control or the metal alloy groups (Figure 5g–i).

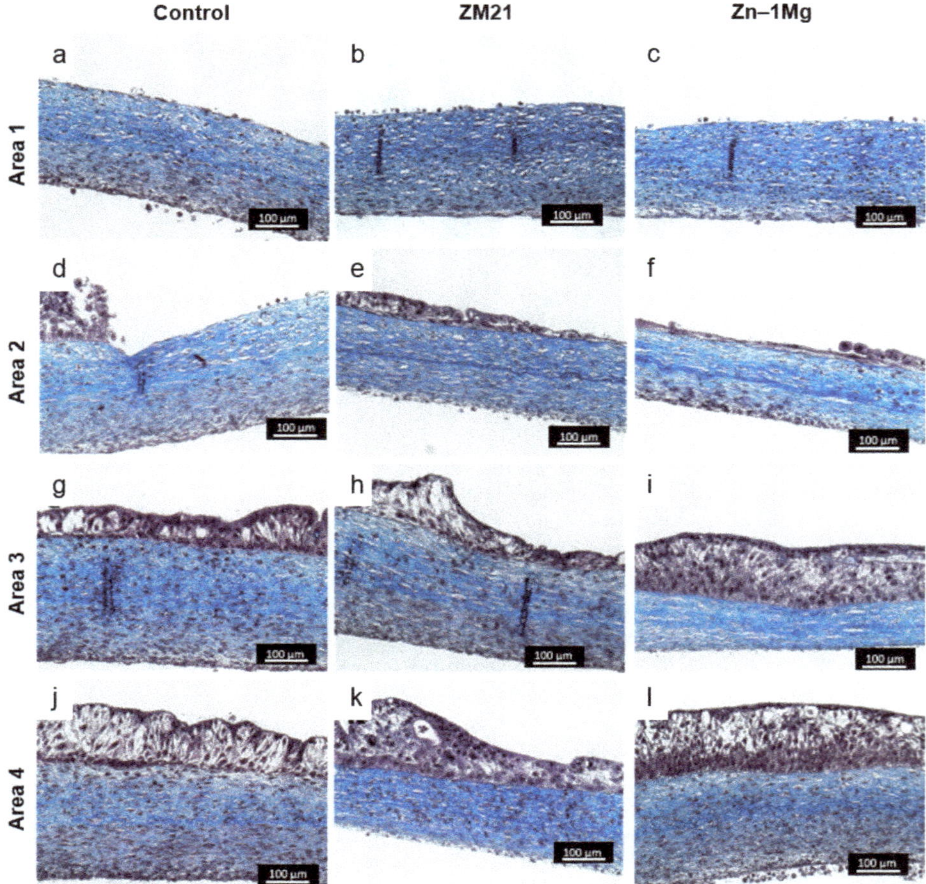

Figure 4. Histological analysis of UW tissue with different metal disks, stained with Masson's Trichrome. Nucleus stained as purple and stroma/collagen stained as blue. Tissue sections are divided into four areas: (**1**) under/in contact with metal disks, (**2**) peri-metal site, (**3**) nearest area to peri-metal site, and (**4**) farthest area from peri-metal site. Bar = 100 µm.

Tight junctions (TJ) between umbrella cells represent the main permeability barrier in combination with another protein, uroplakins [54]. TJs restrict paracellular diffusion and movement, and they contribute to maintenance of the surface polarity of cells by restricting the movement of proteins and lipids between membrane compartments. TJs are composed of cytoplasmic proteins, such as ZO-1, linking TJs to the cytoskeleton, and integral transmembrane proteins, such as occludin, junctional adhesion molecule, and claudins [55–57].

Laminins are a group of proteins that are the most important components of the basal membrane. Laminins play a role in the interaction between urothelial cells and the extracellular matrix. In the ureteral mucosa, laminins are expressed by the basal cells and support the assembly of the basal membrane, contributing to maintenance of cell and tissue integrity. Laminins are also involved in cell adhesion, hemidesmosome formation, migration, proliferation, differentiation, and prevention of programmed cell death [58–61]. The presence of two important proteins, ZO-1 and laminin-5, was noted in our UW model with the direct toxicity method. This means that the urothelium function of producing barrier proteins was not interrupted by the metals and their corrosion products (except in the area directly under the disk, where the epithelium was destroyed). The urothelium may also have preserved its permeable barrier function in close proximity to the cells.

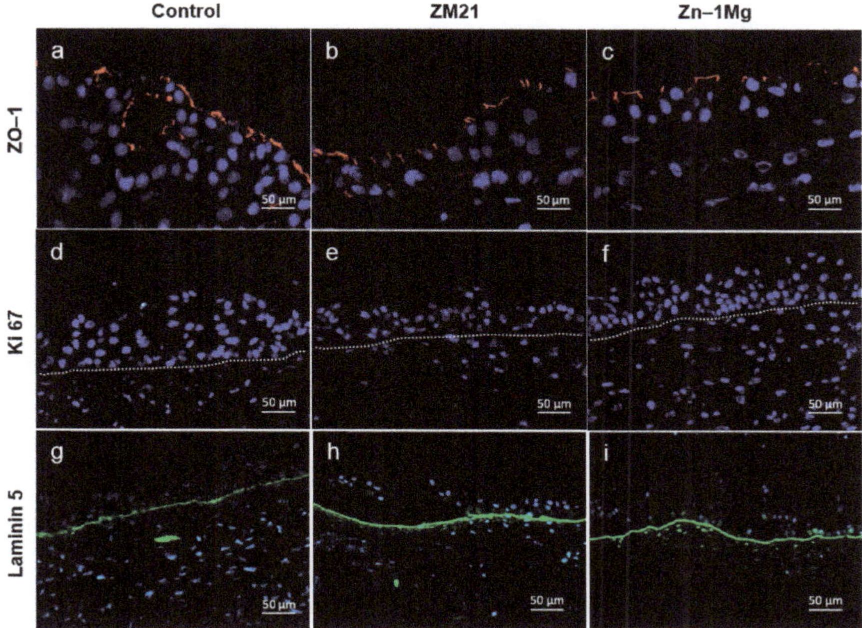

Figure 5. Expression of urothelial differentiation-associated protein and basal lamina in response to contact with metal disk after 7 days of incubation: (**a–c**) immunofluorescence anti-ZO-1 (red), (**d–f**) Ki67 (green), and (**g–i**) anti-laminin-5 (green). Nuclei were stained with Hoechst (blue).

3.4. Scanning Electron Microscopy Analysis

Evidence of corrosion of metals was observed in alloy groups after 7 days of incubation, while the control group showed the same surface as before the incubation (Figure 6a,b). Corrosion of ZM21 was seen as layers of metal flakes all over the surface (Figure 6d,e), while the Zn–1Mg sample had micro-size holes with attached precipitation on some parts of its surface (Figure 6g,h). Urothelial cells and stroma were observed on the surface of each metal disk (Figure 6c,f,i). Attachment of cells is marked by pseudopodia structures in cells with flattened shape. These results confirmed that the lost or removed parts of urothelium from area 1 in the UW model were actually attached to the metal and were removed together with the metal disks. The cells that were attached 3–13 µm in size, indicating that they were basal and intermediate cells [62]. This indicates that the compression of the urothelium probably inhibited the urothelial cells from differentiating well, which occurred similarly in the control and alloy groups. Basal and intermediate cells alone, i.e., non-differentiated urothelial cells, do not have the ability to produce such proteins able to maintain the impermeable barrier. These very encouraging results pave the way for further studies using a more representative metal structure, e.g., a metal mesh, to closely simulate the ureteral stent.

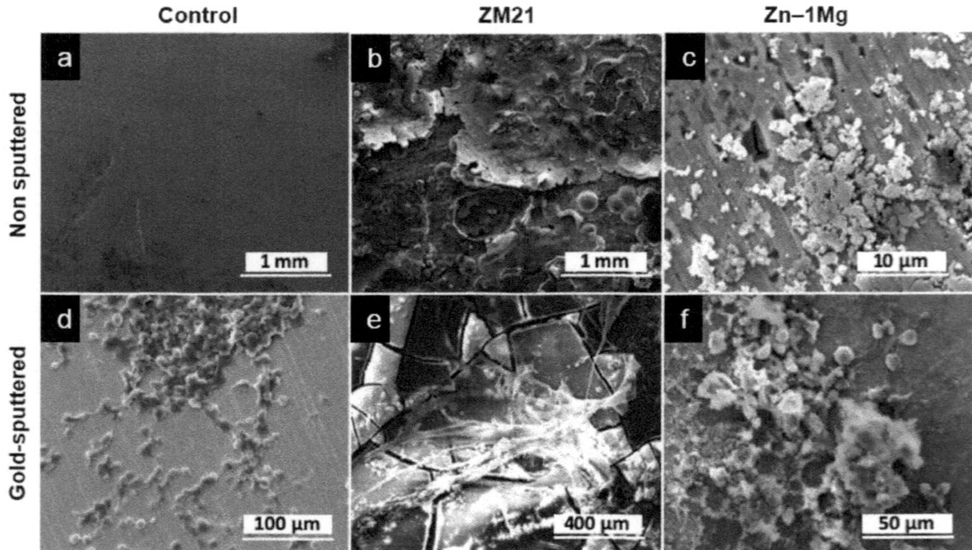

Figure 6. Scanning electron microscope micrographs of the surface of metal disks incubated for 7 days on the UW model: (**a–c**) corrosion layer was observed on the alloys' surface, while the control showed no change, (**d–f**), urothelial cells and stroma collagen layer attachments were observed on the surface of each disk.

4. Conclusions

To our knowledge, this is the first study reporting a biological assessment of absorbable metals in view of ureteral stent application using normal human urothelial cells (NHUCs) and a metal direct toxicity test on a flat 3D ureteral wall tissue model that mimics the in vivo conditions. The NHUCs exhibited a survival mode in response to the toxicity of the magnesium and zinc extracts in monolayer culture, shown by reduced viability and unhealthy appearance, but recovered after some days. This result suggests that the cells could rapidly colonize damaged areas to restore mature urothelium, restoring the barrier function to block urine extravasation in the body. The direct toxicity test in the 3D ureteral wall model showed that the corrosion of metals had an effect on the morphology of cell layers, but the urothelium was still able to maintain its function in producing the barrier protein and may have had a permeable barrier function in close proximity to the cells. This indicates that the tissue was healthy despite the presence of the metal disks and their corrosion products after 7 days of exposure. The attached cell layers on the surface of metal disks may indicate that the metal specimens were non-toxic. The zinc alloy group seemed to be more cytocompatible toward human urothelial cells compared to magnesium alloys, and could be further exploited for biodegradable ureteral stent applications. In the future, in vitro test using a flat 3D ureteral wall tissue model could be used to partially replace animal testing in the initial steps of optimization of absorbable metal stents.

Author Contributions: Conceptualization, D.P., H.H., S.B., and S.C.; Data Curation, D.P., and S.C.; Formal Analysis, D.P., H.H., S.B., and S.C.; Methodology, D.P., H.H., S.B., and S.C.; Supervision, H.H. and S.B.; Writing—Original Draft Preparation, D.P.; Writing—Review & Editing, D.P., H.H., S.B., and S.C.

Funding: This work is supported by the Natural Sciences and Engineering Research Council of Canada (NSERC) and the Fonds de recherche du Québec-Santé (FRQS).

Acknowledgments: The authors would like to thank Geneviève Bernard for the help in preparing and conducting the 3D assay, Agung Purnama for the help in the SEM experiment, Sébastien Champagne for the help in corrosion test, Ehsan Mostaed and Maurizio Vedani for providing us with the metals.

Conflicts of Interest: The authors declare no conflict of interest.

References

1. Frazier, M.S. *Essentials of Human Diseases and Conditions*, 3rd ed.; Elsevier Saunders: St. Louis, MO, USA, 2004; ISBN 9780323228367.
2. Becker, A.; Baum, M. Obstructive uropathy. *Early Hum. Dev.* **2006**, *82*, 15–22. [CrossRef] [PubMed]
3. Dyer, R.B.; Chen, M.Y.; Zagoria, R.J.; Regan, J.D.; Hood, C.G.; Kavanagh, P.V. Complications of Ureteral Stent Placement. *Radiographics* **2002**, *22*, 1005–1022. [CrossRef] [PubMed]
4. Brotherhood, H.; Lange, D.; Chew, B.H. Advances in ureteral stents. *Transl. Androl. Urol.* **2014**, *3*, 314–319. [PubMed]
5. Lange, D.; Bidnur, S.; Hoag, N.; Chew, B.H. Ureteral stent-associated complications—where we are and where we are going. *Nat. Rev. Urol.* **2015**, *12*, 17–25. [CrossRef]
6. Auge, B.K.; Sarvis, J.A.; L'Esperance, J.O.; Preminger, G.M. Practice Patterns of Ureteral Stenting after Routine Ureteroscopic Stone Surgery: A Survey of Practicing Urologists. *J. Endourol.* **2007**, *21*, 1287–1292. [CrossRef]
7. Lange, D.; Chew, B.H. Update on ureteral stent technology. *Ther. Adv. Urol.* **2009**, *1*, 143–148. [CrossRef]
8. Talja, M.; Välimaa, T.; Tammela, T.; Petas, A.; Törmälä, P. Bioabsorbable and Biodegradable Stents in Urology. *J. Endourol.* **1997**, *11*, 391–397. [CrossRef]
9. Wang, X.; Shan, H.; Wang, J.; Hou, Y.; Ding, J.; Chen, Q.; Guan, J.; Wang, C.; Chen, X. Characterization of nanostructured ureteral stent with gradient degradation in a porcine model. *Int. J. Nanomed.* **2015**, *10*, 3055–3064. [CrossRef]
10. Pedro, R.N.; Hendlin, K.; Kriedberg, C.; Monga, M. Wire-Based Ureteral Stents: Impact on Tensile Strength and Compression. *Urology* **2007**, *70*, 1057–1059. [CrossRef]
11. Hendlin, K.; Korman, E.; Monga, M. New Metallic Ureteral Stents: Improved Tensile Strength and Resistance to Extrinsic Compression. *J. Endourol.* **2012**, *26*, 271–274. [CrossRef]
12. Hermawan, H. Updates on the research and development of absorbable metals for biomedical applications. *Prog. Biomater.* **2018**, *7*, 93–110. [CrossRef] [PubMed]
13. Francis, A.; Yang, Y.; Virtanen, S.; Boccaccini, A.R. Iron and iron-based alloys for temporary cardiovascular applications. *J. Mater. Sci. Mater. Electron.* **2015**, *26*, 138. [CrossRef] [PubMed]
14. Bosiers, M. AMS INSIGHT—Absorbable Metal Stent Implantation for Treatment of Below-the-Knee Critical Limb Ischemia: 6-Month Analysis. *J. Vasc. Surg.* **2009**, *50*, 1241. [CrossRef]
15. Murni, N.; Dambatta, M.; Yeap, S.K.; Froemming, G.; Hermawan, H.; Froemming, G. Cytotoxicity evaluation of biodegradable Zn–3Mg alloy toward normal human osteoblast cells. *Mater. Sci. Eng. C* **2015**, *49*, 560–566. [CrossRef]
16. Kubásek, J.; Vojtěch, D.; Jablonská, E.; Pospíšilová, I.; Lipov, J.; Ruml, T. Structure, mechanical characteristics and in vitro degradation, cytotoxicity, genotoxicity and mutagenicity of novel biodegradable Zn–Mg alloys. *Mater. Sci. Eng. C* **2016**, *58*, 24–35. [CrossRef]
17. Bowen, P.K.; Shearier, E.R.; Zhao, S.; Guillory, R.J.; Zhao, F.; Goldman, J.; Drelich, J.W. Biodegradable Metals for Cardiovascular Stents: From Clinical Concerns to Recent Zn-Alloys. *Adv. Health Mater.* **2016**, *5*, 1121–1140. [CrossRef]
18. Lock, J.Y.; Wyatt, E.; Upadhyayula, S.; Whall, A.; Nuñez, V.; Vullev, V.I.; Liu, H. Degradation and antibacterial properties of magnesium alloys in artificial urine for potential resorbable ureteral stent applications. *J. Biomed. Mater. Res. Part A* **2014**, *102*, 781–792. [CrossRef]
19. Zhang, S.; Zheng, Y.; Zhang, L.; Bi, Y.; Li, J.; Liu, J.; Yu, Y.; Guo, H.; Li, Y. In vitro and in vivo corrosion and histocompatibility of pure Mg and a Mg-6Zn alloy as urinary implants in rat model. *Mater. Sci. Eng. C* **2016**, *68*, 414–422. [CrossRef]
20. Tian, Q.; Deo, M.; Rivera-Castaneda, L.; Liu, H. Cytocompatibility of Magnesium Alloys with Human Urothelial Cells: A Comparison of Three Culture Methodologies. *ACS Biomater. Sci. Eng.* **2016**, *2*, 1559–1571. [CrossRef]
21. Zhang, S.; Bi, Y.; Li, J.; Wang, Z.; Yan, J.; Song, J.; Sheng, H.; Guo, H.; Li, Y. Biodegradation behavior of magnesium and ZK60 alloy in artificial urine and rat models. *Bioact. Mater.* **2017**, *2*, 53–62. [CrossRef]
22. Lan, P.; Zhou, F.; Pu, Z.; Shen, C.; Liu, X.; Fan, B.; Li, X.; Wang, H.; Xiao, X.; Zhao, S.; et al. Mechanical properties, in vitro degradation behavior, hemocompatibility and cytotoxicity evaluation of Zn–1.2Mg alloy for biodegradable implants. *RSC Adv.* **2016**, *6*, 86410–86419.

23. Mostaed, E.; Sikora-Jasinska, M.; Mostaed, A.; Loffredo, S.; Demir, A.; Previtali, B.; Mantovani, D.; Beanland, R.; Vedani, M. Novel Zn-based alloys for biodegradable stent applications: Design, development and in vitro degradation. *J. Mech. Behav. Biomed. Mater.* **2016**, *60*, 581–602. [CrossRef] [PubMed]
24. Champagne, S.; Mostaed, E.; Safizadeh, F.; Ghali, E.; Vedani, M.; Hermawan, H. In Vitro Degradation of Absorbable Zinc Alloys in Artificial Urine. *Materials* **2019**, *12*, 295. [CrossRef] [PubMed]
25. Matias, T.; Roche, V.; Nogueira, R.; Asato, G.; Kiminami, C.; Bolfarini, C.; Botta, W.; Jorge, A.; Kiminami, C.; Botta, W. Mg-Zn-Ca amorphous alloys for application as temporary implant: Effect of Zn content on the mechanical and corrosion properties. *Mater. Des.* **2016**, *110*, 188–195. [CrossRef]
26. Wang, C.; Yang, H.; Li, X.; Zheng, Y. In Vitro Evaluation of the Feasibility of Commercial Zn Alloys as Biodegradable Metals. *J. Mater. Sci. Technol.* **2016**, *32*, 909–918. [CrossRef]
27. Wang, X.; Li, Y.; Hou, Y.; Bian, H.; Koizumi, Y.; Chiba, A. Effects of surface friction treatment on the in vitro release of constituent metals from the biomedical Co–29Cr–6Mo–0.16N alloy. *Mater. Sci. Eng. C* **2016**, *64*, 260–268. [CrossRef] [PubMed]
28. Chabaud, S.; Marcoux, T.-L.; Deschênes-Rompré, M.-P.; Rousseau, A.; Morissette, A.; Bouhout, S.; Bernard, G.; Bolduc, S. Lysophosphatidic acid enhances collagen deposition and matrix thickening in engineered tissue. *J. Tissue Eng. Regen. Med.* **2015**, *9*, E65–E75. [CrossRef]
29. Chabaud, S.; Rousseau, A.; Marcoux, T.-L.; Bolduc, S. Inexpensive production of near-native engineered stromas: Inexpensive production of engineered tissues. *J. Tissue Eng. Regen. Med.* **2017**, *11*, 1377–1389. [CrossRef] [PubMed]
30. Bouhout, S.; Émilie;, P.; Gauvin, R.; Bernard, G.; Ouellet, G.; Cattan, V.; Bolduc, S. In Vitro Reconstruction of an Autologous, Watertight, and Resistant Vesical Equivalent. *Tissue Eng. Part A* **2010**, *16*, 1539–1548. [CrossRef] [PubMed]
31. Imbeault, A.; Bernard, G.; Rousseau, A.; Morissette, A.; Chabaud, S.; Bouhout, S.; Bolduc, S. An endothelialized urothelial cell-seeded tubular graft for urethral replacement. *Can. Urol. Assoc. J.* **2013**, *7*, E4–E9. [CrossRef] [PubMed]
32. Chabaud, S.; Saba, I.; Baratange, C.; Boiroux, B.; Leclerc, M.; Rousseau, A.; Bouhout, S.; Bolduc, S. Urothelial cell expansion and differentiation are improved by exposure to hypoxia. *J. Tissue Eng. Regen. Med.* **2017**, *11*, 3090–3099. [CrossRef] [PubMed]
33. Bureau, M.; Pelletier, J.; Rousseau, A.; Bernard, G.; Chabaud, S.; Bolduc, S. Demonstration of the direct impact of ketamine on urothelium using a tissue engineered bladder model. *Can. Urol. Assoc. J.* **2015**, *9*, E613–E617. [CrossRef]
34. Goulet, C.R.; Bernard, G.; Chabaud, S.; Couture, A.; Langlois, A.; Neveu, B.; Pouliot, F.; Bolduc, S. Tissue-engineered human 3D model of bladder cancer for invasion study and drug discovery. *Biomaterials* **2017**, *145*, 233–241. [CrossRef]
35. Heckman, C.; Kanagasundaram, S.; Cayer, M.; Paige, J. Preparation of cultured cells for scanning electron microscope. *Protoc. Exch.* **2007**. [CrossRef]
36. Witte, F.; Eliezer, A. Biodegradable Metals. In *Degradation of Implant Materials*; Springer Science and Business Media LLC: Berlin, Germany, 2012; pp. 93–109.
37. Witecka, A.; Bogucka, A.; Yamamoto, A.; Máthis, K.; Krajňák, T.; Jaroszewicz, J.; Swieszkowski, W. In vitro degradation of ZM21 magnesium alloy in simulated body fluids. *Mater. Sci. Eng. C* **2016**, *65*, 59–69. [CrossRef] [PubMed]
38. Banerjee, P.C.; Al-Saadi, S.; Choudhary, L.; Harandi, S.E.; Singh, R. Magnesium implants: Prospects and challenges. *Materials* **2019**, *12*, 136. [CrossRef]
39. Dambatta, M.; Murni, N.; Izman, S.; Kurniawan, D.; Froemming, G.; Hermawan, H.; Froemming, G. In vitro degradation and cell viability assessment of Zn–3Mg alloy for biodegradable bone implants. *Proc. Inst. Mech. Eng. Part H J. Eng. Med.* **2015**, *229*, 335–342. [CrossRef] [PubMed]
40. Kruse, C.R.; Singh, M.; Targosinski, S.; Sinha, I.; Sørensen, J.A.; Eriksson, E.; Nuutila, K. The effect of pH on cell viability, cell migration, cell proliferation, wound closure and wound re-epithelialization: In vitro and in vivo study. *Wound Repair Regen.* **2017**, *25*, 260–269. [CrossRef]
41. Gu, X.; Wang, F.; Xie, X.; Zheng, M.; Li, P.; Zheng, Y.; Qin, L.; Fan, Y. In vitro and in vivo studies on as-extruded Mg- 5.25 wt.%Zn-0.6 wt.%Ca alloy as biodegradable metal. *Sci. China Mater.* **2018**, *61*, 619–628. [CrossRef]

42. Galow, A.-M.; Rebl, A.; Koczan, D.; Bonk, S.M.; Baumann, W.; Gimsa, J. Increased osteoblast viability at alkaline pH in vitro provides a new perspective on bone regeneration. *Biochem. Biophys. Rep.* **2017**, *10*, 17–25. [CrossRef]
43. Alguacil, J.; Kogevinas, M.; Silverman, D.T.; Malats, N.; Real, F.X.; García-Closas, M.; Tardón, A.; Rivas, M.; Torà, M.; García-Closas, R.; et al. Urinary pH, cigarette smoking and bladder cancer risk. *Carcinogenesis* **2011**, *32*, 843–847. [CrossRef] [PubMed]
44. Yi, J.-H.; Shin, H.-J.; Kim, S.-M.; Han, S.-W.; Kim, H.-J.; Oh, M.-S. Does the Exposure of Urine Samples to Air Affect Diagnostic Tests for Urine Acidification? *Clin. J. Am. Soc. Nephrol.* **2012**, *7*, 1211–1216. [CrossRef] [PubMed]
45. Yaroshenko, I.; Kirsanov, D.; Kartsova, L.; Sidorova, A.; Borisova, I.; Legin, A. Determination of urine ionic composition with potentiometric multisensor system. *Talanta* **2015**, *131*, 556–561. [CrossRef] [PubMed]
46. Bourdeau, P. ; *Scientific Committee on Problems of the Environment. Short-Term Toxicity Tests for Non-Genotoxic Effects*; Wiley: Hoboken, NJ, USA, 1990; Available online: https://apps.who.int/iris/handle/10665/38957 (accessed on 30 June 2019).
47. Hiebl, B.; Peters, S.; Gemeinhardt, O.; Niehues, S.M.; Jung, F. Impact of serum in cell culture media on in vitro lactate dehydrogenase (LDH) release determination. *J. Cell. Biotechnol.* **2017**, *3*, 9–13. [CrossRef]
48. Hsiao, I.-L.; Huang, Y.-J. Effects of serum on cytotoxicity of nano- and micro-sized ZnO particles. *J. Nanoparticle Res.* **2013**, *15*, 1829. [CrossRef]
49. Sumitran-Holgersson, P.C.A.S. Cytokeratins of the Liver and Intestine Epithelial Cells During Development and Disease. In *Cytokeratins—Tools in Oncology*; Hamilton, G., Ed.; InTech: London, UK, 2012; pp. 15–32.
50. Ku, N.-O.; Toivola, D.M.; Strnad, P.; Omary, M.B. Cytoskeletal keratin glycosylation protects epithelial tissue from injury. *Nat. Cell Biol.* **2010**, *12*, 876–885. [CrossRef]
51. Karantza, V. Keratins in health and cancer: More than mere epithelial cell markers. *Oncogene* **2011**, *30*, 127–138. [CrossRef]
52. Birder, L.; Andersson, K.-E. Urothelial signaling. *Physiol. Rev.* **2013**, *93*, 653–680. [CrossRef]
53. Rousseau, A.; Fradette, J.; Bernard, G.; Gauvin, R.; Laterreur, V.; Bolduc, S. Adipose-derived stromal cells for the reconstruction of a human vesical equivalent. *J. Tissue Eng. Regen. Med.* **2015**, *9*, E135–E143. [CrossRef]
54. Hu, P.; Meyers, S.; Liang, F.-X.; Deng, F.-M.; Kachar, B.; Zeidel, M.L.; Sun, T.-T. Role of membrane proteins in permeability barrier function: Uroplakin ablation elevates urothelial permeability. *Am. J. Physiol. Physiol.* **2002**, *283*, F1200–F1207. [CrossRef]
55. Varley, C.L.; Garthwaite, M.A.; Cross, W.; Hinley, J.; Trejdosiewicz, L.K.; Southgate, J. PPARgamma-regulated tight junction development during human urothelial cytodifferentiation. *J. Cell. Physiol.* **2006**, *208*, 407–417. [CrossRef] [PubMed]
56. Smith, N.J.; Hinley, J.; Varley, C.L.; Eardley, I.; Trejdosiewicz, L.K.; Southgate, J. The human urothelial tight junction: Claudin 3 and the ZO-1alpha+ switch. *Bladder* **2015**, *2*, e9. [CrossRef] [PubMed]
57. Rickard, A.; Dorokhov, N.; Ryerse, J.; Klumpp, D.J.; McHowat, J. Characterization of tight junction proteins in cultured human urothelial cells. *In Vitro Cell. Dev. Biol. Anim.* **2008**, *44*, 261–267. [CrossRef] [PubMed]
58. Brunner, A.; Tzankov, A. The Role of Structural Extracellular Matrix Proteins in Urothelial Bladder Cancer (Review). *Biomark. Insights* **2007**, *2*, 418–427. [CrossRef]
59. Florea, F.; Koch, M.; Hashimoto, T.; Sitaru, C. Autoimmunity against laminins. *Clin. Immunol.* **2016**, *170*, 39–52. [CrossRef]
60. Hattori, K.; Mabuchi, R.; Fujiwara, H.; Sanzen, N.; Sekiguchi, K.; Kawai, K.; Akaza, H. Laminin Expression Patterns in Human Ureteral Tissue. *J. Urol.* **2003**, *170*, 2040–2043. [CrossRef]
61. Southgate, J.; Harnden, P.; Selby, P.J.; Thomas, D.F.M.; Trejdosiewicz, L.K. Urothelial Tissue Regulation. In *Advances in Experimental Medicine and Biology*; Springer Science and Business Media LLC: Berlin, Germany, 1999; Volume 462, pp. 19–30.
62. Khandelwal, P.; Abraham, S.N.; Apodaca, G. Cell biology and physiology of the uroepithelium. *Am. J. Physiol. Physiol.* **2009**, *297*, F1477–F1501. [CrossRef]

Article

Microstructure and Properties of Nano-Hydroxyapatite Reinforced WE43 Alloy Fabricated by Friction Stir Processing

Genghua Cao [1], Lu Zhang [2], Datong Zhang [2], Yixiong Liu [1], Jixiang Gao [1], Weihua Li [1] and Zhenxing Zheng [1,*]

1 School of Mechanical and Electronic Engineering, Guangdong Polytechnic Normal University, Guangzhou 510635, China; cghcaogenghua@126.com (G.C.); liuyixiong@gpnu.edu.cn (Y.L.); gjx205@163.com (J.G.); lwh927@163.com (W.L.)

2 Guangdong Key Laboratory for Advanced Metallic Materials processing, South China University of Technology, Guangzhou 510641, China; lewzl@live.com (L.Z.); dtzhang@scut.edu.cn (D.Z.)

* Correspondence: zhengzhenxing@sina.com

Received: 10 July 2019; Accepted: 6 September 2019; Published: 16 September 2019

Abstract: This research mainly focuses on the successful fabrication of nano-hydroxyapatite (nHA) reinforced WE43 alloy by two-pass friction stir processing (FSP). Microstructure evolution, mechanical properties, and in vitro corrosion behavior of FSPed WE43/nHA composite and FSPed WE43 alloy were studied. The results show that nHA particles are effectively dispersed in the processing zone, and the well-dispersed nHA particles can enhance the grain refine effect of FSP. The average grain sizes of FSPed WE43 alloy and WE43/nHA composite are 5.7 and 3.3 μm, respectively. However, a slight deterioration in tensile strength and yield strength is observed on the WE43/nHA composite, compared to the FSPed WE43 alloy, which is attributed to the locally agglomerated nHA particles and the poor quality of interfacial bonding between nHA particles and matrix. The electrochemical test and in vitro immersion test results reveal that the corrosion resistance of the WE43 alloy is greatly improved after FSP. With the addition of nHA particles, the corrosion resistance of the WE43/nHA composite shows an even greater improvement.

Keywords: WE43/HA composite; friction stir processing; microstructure; mechanical properties; corrosion behavior

1. Introduction

Magnesium and its alloys have several advantages when compared with traditional metal biomedical materials. The Young's modulus and density of magnesium and its alloys are similar to that of natural bone, which can effectively avoid the stress shielding effect [1,2]. Magnesium is biodegradable in vivo, and the corrosion products have proven to be nontoxic. In addition, magnesium-based biomedical materials have been widely reported to positively stimulate the formation of new bone, which is favorable for bone fracture healing [3]. Therefore, magnesium alloys have great potential in applications as biodegradable metal materials [4,5]. However, biomedical magnesium alloys face the urgent issue of controlling corrosion behavior by avoiding local corrosion and controlling corrosion rates, in order to meet the safety and mechanical property requirements for biodegradable metal materials [6].

Hydroxyapatite (HA), the main inorganic component of human bone tissue and teeth, has emerged as a promising bioceramic material for its outstanding biocompatibility and bioactivity [7,8]. However, due to its brittleness and poor strength, HA in biomedical applications is currently limited to non-load bearing parts or low load-bearing parts. In this case, introducing HA particles into magnesium alloys is considered an effective method to improve the corrosion rate and biocompatibility of magnesium alloys. At present, several processes such as hot extrusion, stirred casting, powder metallurgy, and other methods are performed to prepare HA magnesium matrix composites with uniform corrosion behavior, good mechanical properties, and biocompatibility [9–11].

Friction stir processing (FSP) is an emerging solid-state processing technology for preparing fine-grained metal materials [12]. In recent years, due to its slight interface reaction between matrix and reinforced particles during processing, FSP has been used for metal matrix composite preparation [13–15]. For instance, nano-hydroxyapatite (nHA) particles have successfully been added to pure Mg substrate by multi-pass FSP, and the Mg/nHA composite shows preferable corrosion resistance in simulated body fluid (SBF) or Dulbecco's phosphate buffered saline compared to the substrate. However, the investigations of mechanical properties are not mentioned in these papers [14,15]. As an implant material, the material should have reasonable mechanical properties in order to meet the clinical requirements of implantation, so investigation of these mechanical properties is also important. WE43 magnesium alloy with high strength and low cytotoxicity is suitable for biomedical applications. Therefore, in this research, the casted WE43 alloy was used as matrix, nano-sized HA particles as the reinforcing phase, and FSP was conducted to prepare WE43/nHA composites. The microstructure evolution during FSP, the effects of rotation speed on the distribution of nHA particles, and the effects of dispersed nHA particles on mechanical properties and corrosion behavior of the composites were studied.

2. Experimental Procedure

2.1. Raw Materials

Commercially available WE43 magnesium alloy (as-cast) sheets of size 150 mm × 30 mm × 6 mm (length × width × height) were used as the base metal (BM) in this study, and the chemical composition of the BM is shown in Table 1. Nano-sized HA powders (nHA) with a purity of >99% used in this study were purchased from Shanxi Baiwei Biotechnology Co., Ltd. (Xi'an, China). The TEM morphology of nHA particles is shown in Figure 1—the particles were of acicular morphology 20–30 nm in width and 60–120 nm in length.

Table 1. Composites of as-cast base metal (wt. %).

Mg	Y	Nd	Gd	Zr	Ni	Ca	Mn	Si	Zn
Bal.	3.34	2.04	1.27	0.39	0.02	0.02	0.02	0.01	0.01

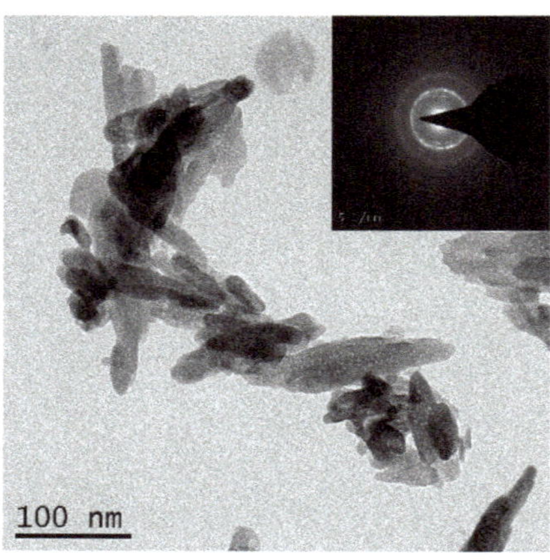

Figure 1. TEM morphology of nano-hydroxyapatite (nHA) particles.

2.2. Processing

Figure 2 presents the schematic illustration of the processing steps. To prepare the WE43/nHA composite, FSP was performed on BM sheets with a groove size of 1 mm × 4 mm (width × depth). After nHA particles were filled into the groove, a cylindrical and pin-less FSP tool was pressed down slowly until the shoulder contacted with the material, and it was then processed at a rotation speed of 600 rpm and a traverse speed of 60 mm/min along the groove direction. After processing, a metal sealing layer was formed above the groove, which could avoid the nHA particles escaping from the groove during FSP (Figure 2c). In the next processing step, cylindrical FSP tools with a shoulder diameter of 15 mm, consisting of a tapered cylindrical pin with a diameter varying from 2 to 5 mm over the length of 5 mm, were used to perform FSP on the sealed specimens.

A two-pass FSP with optimized processing parameters, including a rotation speed of 1000 rpm and a traverse speed of 60 mm/min, was employed to obtain WE43/nHA composites with a fine and uniform microstructure. The plunge depth of 0.8 mm and tilt angle of 2.5° relative to the normal direction of FSP plane were kept constant. The composite specimens obtained by FSP were coded as WE43/nHA. The same processing parameters and conditions were applied for conducting FSP on BM without the addition of nHA particles, with the obtained specimens being coded as FSP-WE43.

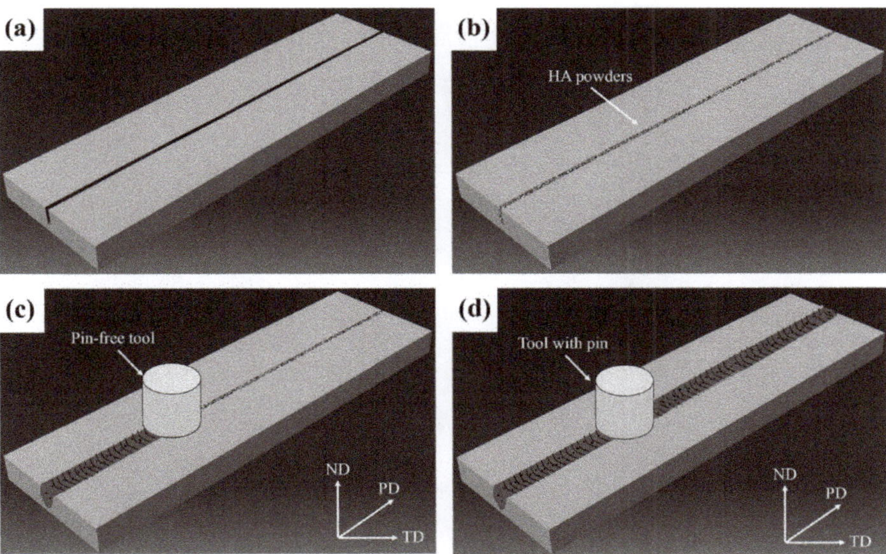

Figure 2. Schematic illustration of the processing steps. (**a**) Step 1; (**b**) Step 2; (**c**) Step 3; (**d**) Step 4.

2.3. Characterization of Microstructure and Phase Composition

The specimens used for microstructure observation were mechanically grinded with emery papers (up to #5000 grade) and polished on a polishing machine. Further, the polished specimens were etched by picric acid solution (picric acid 5 g, alcohol 80 mL, acetic acid 10 mL, and deionized water 10 mL). Optical microscope (DM15000M, Leica, Wizlar, Germany) was used to observe microstructure at lower magnification. The distribution of second phases and reinforcements, as well as the fracture morphology, were observed by a scanning electron microscope (Nova Nano 430, FEI, Hillsboro, OR, USA). The morphology of the nHA particles was characterized by a transmission electron microscopy (JEM-2100F, JEOL, Tokyo, Japan).

2.4. Mechanical Properties Testing

The microhardness test was conducted on a HVS-1000 digital Vickers microhardness tester (YouHong, Corp., Shanghai, China) with the application of a load of 0.98 N and a loading cycle of 10 s. The indention interval was selected to be 0.5 mm in stirred zone (SZ), and every indentation was measured three times and the average value was calculated. Tensile specimens were machined parallel to the processing direction with the gauge being completely within the stirred zone, the shape and dimension of a tensile specimen is shown in Figure 3. The tensile test was carried out on a SANS CMT5105 universal tensile testing machine (MTS, Eden Prairie, MN, USA) with a strain rate of 2×10^{-3} s^{-1}. At least five specimens were tested to evaluate the average property values.

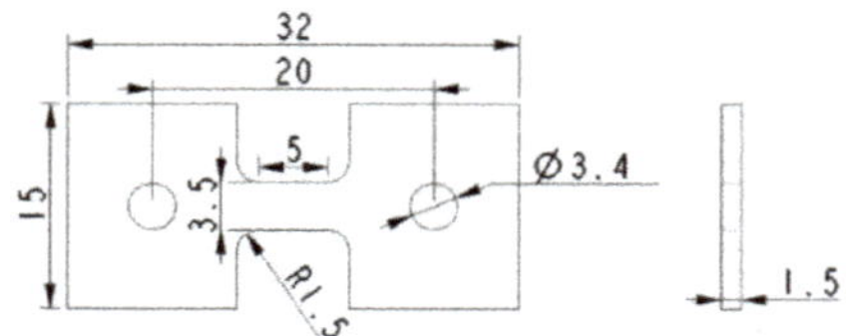

Figure 3. Shape and dimension of tensile specimen.

2.5. Corrosion Behavior

2.5.1. Electrochemical Test

Potentiodynamic polarization curve tests were performed on an electrochemical workstation (Vertex.5A. EIX, IVIUM, Eindhoven, the Netherlands) in SBF solution (8.035 g/L NaCl, 0.355 g/L $NaHCO_3$, 0.225 g/L KCl, 0.231 g/L $K_2HPO_4{\cdot}3H_2O$, 0.311 g/L $MgCl_2{\cdot}6H_2O$, 0.292 g/L $CaCl_2$, 0.072 g/L Na_2SO4, and 6.118 g/L Tris $(HOCH2)_3CNH_2$). The reference electrode was a saturated Ag/AgCl electrode and a platinum electrode was used as the counter electrode. One square cm area of the specimens was used as the working electrode. Specimens were exposed to the SBF solution for 30 min prior to the beginning of the experiments to establish open circuit potential. The potentiodynamic polarization was done between the potentials −2.5 and 0.5 V with a scanning rate of 5 mV/s.

2.5.2. Immersion Test

The immersion test was performed per ASTM-G31-72 in SBF at 37 °C for 24, 48, and 72 h, respectively. Weight loss specimens of size 6 mm × 4 mm × 2 mm (length × width × height) were cut from the SZ of FSP-WE43 and WE43/nHA composite samples and BM as well, for the purpose of measuring the corrosion rate of specimens in SBF. After immersion, the corrosion products were removed by chromic acid (200 g/L CrO_3, 10 g/L $AgNO_3$, and 20 g/L $Ba(NO_3)_2$) and then ultrasonically cleaned in distilled water and ethanol, respectively. The weight of specimens was measured before and after immersion. The corrosion rate was calculated by the following equation:

$$CR = W/At\rho \tag{1}$$

where CR is the corrosion rate (mm/year); W represents the weight loss (g); A refers to the surface area (cm^2); t is the immersion time; and ρ is the standard density of WE43. A density of 1.83 g/cm^3 was used for all specimens in this study.

The corrosion morphology of specimens was observed by the SEM mentioned above.

3. Results and Discussion

3.1. Microstructure Evolution

Figure 4 shows the optical microstructure of BM and the stir zone of FSP-WE43 and WE43/nHA specimens. The average grain size of BM is measured ~50.9 μm. Grain refinement is achieved up to ~5.7 and ~3.3 μm in the FSP-WE43 specimen and WE43/nHA specimen, respectively. During FSP, materials in the stir zone will undergo dynamic recrystallization and coarse second phases will break into small particles, resulting from the severe plastic deformation caused by the FSP tool and thermal effect caused by friction [13,16]. This is the main reason for the apparent refinement of grains after FSP. During FSP of magnesium, the peak temperature of SZ is reported lower than 550 °C, at which temperature the nHA particles remain stable [17–19]. The incorporated insoluble nHA particles act to stimulate nucleation and impede the migration of grain boundaries [20,21]. As a result, further grain refinement is achieved on WE43/nHA composites.

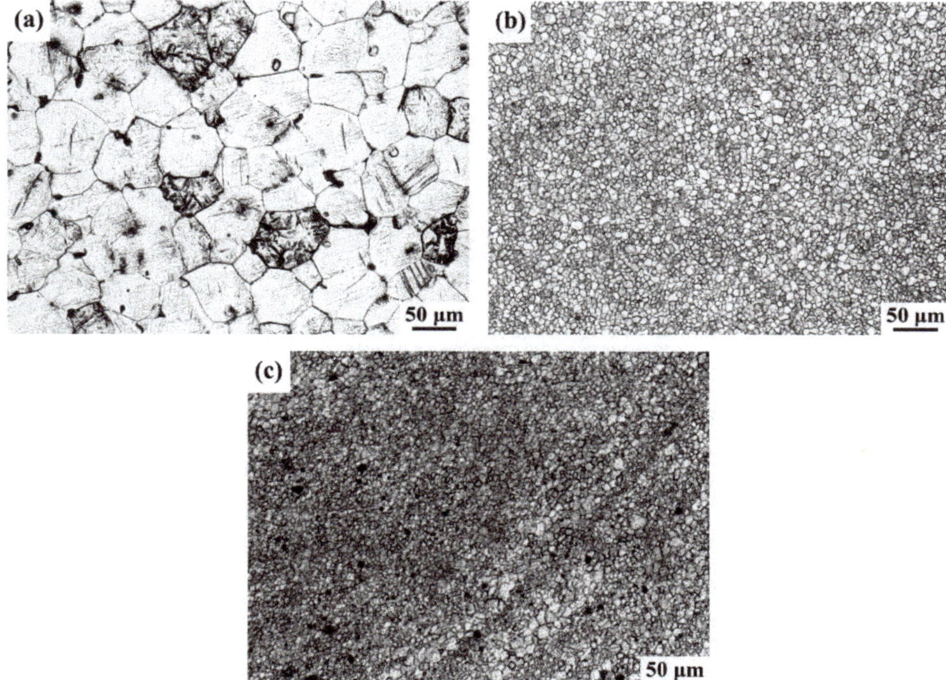

Figure 4. Optical image of (**a**) base metal (BM); (**b**) friction stir processing (FSP)-WE43; and (**c**) WE43/nHA specimens.

3.2. Distribution of HA Particles

Figure 5 shows the distribution of the second phase particles and its corresponding EDS analysis in the WE43/nHA specimen. nHA particles are found well dispersed on the matrix after FSP, joined by only a few clusters with a diameter of less than 10 μm (Figure 5a). The high-angle annular dark field (HAADF) image and the corresponding EDS analyses of the stir zone (Figure 5b,c) confirm that nHA particles are successfully added into the WE43 alloy matrix and most nHA particles remain at nanoscale.

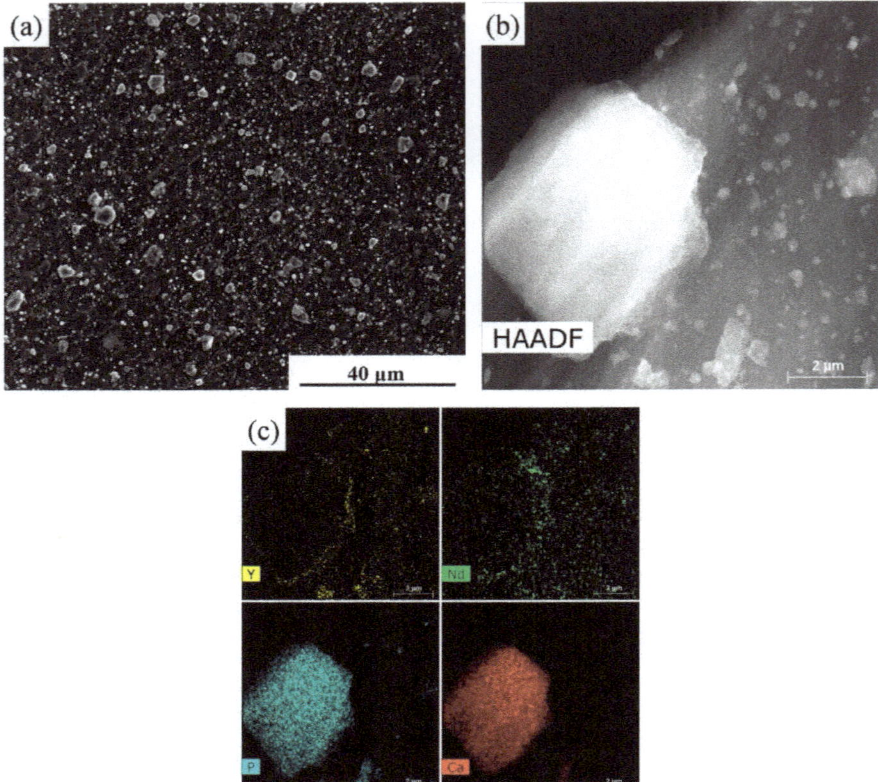

Figure 5. (**a**) SEM image; (**b**) TEM image; and (**c**) EDS analysis of the WE43/nHA specimen.

3.3. Mechanical Properties

3.3.1. Microhardness

The microhardness distribution curves of FSP-WE43 and WE43/nHA specimens are plotted in Figure 6. After FSP, the Vickers microhardness value of the SZ is significantly increased in FSP-WE43 and WE43/nHA specimens. The microhardness value at the SZ of the FSP-WE43 and WE43/nHA specimens is relatively stable, while the microhardness value of the base metal region fluctuates. The region with a higher microhardness value is about 5 mm in width, which is approximately equal to the diameter of the pin on FSP tool. The mean microhardness value of WE43 substrate is ~62.6 HV, which increases to ~79.6 HV by FSP without the addition of nHA powder. In the case of introducing nHA powder during FSP, the mean microhardness value is improved up to ~85.2 HV.

The grain size of as-cast WE43 alloy is coarse and the microstructure is ununiform, so the microhardness value of the BM zone is lower and fluctuates. After FSP, the grains in the stir zone are remarkably refined and grain boundary strengthening is considered the main reason for the significant increase of microhardness value in the stir zone. For the WE43/nHA specimen, the distribution of microhardness values is related to the dispersion of nHA particles, and a relatively stable microhardness value fluctuation for the WE43/nHA specimen indicates that nHA particles are dispersed uniformly on WE43 substrate, which is consistent with the microstructure observation.

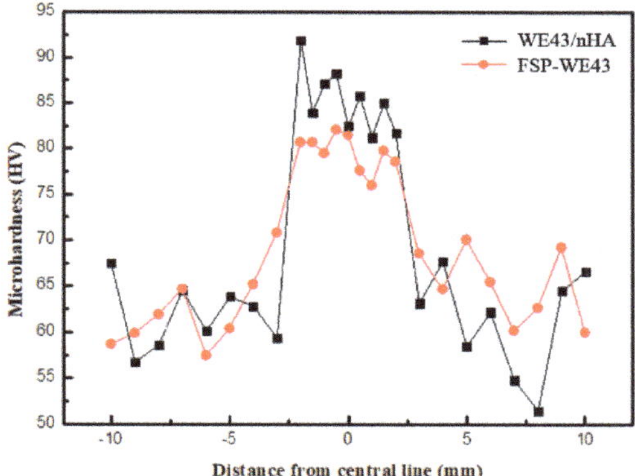

Figure 6. Microhardness distribution curves of WE43/nHA and FSP-WE43 specimens.

3.3.2. Tensile Properties

Figure 7 shows a comparison of the tensile properties of the base metal and processed WE43 (with or without adding nHA particles) specimens. As plotted in Figure 7, the ultimate tensile strength (UTS), yield strength (YS), and elongation of specimens after processing are improved in different degrees compared with the BM. For BM, the YS and UTS are measured as only ~153.3 and ~193.2 MPa, respectively. After FSP, the YS and UTS of the FSP-WE43 specimen are improved up to ~198.7 MPa and ~255.4 MPa, respectively. Compared with the FSP-WE43 specimen, a slight decline in strength is observed on the WE43/nHA specimen, with the YS and UTS being measured as ~185.1 and ~232.3 MPa, respectively. Furthermore, the value of elongation after fracture is greatly increased by FSP as well. The elongation of BM is only ~5.2%, while in the FSP-WE43 and WE43/nHA specimen it is 20.2% and 10.1%, respectively.

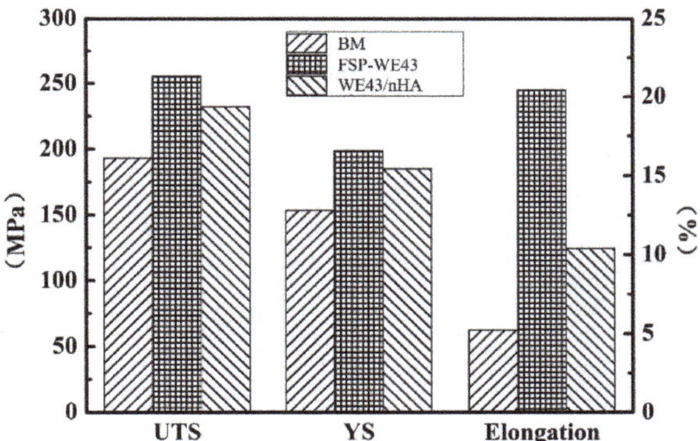

Figure 7. Tensile properties of BM, FSP-WE43, and WE43/nHA specimens. UTS: ultimate tensile strength; YS: yield strength.

3.3.3. Fractographic Studies

Figure 8 shows the fractographic images of tensile specimens. The fractographic of BM (Figure 8a) consists of a large amount of cleavage facets and voids defects, which indicates that BM fails in a brittle way. Figure 8b shows the fractographic image of an FSP-WE43 specimen, in which a large amount of fine equal-sized equiaxed dimples can be observed. The fracture morphology exhibits plastic fracture characteristics. This confirms the high ductility of the FSP-WE43 specimen as shown in Figure 7. The fractographic images of the WE43/nHA specimen are represented in Figure 8c,d. Fine dimples can also be observed in the WE43/nHA specimen and partially agglomerated nHA particles can be seen (as shown by white circles). By observing agglomerated nHA particles at higher magnification, several cracks across the nHA cluster is found (Figure 8d). The loose nHA clusters can be the initiation source of cracks during failure, which is the main reason for the decline of elongation in the WE43/nHA specimen compared with that in the FSP-WE43 specimen.

In general, coarse grains and brittle $Mg_{12}Nd$ networks in as-cast WE43 magnesium alloy cause its poor tensile properties [22]. As shown in Figure 4, remarkable grain refinement is achieved and coarse $Mg_{12}Nd$ phases are broken into fine particles after FSP. Under the combined effects of grain boundary strengthening and dispersion strengthening, the tensile strength of processed specimens (with or without addition of nHA particles) are significantly improved. In addition, FSP eliminates the voids defects in as-cast WE43, which is beneficial to the improvement of strength and ductility. However, the localized nHA agglomerates reduce the tensile properties of the composites to a certain extent compared with those in the FSP-WE43 alloy, while the tensile properties of the WE43/nHA specimen are still improved compared with those in the BM.

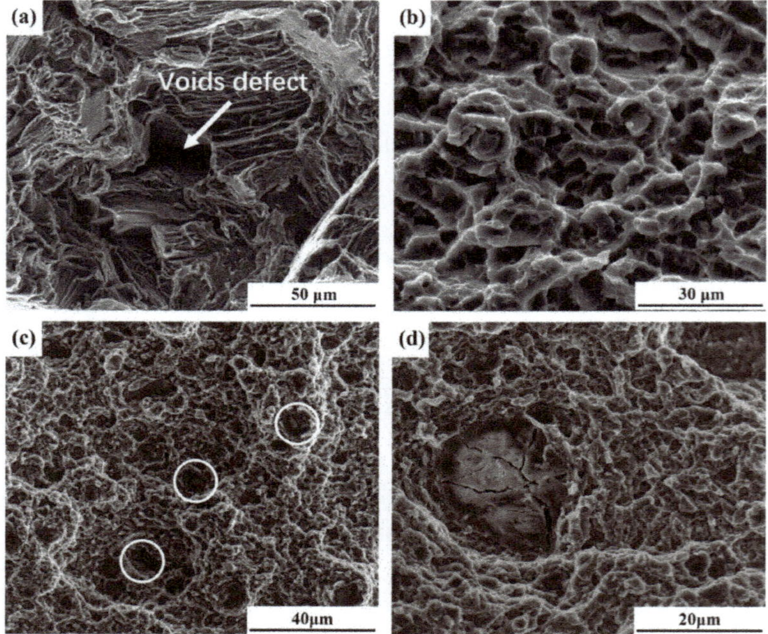

Figure 8. Fractographic images of (**a**) BM, (**b**) FSP-WE43, and (**c**) and (**d**) WE43/nHA specimens.

3.4. Corrosion Behavior

3.4.1. Electrochemical Test

The potentiodynamic polarization curves of as-cast WE43 and processed WE43 specimens are demonstrated in Figure 9. The corrosion potentials (E_{corr}) and corrosion current density (i_{corr}) of BM are measured as −1.691 mV (vs. Ag/AgCl) and 109.6 $\mu A/cm^2$, respectively. For the processed specimens, the E_{corr} of the WE43/nHA specimen (−1.661 mV) shifts toward the positive side and the i_{corr} (46.7 $\mu A/cm^2$) is lower than that of BM. The E_{corr} and i_{corr} values of the FSP-WE43 sample are measured as −1.678 mV and 53.7 $\mu A/cm^2$, respectively. The highest E_{corr} value and lowest i_{corr} value indicate that the WE43/nHA composite has the best corrosion resistance in this study.

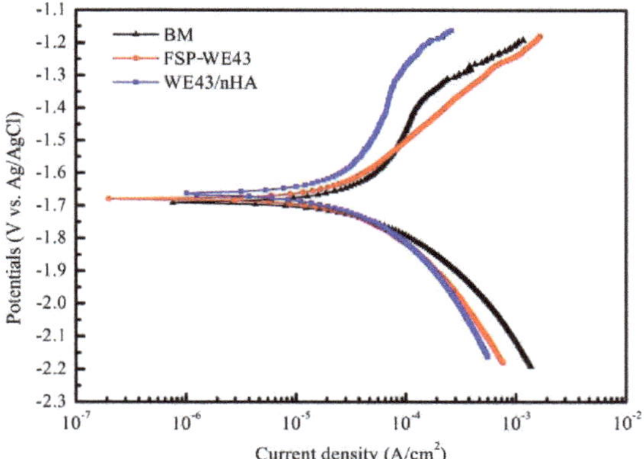

Figure 9. Potentiodynamic polarization curves of BM, FSP-WE43, and WE43/nHA specimens in simulated body fluid (SBF).

As discussed in Section 3.1, grain size is significantly refined by FSP. For magnesium alloy, fine grains are proved beneficial to the formation of a passive layer, as the result of increasing the number of grain boundaries per unit volume and reducing the galvanic couple between grain boundary and grain interior [23,24]. This is the main reason for the increase of corrosion resistance on processed WE43 specimens. Moreover, tiny dispersed nHA particles can also contribute to uniform corrosion behavior [25]. Therefore, the combined effects of grain refinement and dispersion of nHA particles lead to the improvement of corrosion resistance on WE43/nHA composites.

3.4.2. Degradation in Immersion Test

Figure 10 shows the corrosion weight loss curve of BM, FSP-WE43, and WE43/nHA specimens. During the immersion period of 120 h, the corrosion weight loss of BM increases rapidly. After immersion for 120 h, BM specimens are almost completely degraded in the SBF solution. The corrosion weight loss rates of the FSP-WE43 specimen and WE43/nHA specimen are relatively stable. In the first 72 h, the weight loss rates of the FSP-WE43 and WE43/nHA specimen are about the same. After immersion for 72 h, the weight loss rate of the FSP-WE43 specimen increases rapidly, while that of the WE43/nHA specimen maintains a relatively stable value. The corrosion rates are calculated according to Equation (1). After immersion for 120 h, the corrosion rate of the BM specimen is 26.8 mm/year, and the corrosion rates of the FSP-WE43 and the WE43/nHA specimen are 8.1 mm/year and 3.9 mm/year, respectively. Obviously, the FSP process has greatly improved the corrosion resistance of the casted WE43 alloy, and the addition of nHA particles has further improved the corrosion resistance of the material.

Corrosion morphologies (with corrosion products) of specimens after immersion in SBF for 72 h are shown in Figure 11. The SEM images of corrosion morphology show that BM experiences severe localized corrosion after immersion for 72 h and the accumulation of thick corrosion products occur locally, which are reported to be $Mg(OH)_2$ and calcium phosphate bio-minerals [26,27]. For the FSP-WE43 specimen, uniform protective films are observed in most areas, while a small number of protective films fall off locally, which may decrease the protective effect on corrosion attack. For the WE43/nHA specimen, a dense and uniform protective layer generates on the composite surface. These results indicate that the corrosion morphology is changed from local corrosion in as-cast WE43 alloy to uniform corrosion in FSP-WE43 alloy, which is attributed to the fine-grained and homogeneous microstructure by FSP. With the addition of dispersed nHA particles, the uniform corrosion morphology on WE43/nHA composite is more obvious.

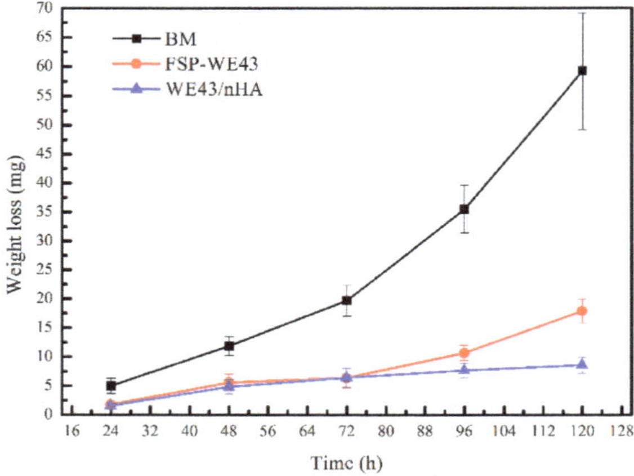

Figure 10. The weight loss and corrosion rate curves of immersion specimens.

Figure 11. Corrosion morphologies of (**a–c**) BM, (**d–f**) FSP-WE43, and (**g–i**) WE43/nHA specimens after immersion in SBF for 72 h at 37 °C (with corrosion products).

Corrosion morphologies (without corrosion products) of specimens after immersion in SBF for 72 h are shown in Figure 12. It can be seen from Figure 12a that the BM specimen experiences extremely severe corrosion attack and a large amount of material is dissolved in SBF. Deep and large etch pits can be observed (Figure 12b), proving that the material has been eroded by SBF and the corrosion products cannot prevent further corrosion. However, the FSP-WE43 specimen still keeps a relatively complete surface morphology, although parts of materials are dissolved in SBF (Figure 12c,d). For the composite specimen, the original shape is almost maintained after immersion in SBF for three days and only shallow corrosion pits can be observed locally (Figure 12e,f). The corrosion morphology observation indicates that the corrosion resistance of the WE43/nHA composite is superior to that of the FSP-WE43 alloy and much superior to that of the as-cast WE43 alloy, which is in accord with the electrochemical test results.

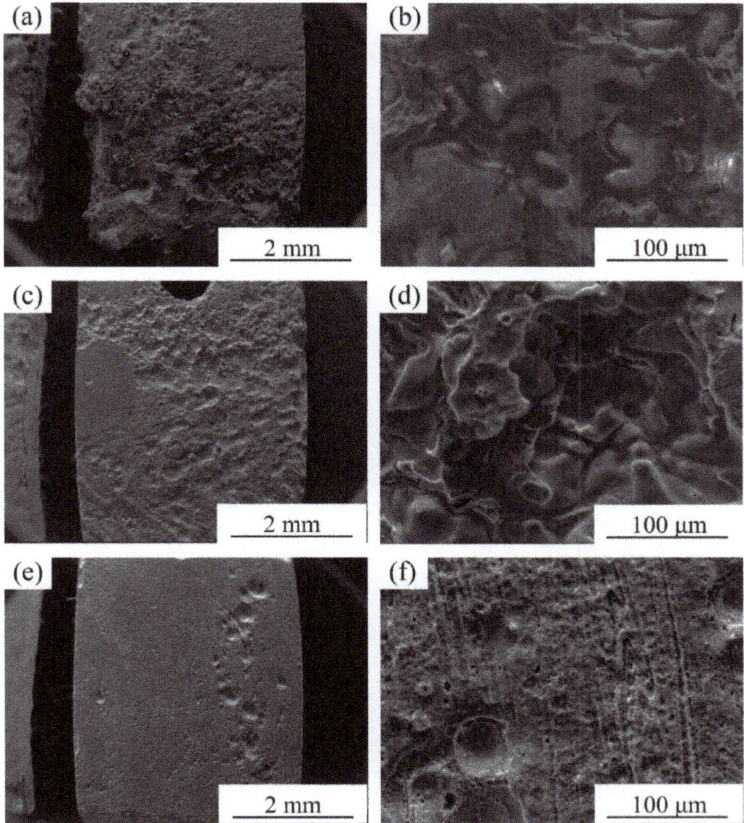

Figure 12. Corrosion morphologies of (**a**,**b**) BM, (**c**,**d**) FSP-WE43, and (**e**,**f**) WE43/nHA specimens after immersion in SBF for 72 h at 37 °C (without corrosion products).

4. Conclusions

Fine-grained WE43/nHA composite was successfully prepared through friction stir processing. Microstructure evolution and mechanical properties as well as in vitro corrosion behavior of WE43/nHA composite were studied. The main findings are summarized as follows:

1. After friction stir processing, nHA particles disperse uniformly on WE43 matrix, and the dispersed nHA particles enhance the grain refinement effect during processing.
2. The tensile properties of the WE43/nHA composite are significantly improved compared with those of the casted WE43 alloy, while experiencing a slight deterioration compared with the tensile properties of the FSP-WE43 alloy, which are the result of the locally agglomerated nHA particles and the poor quality of interfacial bonding between nHA particles and matrix.
3. Due to the grain refinement and dispersed nHA particles, the corrosion resistance of the WE43/nHA composite is superior to that of the FSP-WE43 alloy and much superior to that of the as-cast WE43 alloy.

Author Contributions: Methodology G.C. and D.Z.; Supervision Z.Z. and J.G.; resources Z.Z. and D.Z.; Analysis G.C.; W.L. and Y.L.; Writing—original draft G.C. and L.Z.; Writing—reviewing and editing G.C. and J.G.

Funding: This research is funded by the Natural Science Foundation of Guangdong Province (2017A030310630), Guangdong Science and Technology Planning Project (2017A070715012), and Guangdong education department project (2017GCZX003, 2017KTSCX115).

Conflicts of Interest: The authors declare no conflict of interest.

References

1. Agarwal, S.; Curtin, J.; Duffy, B.; Jaiswal, S. Biodegradable Magnesium Alloys for Orthopaedic Applications: A Review on Corrosion, Biocompatibility and Surface Modifications. *Mater. Sci. Eng. C* **2016**, *68*, 948–963. [CrossRef] [PubMed]
2. Gu, X.N.; Zheng, Y.F. A review on magnesium alloys as biodegradable materials. *Front. Mater. Sci.* **2010**, *4*, 111–115. [CrossRef]
3. Staiger, M.P.; Pietak, A.M.; Huadmai, J.; Dias, J. Magnesium and its Alloys as Orthopedic Biomaterials. *Biomaterials* **2006**, *27*, 1728–1734. [CrossRef] [PubMed]
4. Zhao, D.; Witte, F.; Lu, F.; Wang, J.; Li, J.; Qin, L. Current status on clinical applications of magnesium-based orthopaedic implants: A review from clinical translational perspective. *Biomaterials* **2017**, *112*, 287–302. [CrossRef] [PubMed]
5. Gu, X.N.; Zheng, Y.F.; Cheng, Y.; Zhong, S.P.; Xi, T.F. In vitro corrosion and biocompatibility of binary magnesium alloys. *Biomaterials* **2009**, *30*, 484–498. [CrossRef] [PubMed]
6. Zhao, Y.; Wu, G.S.; Jiang, J.J.; Wong, H.M.; Yeung, K.W.K.; Chu, P.K. Improved corrosion resistance and cytocompatibility of magnesium alloy by two-stage cooling in thermal treatment. *Corros. Sci.* **2012**, *59*, 360–365. [CrossRef]
7. Suchanek, W.; Yoshimura, M. Processing and properties of hydroxyapatite-based biomaterials for use as hard tissue replacement implants. *J. Mater. Res.* **1998**, *13*, 94–117. [CrossRef]
8. Karageorgiou, V.; Kaplan, D. Porosity of 3D biomaterial scaffolds and osteogenesis. *Biomaterials* **2005**, *26*, 5474–5491. [CrossRef]
9. Campo, R.D.; Savoini, B.; Muñoz, A.; Monge, M.A.; Garcés, G. Mechanical properties and corrosion behavior of Mg-HAP composites. *J. Mech. Behav. Biomed.* **2014**, *39*, 238–246. [CrossRef] [PubMed]
10. Khalajabadi, S.Z.; Kadir, M.R.A.; Izman, S.; Ebrahimi-Kahrizsangi, R. Fabrication, bio-corrosion behavior and mechanical properties of a Mg/HA/MgO nanocomposite for biomedical applications. *Mater. Des.* **2015**, *88*, 1223–1233. [CrossRef]
11. Xiong, G.Y.; Nie, Y.J.; Ji, D.H.; Li, J.; Li, C.Z.; Li, W.; Zhu, Y.; Luo, H.L.; Wan, Y.Z. Characterization of biomedical hydroxyapatite/magnesium composites prepared by powder metallurgy assisted with microwave sintering. *Curr. Appl. Phys.* **2016**, *16*, 830–836. [CrossRef]
12. Mishra, R.S.; Ma, Z.Y. Friction Stir Welding and Processing. *Mater. Sci. Eng. R* **2005**, *50*, 1–78. [CrossRef]
13. Rahmati, R.; Khodabakhshi, F. Microstructural evolution and mechanical properties of a friction-stir processed Ti-hydroxyapatite (HA) nanocomposite. *J. Mech. Behav. Biomed.* **2018**, *88*, 127–139. [CrossRef] [PubMed]
14. Sunil, B.R.; Kumar, T.S.S.; Uday, C.; Nandakumar, V.; Doble, M. Friction stir processing of magnesium-nanohydroxyapatite composites with controlled in vitro degradation behavior. *Mater. Sci. Eng. C* **2014**, *39*, 315–324. [CrossRef] [PubMed]
15. Ahmadkhaniha, D.; Fedel, M.; Sohi, M.H.; Hanzaki, A.Z.; Deflorian, F. Corrosion behavior of magnesium and magnesium–hydroxyapatite composite fabricated by friction stir processing in Dulbecco's phosphate buffered saline. *Corros. Sci.* **2016**, *104*, 319–329. [CrossRef]
16. Khodabakhshi, F.; Simchi, A.; Kokabi, A.H. Surface modifications of an aluminum-magnesium alloy through reactive stir friction processing with titanium oxide nanoparticles for enhanced sliding wear resistance. *Surf. Coat. Technol.* **2017**, *309*, 114–123. [CrossRef]
17. Chai, F.; Zhang, D.T.; Li, Y.Y. Effect of Thermal History on Microstructures and Mechanical Properties of AZ31 Magnesium Alloy Prepared by Friction Stir Processing. *Materials* **2014**, *7*, 1573–1589. [CrossRef]
18. Wang, Y.B.; Huang, Y.X.; Meng, X.C.; Wan, L.; Feng, J.C. Microstructural evolution and mechanical properties of Mg-Zn-Y-Zr alloy during friction stir processing. *J. Alloy. Compd.* **2017**, *696*, 875–883. [CrossRef]

19. Rameshbabu, N.; Rao, K.P.; Kumar, T.S.S. Acclerated microwave processing of nanocrystalline hydroxyapatite. *J. Mater. Sci.* **2005**, *40*, 6319–6323. [CrossRef]
20. Khodabakhshi, F.; Simchi, A.; Kokabi, A.H.; Gerlich, A.P. Friction stir processing of an aluminum-magnesium alloy with pre-placing elemental titanium powder: In-situ formation of an Al3Ti-reinforced nanocomposite and materials characterization. *Mater. Charact.* **2015**, *108*, 102–114. [CrossRef]
21. Ammouri, A.H.; Kridli, G.; Ayoub, G.; Hamade, R.F. Relating grain size to the Zener-Hollomon parameter for twin-roll-cast AZ31B alloy refined by friction stir processing. *J. Mater. Process. Technol.* **2015**, *222*, 301–306. [CrossRef]
22. Cao, G.H.; Zhang, D.T.; Zhang, W.; Qiu, C. Microstructure evolution and mechanical properties of Mg-Nd-Y alloy in different friction stir processing conditions. *J. Alloy. Compd.* **2015**, *636*, 12–19. [CrossRef]
23. Argade, G.R.; Panigrahi, S.K.; Mishra, R.S. Effects of grain size on the corrosion resistance of wrought magnesium alloys containing neodymium. *Corros. Sci.* **2012**, *58*, 145–151. [CrossRef]
24. Argade, G.R.; Kandasamy, K.; Panigrahi, S.K.; Mishra, R.S. Corrosion behavior of a friction stir processed rare-earth added magnesium alloy. *Corros. Sci.* **2012**, *58*, 321–326. [CrossRef]
25. Witte, F.; Feyerabend, F.; Maier, P.; Fischer, J.; Störmer, M.; Blawert, C.; Dietzel, W.; Hort, N. Biodegradable magnesium-hydroxyapatite metal matrix composites. *Biomaterials* **2007**, *28*, 2163–2174. [CrossRef] [PubMed]
26. Jamesh, M.I.; Wu, G.S.; Zhao, Y.; McKenzie, D.R.; Bilek, M.M.M.; Chu, P.K. Electrochemical corrosion behavior of biodegradable Mg-Y-RE and Mg-Zn-Zr alloys in Ringer's solution and simulated body fluid. *Corros. Sci.* **2015**, *91*, 160–184. [CrossRef]
27. Chu, P.W.; Marquis, E.A. Linking the microstructure of a heat-treated WE43 Mg alloy with its corrosion behavior. *Corros. Sci.* **2015**, *101*, 94–104. [CrossRef]

Article

Biomechanical Loading Comparison between Titanium and Unsintered Hydroxyapatite/Poly-L-Lactide Plate System for Fixation of Mandibular Subcondylar Fractures

Shintaro Sukegawa [1,4,*], Takahiro Kanno [2], Norio Yamamoto [3], Keisuke Nakano [4], Kiyofumi Takabatake [4], Hotaka Kawai [4], Hitoshi Nagatsuka [4] and Yoshihiko Furuki [1]

1 Department of Oral and Maxillofacial Surgery, Kagawa Prefectural Central Hospital, 1-2-1, Asahi-machi, Takamatsu, Kagawa 760-8557, Japan; furukiy@ma.pikara.ne.jp
2 Department of Oral and Maxillofacial Surgery, Shimane University Faculty of Medicine, Shimane 693-8501, Japan; tkanno@med.shimane-u.ac.jp
3 Department of Orthopaedic Surgery, Kagawa Prefectural Central Hospital, Takamatsu, Kagawa 761-0396, Japan; lovescaffe@yahoo.co.jp
4 Department of Oral Pathology and Medicine, Okayama University Graduate School of Medicine, Dentistry and Pharmaceutical Sciences, Okayama 7008530, Japan; pir19btp@okayama-u.ac.jp (K.N.); gmd422094@s.okayama-u.ac.jp (K.T.); de18018@s.okayama-u.ac.jp (H.K.); jin@okayama-u.ac.jp (H.N.)
* Correspondence: gouwan19@gmail.com; Tel.: +81-87-811-3333

Received: 5 April 2019; Accepted: 8 May 2019; Published: 13 May 2019

Abstract: Osteosynthesis absorbable materials made of uncalcined and unsintered hydroxyapatite (u-HA) particles, poly-L-lactide (PLLA), and u-HA/PLLA are bioresorbable, and these plate systems have feasible bioactive osteoconductive capacities. However, their strength and stability for fixation in mandibular subcondylar fractures remain unclear. This in vitro study aimed to assess the biomechanical strength of u-HA/PLLA bioresorbable plate systems after internal fixation of mandibular subcondylar fractures. Tensile and shear strength were measured for each u-HA/PLLA and titanium plate system. To evaluate biomechanical behavior, 20 hemimandible replicas were divided into 10 groups, each comprising a titanium plate and a bioresorbable plate. A linear load was applied anteroposteriorly and lateromedially to each group to simulate the muscular forces in mandibular condylar fractures. All samples were analyzed for each displacement load and the displacement obtained by the maximum load. Tensile and shear strength of the u-HA/PLLA plate were each approximately 45% of those of the titanium plates. Mechanical resistance was worst in the u-HA/PLLA plate initially loaded anteroposteriorly. Titanium plates showed the best mechanical resistance during lateromedial loading. Notably, both plates showed similar resistance when a lateromedially load was applied. In the biomechanical evaluation of mandibular condylar fracture treatment, the u-HA/PLLA plates had sufficiently high resistance in the two-plate fixation method.

Keywords: mandibular condylar fracture; unsintered hydroxyapatite/poly-L-lactide composite plate; bioactive resorbable plate; biomechanical loading evaluation; fracture fixation

---

1. Introduction

Mandibular condylar fractures constitute 25%–35% of all mandibular fractures [1] and are the most common form of mandible fracture.

The selection between closed (conservative treatment) and open (surgical treatment) techniques are linked to the type of fracture, the patient's age, and the degree of functional impairment caused by the fracture. Growing patients with certain undisplaced mandibular condylar fractures can undergo closed treatment, but severely displaced fractures are indications for open surgical reduction and fixation with internal rigid devices, generally with plates and screws [2]. In adult patients, the condylar head can be adjusted for functional adaptation of the condyle without restoration of the anatomy caused by displaced base fractures with a loss of ramus height [1]. Therefore, high condylar fractures with little bone available for fixation are mostly treated non-surgically. On the other hand, lower condylar fractures that affect the subcondyle, condylar base, and condylar neck are generally treated by open reduction and internal fixation [3,4]. The surgical treatment of mandibular condylar fractures involves fixation of the fractured stumps with the use of plates and screws. There are several types of fixation for fractures of the mandibular condyle [5]; the most common fixation technique is to use two four-hole straight plates [6]. In general, a titanium metal plate system has been used as a standard osteosynthesis material.

However, titanium metal plates have been reported to produce long-term problems related to implantation in the human body, such as intracranial migration, growth retardation of the craniofacial skeleton, and hypersensitivity to cold. In recent years, as an alternative to metal plates, resorbable plates, which do not require plate removal, have been used as osteosynthesis material in maxillofacial surgery [7]. Bioresorbable osteosynthesis devices have been evolving, with improvements in bioresorbability and marked bioactivity with new material compositions, for different and better in situ behavior. Currently, many osteosynthesis absorbable materials are commercially available; these include OSTEOTRANS-MX (known in Japan as Super FIXSORB MX®; Teijin Medical Technologies Co., Ltd. Osaka, Japan). These bioactive and bioresorbable devices are made from composites of uncalcined and unsintered hydroxyapatite (u-HA) particles and poly-L-lactide (PLLA). This u-HA/PLLA osteosynthesis material is an excellent bioactive device that has an osteoconductive ability in the short-term and absorbs safely and reliably in the long-term [8,9]. However, both the strength and stability of this plate system for fixation to a mandibular subcondylar fracture are still unclear. Thus, the aim of this in vitro study was to assess the biomechanical strength of u-HA/PLLA bioactive resorbable plate systems after internal fixation of mandibular subcondylar fractures.

2. Materials and Methods

2.1. Materials

We used the bioresorbable Super FIXSORB MX® (Teijin Medical Technologies Co., Ltd., Osaka, Japan) osteosynthesis system. Forged composites of u-HA/PLLA were processed by machining or milling treatments to form various miniscrews and miniplates, which contained 30- and 40-weight fractions of u-HA particles (raw HA, neither calcined nor sintered material), respectively, in composites (hereinafter referred to as u-HA 30 miniscrews and u-HA 40 miniplates).

A 2.0 mm miniplate system (MatrixMANDIBLE™ Adaption Plate; DePuy Synthes, Raynham, Mass.) (Johnson & Johnson, New Brunswick, USA) was used as the titanium metal plate.

2.2. Evaluation of Tensile and Shear Strength of Titanium and Bioresorbable Plate Systems

2.2.1. Sample Preparation

We prepared mechanical strength models by affixing the titanium and u-HA/PLLA plates to polyetherketoneketone (PEKK) plates (thickness 3 mm) with screws. The PEKK plates were held in place by different osteosynthesis systems. Plates and screws were used to form the following groups:

1. A single u-HA/PLLA bioresorbable straight plate, with the plate held in place on each side with two screws (total four screws);
2. A single titanium straight plate, with the plate held in place on each side with two screws (total four screws).

In this study, the bioresorbable osteosynthesis material consisted of 1.4 mm-thick plates and screws (diameter 2 mm; length 6 mm). In comparison, the titanium osteosynthesis material consisted of 1.0 mm-thick plates and screws (diameter 2 mm; length 6 mm). The experimental fixed models were mounted on an Autograph AG-20kNXD test frame (Shimadzu Co., Kyoto, Japan) across the chuck. The maximum stress until the plate or screws were destroyed and the stress at the time of 1 mm movement were measured, and the load was applied at a test speed of 10 mm/min. Two types of strength tests (the tensile and shear strength tests) were performed five times each.

2.2.2. Strength Measurement

Using the method illustrated in Figure 1A according to the Japanese Industrial Standard K7113, we measured tensile strength. The peak value of the load profile attained by the Autograph AG-20kNXD (Shimadzu Co., Kyoto, Japan) was considered the tensile strength. Using the method illustrated in Figure 1B according to the Japanese Industrial Standard K7113, we measured shear strength. The peak value of the load profile attained by the Autograph AG-20kNXD was considered the shear strength.

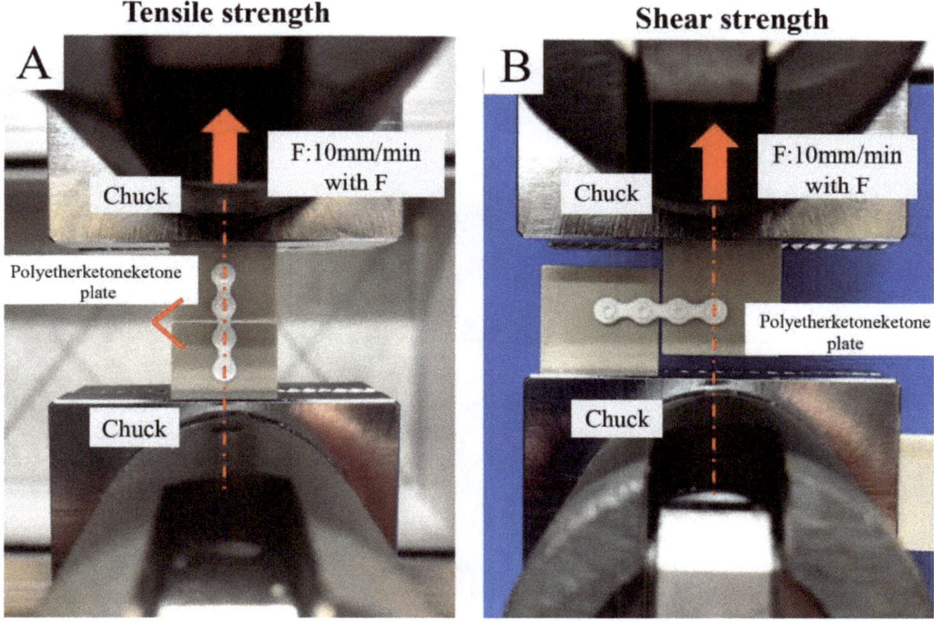

Figure 1. Mechanical strength models were prepared by fixing the plate with screws to the polyetherketoneketone plate. (**A**) Tensile strength; (**B**) shear strength.

2.3. Biomechanical Loading Evaluation

2.3.1. Sample Preparation

We used 20 polyurethane replicas of human hemimandibles (Mandible, Code #8900, SYNBONE AG, Laudquart, Tardisstrasse, Switzerland). Although polyurethane mandible replicates the property of spongy bone [10], this model is useful to obtain preliminary results concerning the stability of the investigated osteosynthesis systems. An ordinary subcondylar fracture model was investigated by using the osteotomy line connecting the mandibular notch to the midpoint of the mandibular ramus' posterior border. We created a fracture in the right condyle of each model, mimicking the experimental method of Meyer et al. [11]. Initially, these fracture models were created with the use of a cutting guide by a computer-controlled program. A partial cut was made in each hemimandibular model with a diamond disk (KG Sorensen, Cotia, São Paulo, Brazil). A silicone mold of the subcondylar area of the osteosynthesis was made on the mandible and drilled to form guide holes, and then, using the mold as a guide, we completed the cut. This method was used to create identical cuts and perforations in all the hemimandibular models, thus guiding the positions for plate fixation according to each study group.

2.3.2. Fixation Method for Mandibular Condylar Fracture

We used two straight miniplates along the ideal line of the mandibular condyle as proposed by Meyer et al. [12]. The technique is to plate below the mandibular notch and below the mandibular posterior border, respectively. This technique with monocortical screws had proved to be the most reliable and functionally stable osteosynthesis procedure for subcondylar fractures. The bone segments were held in place by different commercially available osteosynthesis methods (a titanium miniplate and bioresorbable plate) and monocortical screws. With both types of plates, bone fixation could be performed without plate bending. Thus, we created two conditions (Figure 2A):

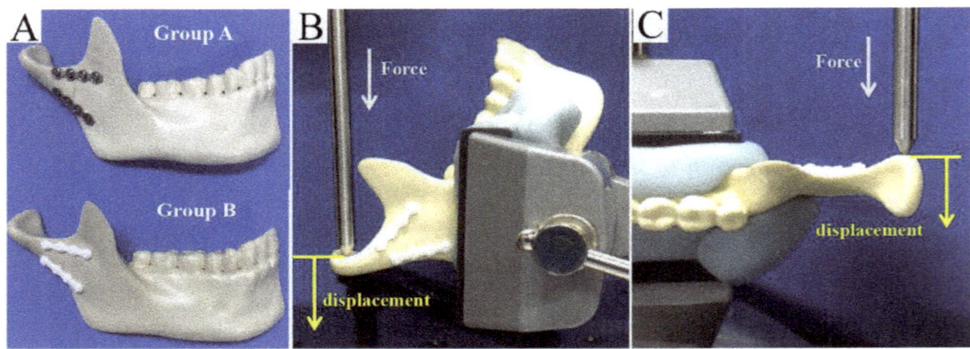

Figure 2. (**A**) The mandibular subcondylar fracture replicas were held in place by double titanium (Ti) straight plates (thickness: 1.0 mm) with four monocortical screws (2.0 mm in diameter and 6 mm long screws; group A) or by double unsintered hydroxyapatite (u-HA)/ poly-L-lactide (PLLA) straight plates (thickness: 1.4 mm), each with four monocortical screws (2.0 mm in diameter and 6 mm long; group B). A linear load was applied at a displacement speed of 1 mm/min. (**B**) Anteroposterior (vertical) linear loading. (**C**) Lateromedial (lateral) linear loading.

(1) In each bone segment in group A, double titanium straight plates (MatrixMANDIBLE™ Adaption Plate, 1.0 mm thick) with four monocortical screws (2.0 mm in diameter and 6 mm long) were installed; and (2) in each bone segment in group B, double-u-HA/PLLA straight plates (Super FIXSORB MX ®, 1.4 mm thick), each with four monocortical screws (2.0 mm in diameter and 6 mm long), were installed.

We used a total of 20 hemimandibular replicas, and 10 were allocated to each of the two groups.

2.3.3. Biomechanical Loading Test

After bone segment fixation, the replicas were mounted on a testing machine (AG-20KNX; Shimazu) (Shimadzu Co., Kyoto, Japan), which was based on a biomechanical cantilever-bending model that simulates masticatory forces. The mandibular body and angle areas of each replica were then stabilized. Adaptation of the polyurethane hemimandible to the machine was guaranteed through a metallic support. In reference to past biomechanical evaluation methods, the replicas were then subjected to linear loading in two directions: from anterior to posterior (vertically; Figure 2B) and from lateral to medial (horizontally; Figure 2C). These forces simulated the muscular forces applied to an actual fractured condyle. The material testing unit created linear displacement at a rate of 1 mm/min, and loading continued until the maximum load was reached. The peak load and displacement for each replica were recorded. All replicas were analyzed for 0.5, 1.0, 1.5, 2.0, 3.0, and 5.0 mm displacements by loading and for the amount of displacement by the maximum load. Means and standard deviations were derived and assessed for statistical significance.

2.4. Statistical Analysis

Data were recorded and entered into an electronic database during the course of the evaluation by means of Microsoft Excel (Microsoft, Inc., Redmond, USA). Means and standard deviations were used for normal data distributions. The database was transferred to JMP version 11.2 for Macintosh computers (SAS Institute, Inc., Cary, USA) for statistical analysis. Then, t-tests for independent samples were performed to investigate whether there were significant differences between the mean values of the groups. The level of significance was set at $p < 0.05$.

3. Results

3.1. Tensile and Shear Strength Evaluation

The results are shown in Figure 3. In the tensile test, the mean maximum test forces were 806.1 N (standard deviation 7.9 N) in group A (titanium plates fixed with titanium screws) and 208.8 N (standard deviation 10.3 N) in group B (u-HA/PLLA bioresorbable plates with bioresorbable screws). The mean test forces at 1 mm displacement were 382.1 N (standard deviation 19.5 N) in group A and 181.8 N (standard deviation 4.2 N) in group B. The mean maximum tensile test force in group B was 25.9% that of group A, and the test force at 1 mm displacement in group B was 47.6% that of group A. These differences were significant ($p < 0.05$) (Figure 3A).

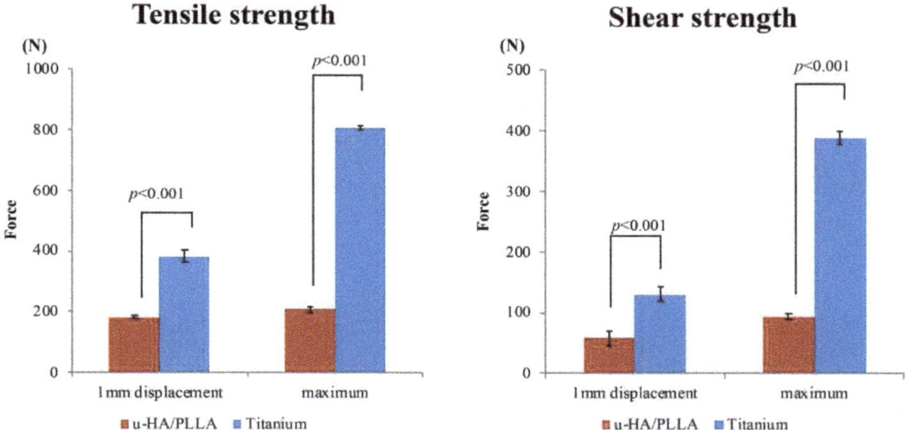

Figure 3. (**A**) Comparison of titanium plates and unsintered hydroxyapatite (u-HA)/poly-L-lactide (PLLA) bioresorbable plates with regard to maximum tensile test force and test force at 1 mm displacement in the tensile test. (**B**) Comparison of titanium plates and u-HA/PLLA bioresorbable plates with regard to maximum tensile test force and test force at 1 mm displacement in the shear test.

In the shear test, the mean maximum test force and the test force at 1 mm displacement were 390.5 N (standard deviation 11.0 N) and 130.6 N (standard deviation 12.2 N), respectively, in group A. In group B, these mean forces were 93.2 N (standard deviation 4.6 N) and 58.1.8 N (standard deviation 12.2 N), respectively. The maximum shear test force in group B was 23.9% that of group A, and the test force at 1 mm displacement in group B was 44.5% that of group A.

The strength of the titanium-based fixation system was obviously higher than that of the u-HA/PLLA bioresorbable system in both the tensile test and the shear test. In the shear test, the titanium plates were significantly stronger ($p < 0.05$) than the u-HA/PLLA bioresorbable plates at maximum shear test force and at 1 mm displacement test force (Figure 3B).

3.2. Biomechanical Loading Evaluation

The results of this experiment show that the mechanical resistances among the osteosynthesis devices remained proportional to the amount of displacement. In the vertical loading test, the titanium plates had a significantly higher load value than did the u-HA/PLLA bioresorbable plates at displacements of 0.5 and 1 mm. In the displacements of 1.5–5 mm, there was no significant difference between the two types of plates (Figure 4). As the load in the anteroposterior direction increased, although the bone fragments fixed by the anterior plate were separated, the position of the posterior bone fragments did not change (Figure 5).

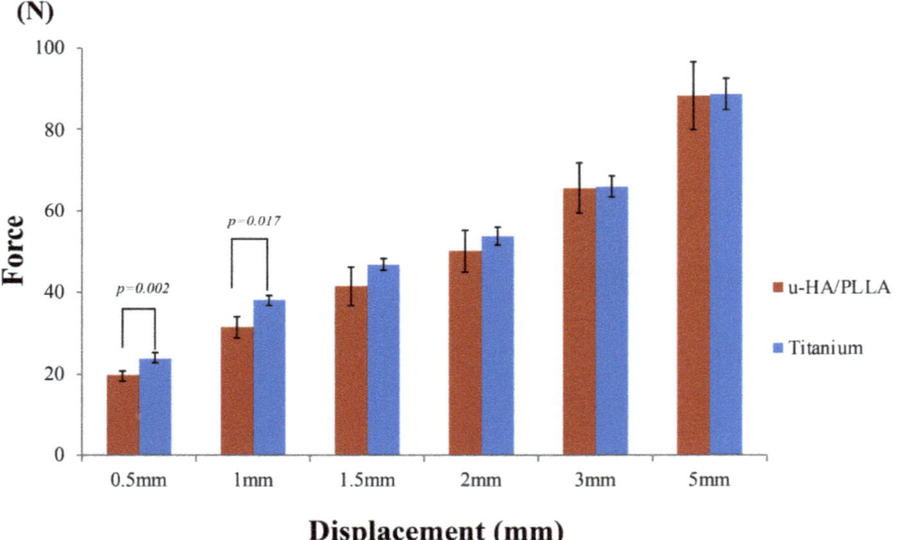

Figure 4. Load values of the titanium plates and unsintered hydroxyapatite (u-HA)/poly-L-lactide (PLLA) bioresorbable plates according to the amount of displacement in the anteroposterior loading test.

In contrast, when the forces were applied to the condyle in the lateromedial direction, there was no significant difference between the u-HA/PLLA bioresorbable plates and the titanium plates at all displacements (Figure 6). As the load in the lateromedial direction increased, there was no excessive load on either titanium or u-HA/PLLA bioresorbable anterior and posterior plates, and both types of plates in both locations were flexed (Figure 7).

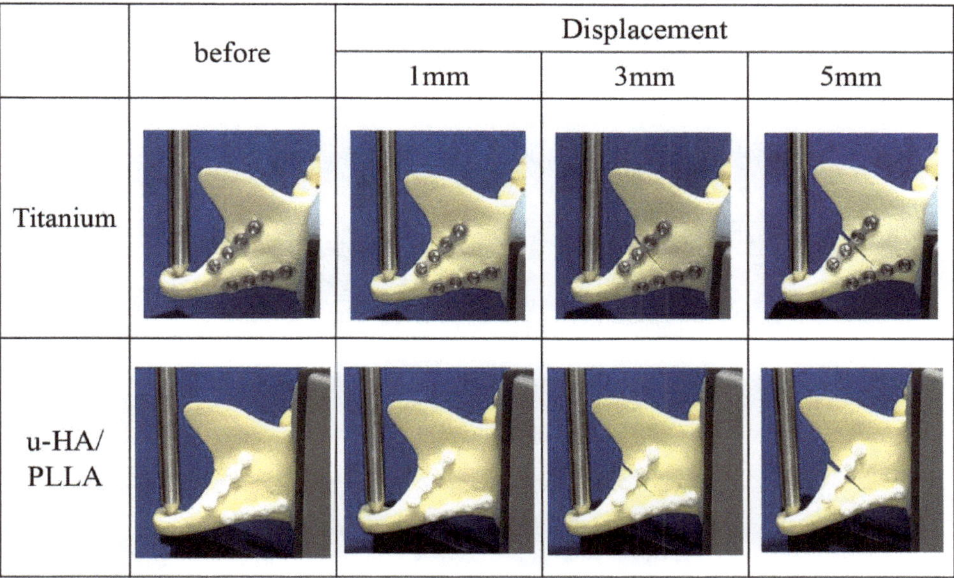

Figure 5. Change in subcondylar fracture segments under anteroposterior loading. Ti, titanium (plates); u-HA/PLLA, unsintered hydroxyapatite/poly-L-lactide (bioresorbable plates).

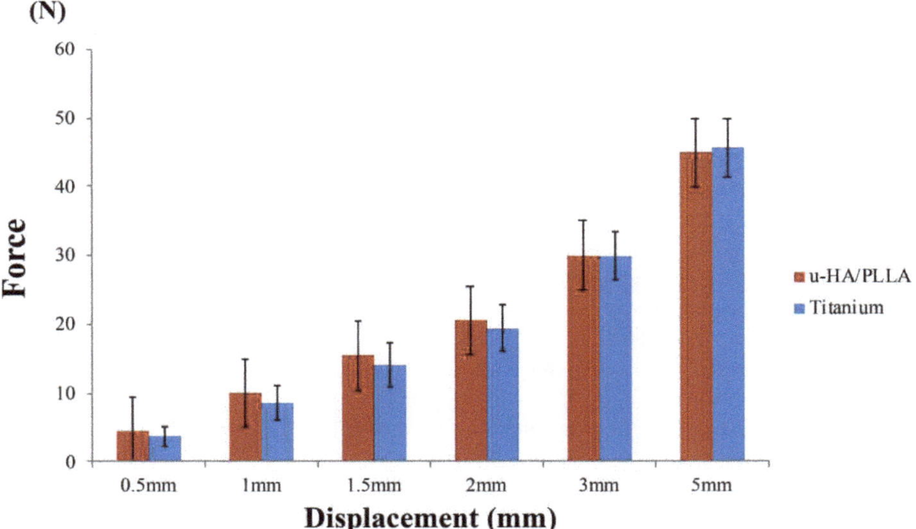

Figure 6. Load values of the titanium and unsintered hydroxyapatite (u-HA)/poly-L-lactide (PLLA) bioresorbable plates according to the amount of displacement in the lateromedial loading test.

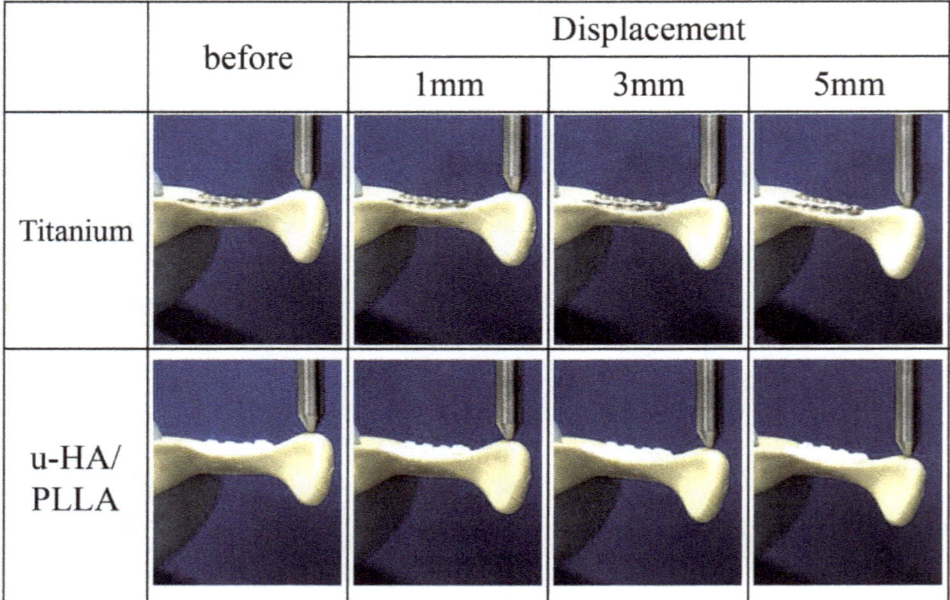

Figure 7. Change in subcondylar fracture segment under lateromedial loading. Ti, titanium (plates); u-HA/PLLA, unsintered hydroxyapatite/poly-L-lactide (bioresorbable plates).

4. Discussion

Bioresorbable materials have been used clinically for osteosynthesis in various fields of maxillofacial surgery such as orthodontic surgery [13], repair of craniomaxillofacial fractures [7], bone augmentation for dental implantation [14], and reconstruction in situations of maxillofacial cysts and tumors. In traumatology, bioresorbable plates are used mainly in the treatment of orbital and zygomatic fractures [15–17] and in the treatment of mandibular fractures [8,18]. However, there are very few clinical studies of the treatment of mandibular subcondylar fractures with resorbable plates, and the number of cases reported is often low [19]. Bioresorbable material, unlike titanium plates, generally cannot resist high mechanical loads. Therefore, for mandibular condylar fracture treatment, it is necessary to know how much load the u-HA/PLLA bioresorbable plates can withstand in comparison with titanium plates. We aimed to clarify the strength of the bioresorbable plate system and the titanium plate system and to examine the possibility of the treatment choice of the osteosynthesis material for mandibular subcondylar fracture.

First, the mechanical strengths of the titanium and u-HA/PLLA bioresorbable plate systems were examined. The strength of the u-HA/PLLA bioresorbable system, in either tension or shear, was only approximately 45% that of the titanium system. The u-HA/PLLA bioresorbable plate strength was improved slightly when the form of the plates was modified. This is comparable to previous research on the strength of titanium plates when the same material of the bioresorbable plate was used. However, our results were similar to those of previous research [20]. These were very important findings. For pure plate strength, these results suggest that the u-HA/PLLA bioresorbable plate system was much weaker than the titanium plate system.

However, the results of our biomechanical loading tests were very interesting. In the loading tests, we reproduced mandibular subcondylar fracture treatment with two straight miniplates along the ideal line of the mandibular condyle. In the lateromedial loading test, there was no significant difference between the u-HA/PLLA bioresorbable plates and the titanium plates. One of the most important clinical indications for surgical treatment of mandibular condylar fracture is the medial displacement of the condyle [21], which is caused by traction of the lateral pterygoid muscle. The primary function of the lateral pterygoid muscle is to pull the head of the condyle out of the mandibular fossa along the articular eminence to make the mandible protrude [22]. The lateral pterygoid muscle allows lateral movement of the mandible. This test simulates the lateral movements of mastication, and the strength of the bioresorbable plate system was sufficient. In contrast, in the anteroposterior loading test, the load value at the initial displacements of 0.5 and 1.0 mm was significantly larger in the titanium plates than in the u-HA/PLLA bioresorbable plates; the load in the latter was approximately 80% that of the titanium plates. The u-HA/PLLA bioresorbable plates were much weaker than the titanium plates. However, proper plate placement for mandibular subcondylar fracture treatment greatly improved the strength of u-HA/PLLA bioresorbable plates.

Champy and Lodde [23] advocated "functionally stable osteosynthesis" or "dynamic osteosynthesis," according to which the plates must be placed along physiological tension lines that appear during the fracture. This concept was based on the theory that stabilization depends mainly on the compression of the fracture site until function is restored. Meyer et al. [11] demonstrated the presence of the compression line on the mandibular posterior border and the tension line on the mandibular notch in the mandibular condylar process. One posterior plate is located on the compression strain lines when placed along the condylar neck axis. This plate enables functionally stable osteosynthesis. The other plate, which enables dynamic osteosynthesis, should be on the traction line. To conform to dynamic osteosynthesis principles, the plate should be placed higher and more obliquely, parallel to the mandibular notch [11]. In short, the plate located at the traction line resists the dynamic force of the occlusion. The key to anteroposterior direction load testing is the presence of plates on the traction line.

In our experiment, as the anteroposterior load increased, the bone opening on the traction line increased. This opening was influenced by the strength of the plate on the traction line and was thought to cause the difference in load values of the titanium plates. Load values at displacements of 1.5 mm or more in the bioresorbable plates did not differ significantly from those in the titanium plate group. Therefore, we believe that the bioresorbable plate can serve clinically in osteosynthesis of mandibular subcondylar fractures. However, the mechanical model used for this study was a polyurethane model, which is softer than the actual mandible [9]. Hence, considering a slight decrease in the biomechanical strength of the bioresorbable model compared with the titanium equivalent, in clinical use of the bioresorbable plate system, early overloading soon after surgery should be avoided. A young patient with strong occlusal force also needs dietary instruction, such as eating soft foods after surgery.

At present, resorbable plate systems are used widely, and various resorbable plate systems have been developed on the basis of PLLA. Like the u-HA/PLLA plate system used in this study, these plate systems composed of various materials, including pure PLLA [15] or polyglycolic acid/PLLA [24], are commercially available. In particular, the u-HA/PLLA bioresorbable plate system that we studied has a very large biological advantage. Resorbable plates composed of pure PLLA are very strong [25], however, the strength of plates is reduced when PLLA is mixed with other materials. Unfortunately, PLLA osteosynthetic devices have several disadvantages, including lower dynamic strength, an inability to bond directly to bones, unstable resorption and decomposition processes, and long replacement times [26–28]. In contrast, u-HA/PLLA osteosynthesis devices significantly improved biological activity. It has been reported that u-HA/PLLA material induced no apparent inflammatory or foreign body reactions after implantation, and bonded directly to the human bone quickly [14]. Clinical studies have also clearly shown the effective osteoconductivity of the u-HA/PLLA plate system in maxillofacial treatment [8,29]. In addition, the reduction in strength due to the use of composite material was overcome by the unique compression molding and machining treatment in this osteosynthesis device [30]. The early bioactivity, such as osteoconductivity and direct bone bonding capacity, can produce early functional improvements in maxillofacial osteosynthesis of mandibular condylar fractures. However, to clarify the relationship between loading and mandibular condylar fracture healing with this plate system, it is necessary to evaluate the tissue over time. Further research reports are expected in the future.

The approach to mandibular condylar fracture is generally an extraoral approach [4]. Therefore, when it is necessary to remove the plate, there is a risk of recurrent facial nerve damage [31]. This is greatly stressful for both the surgeon and the patient. The bioresorbable system applied clinically may be an important option for the treatment of mandibular condylar fractures in the future.

The search for improved methods of subcondylar fracture fixation devices has been the subject of several studies. Mechanical loading tests are used to evaluate the behavior of fixation devices and methods, allowing the study of different osteosynthesis constructs. However, the present study did not involve a real mandible, and it will be necessary to conduct further research on mechanical experiments, as well as studies from various clinical viewpoints.

5. Conclusions

The titanium-based fixation system was much stronger than that based on bioresorbable u-HA/PLLA. In the biomechanical evaluation of mandibular condylar fracture treatment, however, our results showed that the u-HA/PLLA bioresorbable plate had sufficiently high resistance in the two-plate fixation method. The titanium plates were more resistant than the bioresorbable plates in subcondylar fractures when the initial force was applied in the anteroposterior direction; conversely, in the medial direction, there were no statistically significant differences. These results suggest that titanium and bioresorbable plate fixation systems have similar mechanical resistance within the limitation of our mandible model study.

Author Contributions: S.S. participated in the design of the study and experiments, acquisition of data, and drafted the manuscript. T.K. performed and coordinated the study and contributed to the drafting of the manuscript. N.Y. analyzed the data and contributed to the drafting of the manuscript. H.N. and K.N. performed and coordinated the study and contributed to the drafting of the manuscript. K.T. and H.K. made many contributions to the experiment and the correction of manuscript of this paper. Y.F. conceived the study and participated in its design. All authors approved the final version of the manuscript prior to submission.

Funding: This work was supported by JSPS KAKENHI Grant Number 20837661.

Acknowledgments: We declare no financial support from Teijin Co., Ltd., Osaka, Japan. Some analyses, such as those of tensile and shear strength and biomechanical loading evaluations, were performed at the Teijin Medical Institute, Okayama, Japan.

Conflicts of Interest: The authors declare no conflict of interest.

References

1. Ellis, E.; Throckmorton, G.S. Treatment of mandibular condylar process fractures: Biological considerations. *J. Oral Maxillofac. Surg.* **2005**, *63*, 115–134. [CrossRef]
2. Fernández-Olarte, H.; Gómez-Delgado, A.; López-Dávila, D.; Rangel-Perdomo, R.; Lafaurie, G.I.; Chambrone, L. Is the Mandibular Growth Affected by Internal Rigid Fixation? A Systematic Review. *J. Maxillofac. Oral Surg.* **2017**, *16*, 277–283. [CrossRef]
3. Kanno, T.; Sukegawa, S.; Fujioka, M.; Takabatake, K.; Furuki, Y. Transoral Open Reduction with Rigid Internal Fixation for Subcondylar Fractures of the Mandible Using a Small Angulated Screwdriver System: Is Endoscopic Assistance Necessary? *J. Oral Maxillofac. Surg.* **2011**, *69*, e372–e384. [CrossRef] [PubMed]
4. Kanno, T.; Sukegawa, S.; Tatsumi, H.; Nariai, Y.; Ishibashi, H.; Furuki, Y.; Sekine, J. The retromandibular transparotid approach for reduction and rigid internal fixation using two locking miniplates in mandibular condylar neck fractures. *Int. J. Oral Maxillofac. Surg.* **2014**, *43*, 177–184. [CrossRef]
5. Sukegawa, S.; Kanno, T.; Katase, N.; Shibata, A.; Takahashi, Y.; Furuki, Y. Which fixation methods are better between three-dimensional anatomical plate and two miniplates for the mandibular subcondylar fracture open treatment? *J. Craniomaxillofac. Surg.* **2019**, *47*, 771–777. [CrossRef]
6. Parascandolo, S.; Spinzia, A.; Parascandolo, S.; Piombino, P.; Califano, L. Two load sharing plates fixation in mandibular condylar fractures: Biomechanical basis. *J. Cranio-Maxillofac. Surg.* **2010**, *38*, 385–390. [CrossRef] [PubMed]

7. Singh, V.; Kshirsagar, R.; Halli, R.; Sane, V.; Chhabaria, G.; Ramanojam, S.; Joshi, S.; Patankar, A. Evaluation of bioresorbable plates in condylar fracture fixation: A case series. *Int. J. Oral Maxillofac. Surg.* **2013**, *42*, 1503–1505. [CrossRef]
8. Sukegawa, S.; Kanno, T.; Katase, N.; Shibata, A.; Takahashi, Y.; Furuki, Y. Clinical Evaluation of an Unsintered Hydroxyapatite/Poly-L-Lactide Osteoconductive Composite Device for the Internal Fixation of Maxillofacial Fractures. *J. Craniofac. Surg.* **2016**, *27*, 1391–1397. [CrossRef] [PubMed]
9. Sukegawa, S.; Kanno, T.; Kawai, H.; Shibata, A.; Takahashi, Y.; Nagatsuka, H.; Furuki, Y. Long-Term Bioresorption of Bone Fixation Devices Made from Composites of Unsintered Hydroxyapatite Particles and Poly-L-Lactide. *J. Hard Tissue Biol.* **2015**, *24*, 219–224. [CrossRef]
10. De Santis, R.; Sarracino, F.; Mollica, F.; Netti, P.A.; Ambrosio, L.; Nicolais, L. Continuous fiber reinforced polymers as connective tissue replacement. *Compos. Sci. Technol.* **2004**, *64*, 861–871. [CrossRef]
11. Meyer, C.; Serhir, L.; Boutemi, P. Experimental evaluation of three osteosynthesis devices used for stabilizing condylar fractures of the mandible. *J. Cranio-Maxillofac. Surg.* **2006**, *34*, 173–181. [CrossRef] [PubMed]
12. Meyer, C.; Kahn, J.-L.; Lambert, A.; Boutemy, P.; Wilk, A. Development of a static simulator of the mandible. *J. Cranio-Maxillofac. Surg.* **2000**, *28*, 278–286. [CrossRef] [PubMed]
13. Ballon, A.; Laudemann, K.; Sader, R.; Landes, C.A. Segmental stability of resorbable P(L/DL)LA-TMC osteosynthesis versus titanium miniplates in orthognatic surgery. *J. Cranio-Maxillofac. Surg.* **2012**, *40*, e408–e414. [CrossRef]
14. Sukegawa, S.; Kawai, H.; Nakano, K.; Kanno, T.; Takabatake, K.; Nagatsuka, H.; Furuki, Y. Feasible Advantage of Bioactive/Bioresorbable Devices Made of Forged Composites of Hydroxyapatite Particles and Poly-L-lactide in Alveolar Bone Augmentation: A Preliminary Study. *Int. J. Med. Sci.* **2019**, *16*, 311–317. [CrossRef] [PubMed]
15. Sukegawa, S.; Kanno, T.; Nagano, D.; Shibata, A.; Sukegawa-Takahashi, Y.; Furuki, Y. The Clinical Feasibility of Newly Developed Thin Flat-Type Bioresorbable Osteosynthesis Devices for the Internal Fixation of Zygomatic Fractures: Is There a Difference in Healing Between Bioresorbable Materials and Titanium Osteosynthesis? *J. Craniofac. Surg.* **2016**, *27*, 2124–2129. [CrossRef] [PubMed]
16. Dong, Q.N.; Karino, M.; Koike, T.; Ide, T.; Okuma, S.; Kaneko, I.; Osako, R.; Kanno, T. Navigation-Assisted Isolated Medial Orbital Wall Fracture Reconstruction Using an U-HA/PLLA Sheet via a Transcaruncular Approach. *J. Investig. Surg.* **2019**, 1–9. [CrossRef] [PubMed]
17. Sukegawa, S.; Kanno, T.; Tanaka, S.; Matsumoto, K.; Sukegawa-Takahashi, Y.; Masui, M.; Koyama, Y.; Furuki, Y. Precision of Post-Traumatic Orbital Reconstruction Using Unsintered Hydroxyapatite Particles/Poly-L-Lactide Composite Bioactive/Resorbable Mesh Plate with and without Navigation: A Retrospective Study. *J. Hard Tissue Biol.* **2017**, *26*, 274–280. [CrossRef]
18. Leno, M.B.; Liu, S.Y.; Chen, C.-T.; Liao, H.-T. Comparison of functional outcomes and patient-reported satisfaction between titanium and absorbable plates and screws for fixation of mandibular fractures: A one-year prospective study. *J. Cranio-Maxillofac. Surg.* **2017**, *45*, 704–709. [CrossRef]
19. Lauer, G.; Pradel, W.; Leonhardt, H.; Loukota, R.; Eckelt, U. Resorbable triangular plate for osteosynthesis of fractures of the condylar neck. *Br. J. Oral Maxillofac. Surg.* **2010**, *48*, 532–535. [CrossRef]
20. Shikinami, Y.; Okuno, M. Bioresorbable devices made of forged composites of hydroxyapatite (HA) particles and poly L-lactide (PLLA). Part II: practical properties of miniscrews and miniplates. *Biomaterials* **2001**, *22*, 3197–3211. [CrossRef]
21. de Souza, G.M.; Rodrigues, D.C.; Celegatti Filho, T.S.; Moreira, R.W.F.; Falci, S.G.M. In vitro comparison of mechanical resistance between two straight plates and a Y-plate for fixation of mandibular condyle fractures. *J. Craniomaxillofac. Surg.* **2018**, *46*, 168–172. [CrossRef] [PubMed]
22. Koolstra, J.H.; Van Eijden, T.M. Dynamics of the human masticatory muscles during a jaw open-close movement. *J. Biomech.* **1997**, *30*, 883–889. [CrossRef]
23. Champy, M.; Lodde, J. Mandibular synthesis. Positioning of the syntheses according to mandibular stress. *Rev. Stomatol. Chir Maxillofac.* **1976**, *77*, 971–976. [PubMed]
24. Sukegawa, S.; Kanno, T.; Matsumoto, K.; Sukegawa-Takahashi, Y.; Masui, M.; Furuki, Y. Complications of a poly-L-lactic acid and polyglycolic acid osteosynthesis device for internal fixation in maxillofacial surgery. *Odontology* **2018**, *106*, 360–368. [CrossRef] [PubMed]
25. Daniels, A.U.; Chang, M.K.O.; Andriano, K.P.; Heller, J. Mechanical properties of biodegradable polymers and composites proposed for internal fixation of bone. *J. Appl. Biomater.* **1990**, *1*, 57–78. [CrossRef]

26. Bergsma, J.E.; de Bruijn, W.C.; Rozema, F.R.; Bos, R.R.; Boering, G. Late degradation tissue response to poly(L-lactide) bone plates and screws. *Biomaterials* **1995**, *16*, 25–31. [CrossRef]
27. Matsusue, Y.; Nakamura, T.; Iida, H.; Shimizu, K. A long-term clinical study on drawn poly-L-lactide implants in orthopaedic surgery. *J. Long. Term. Eff. Med. Implants* **1997**, *7*, 119–137.
28. Matsusue, Y.; Yamamuro, T.; Yoshii, S.; Oka, M.; Ikada, Y.; Hyon, S.-H.; Shikinami, Y. Biodegradable screw fixation of rabbit tibia proximal osteotomies. *J. Appl. Biomater.* **1991**, *2*, 1–12. [CrossRef]
29. Sukegawa, S.; Kanno, T.; Hotaka, K.; Akane, S.; Matsumoto, K.; Sukegawa-Takahashi, Y.; Sakaida, K.; Nagatsuka, H.; Furuki, Y. Surgical Treatment and Dental Implant Rehabilitation after the Resection of an Osseous Dysplasia. *J. Hard Tissue Biol.* **2016**, *25*, 437–441. [CrossRef]
30. Shikinami, Y.; Okuno, M. Bioresorbable devices made of forged composites of hydroxyapatite (HA) particles and poly-L-lactide (PLLA): Part I. Basic characteristics. *Biomaterials* **1999**, *20*, 859–877. [CrossRef]
31. Al-Moraissi, E.A.; Ellis, E.; Neff, A. Does encountering the facial nerve during surgical management of mandibular condylar process fractures increase the risk of facial nerve weakness? A systematic review and meta-regression analysis. *J. Cranio-Maxillofac. Surg.* **2018**, *46*, 1223–1231. [CrossRef] [PubMed]

Article

In Vitro Degradation of Absorbable Zinc Alloys in Artificial Urine

Sébastien Champagne [1,2], Ehsan Mostaed [3], Fariba Safizadeh [1], Edward Ghali [1], Maurizio Vedani [3] and Hendra Hermawan [1,2,*]

1 Department of Mining, Metallurgical and Materials Engineering, Laval University, 1065 avenue de la Médecine, Québec, QC G1V 0A6, Canada; sebastien.champagne.2@ulaval.ca (S.C.); fariba.safizadeh.1@ulaval.ca (F.S.); edward.ghali@gmn.ulaval.ca (E.G.)

2 Research Center of CHU de Québec, 10 rue de l'Espinay, Québec, QC G1L 3L5, Canada

3 Department of Mechanical Engineering, Politecnico di Milano, Via La Masa 1, 20156 Milan, Italy; ehsan.mostaed@polimi.it (E.M.); maurizio.vedani@polimi.it (M.V.)

* Correspondence: hendra.hermawan@gmn.ulaval.ca

Received: 24 December 2018; Accepted: 16 January 2019; Published: 18 January 2019

Abstract: Absorbable metals have potential for making in-demand rigid temporary stents for the treatment of urinary tract obstruction, where polymers have reached their limits. In this work, in vitro degradation behavior of absorbable zinc alloys in artificial urine was studied using electrochemical methods and advanced surface characterization techniques with a comparison to a magnesium alloy. The results showed that pure zinc and its alloys (Zn–0.5Mg, Zn–1Mg, Zn–0.5Al) exhibited slower corrosion than pure magnesium and an Mg–2Zn–1Mn alloy. The corrosion layer was composed mostly of hydroxide, carbonate, and phosphate, without calcium content for the zinc group. Among all tested metals, the Zn–0.5Al alloy exhibited a uniform corrosion layer with low affinity with the ions in artificial urine.

Keywords: absorbable; corrosion; degradation; magnesium; ureteral stent; zinc

1. Introduction

In 2008, up to five in 10,000 Canadian adults suffered from acute urinary tract obstruction, a kidney-related disease contributing to the economy burden of Canada by $3.8 billion [1,2]. It is a blockage that limits the flow of urine out of the body, which can be caused by a variety of factors such as a kidney stone, a swollen prostate, or a tumor [3]. To quickly relieve the kidney of the built-up of urine, urologists will mechanically open the urinary tract using a stent. An ideal ureteral stent will maintain an excellent urine flow to optimize upper tract drainage, be resistant to infection and encrustation (calcification), and be absorbable [4]. The majority of urologists consider the placement of stents after ureteral dilation to be routine. Some studies showed that about 60–85% of urologists maintained the stent for less than seven days [5,6]. This temporary need has brought the concept of biodegradable ureteral stent. Introduced in the late 1980s, their purpose is especially for temporary treatment such as for patients waiting for the effect of a medical therapy [7]. Poly(glycolic acid), poly(lactic acid), and poly(lactic-*co*-glycolic acid) are among the biodegradable polymeric materials used for ureteral stents [4,8]. However, polymeric stents possess a limited ability to resist external compression forces such as those created in a malignant extrinsic ureteral obstruction. This is due to their intrinsic low strength, which often results in their reinforcement by a metal skeleton made of non-degradable alloys [9,10]. At this point, metals that can corrode would constitute the ideal materials for biodegradable ureteral stents.

Nowadays, with the recent development in new biomaterials technology, namely absorbable metals, this ideal ureteral stent should be feasible. Absorbable metals, also known as biodegradable metals, are metals that corrode gradually in vivo with an appropriate host response, then dissolve completely while assisting tissue healing [11]. Iron (Fe), magnesium (Mg), zinc (Zn), and their alloys are among the studied absorbable metals, mostly for cardiovascular and orthopaedic applications. Coronary stents made of Fe and its alloys have shown their safety and efficacy when tested in animals [12], while those made of Mg alloys were clinically tested in humans, showing a continuous favorable safety profile up to 12 months [13]. Zn and its alloys have been more recently proposed as alternatives to Fe and Mg. In recent in vivo studies, pure Zn wire showed a steady corrosion rate for up to 20 months post-implantation in murine artery without causing local toxicity [14], while pure zinc stents corroded in rabbit abdominal aorta without obvious accumulation of corrosion products even after 12 months of implantation [15].

Application of absorbable metals in urology is a novel avenue. In 2014, Lock et al. [16] conducted a study on the degradation and antibacterial properties in artificial urine of Mg, Mg–Y, and Mg–3Al–1Zn (AZ31) that was the first to show the potential of absorbable metals for urological applications. More recent work by Zhang et al. [17,18] reported that implantation of Mg, Mg–6Zn, and Mg–5.4Zn–0.5Zr (ZK60) showed no adverse effect in the surrounding rat's bladder tissue and no sign of toxicity towards the liver and the kidneys. Their degradation tests demonstrated that the pure Mg consistently corroded faster than the alloys in simulated body fluids and artificial urine and in the bladder of rats. However, those previous corrosion studies did not assess the formation of the corrosion layer. In reaction with the ionic content of urine, the surface of the ureteral stent can be easily calcified. Known as encrustation, this is one of the main problems currently faced with respect to ureteral stents [6]. Therefore, this work aims at revealing a detailed in vitro degradation behavior of Zn alloys under simulated urinary tract conditions by using an electrochemical method and advanced surface characterization techniques.

2. Material and Methods

2.1. Material and Specimen Preparation

A group of binary Zn alloys (Zn–0.5Mg, Zn–1Mg, and Zn–0.5Al) was selected based on their excellent mechanical properties as determined in our previous study [19]. Hot rolled pure Zn (99.995%) and pure Mg (99.94%) and commercially available extruded Mg–2Zn–1Mn alloy were used as comparative materials. The Zn alloys were melted at 500 °C in a cylindrical steel mold inside a resistance furnace (Nabertherm, Lilienthal, Germany) and cast as cylindrical billets of 80 mm length and 15 mm diameter. The billets were annealed at 350 °C for 48 h, water-quenched, and extruded to a final diameter of 8 mm as described in detail elsewhere [19]. Specimens for corrosion tests were cut from the extruded metals at a thickness of 4 mm using a slow speed diamond rotating blade (Isomet 1000 Precision Saw, Buehler, Uzwil, Switzerland). The specimens were then polished using SiC paper from #600, to #800, and to #1200 (Carbimet 2 Abrasive Paper, Buehler, Uzwil, Switzerland).

2.2. In Vitro Degradation Testing

Four methods of electrochemical corrosion tests were conducted: open circuit potential (OCP), electrochemical impedance spectroscopy (EIS), potentiodynamic polarization (PDP), and electrochemical noise (EN). The metal specimens were mounted as working electrodes in a waterproof chamber with an exposed area of 0.096 cm^2 (0.35 cm of diameter), and all tests were done in triplicate for statistical analysis. The first three methods were performed one after another on all metal specimens using a three-electrode cell configuration. The metallic samples served as the working electrode, a graphite rod as the counter electrode, and a saturated calomel electrode (SCE) (E = +0.241 V saturated) as the reference electrode. The electrodes were connected to a VersaSTAT3 potentiostat and monitored using the VersaStudio software (Ametek Princeton Applied Research, Oak Ridge, TN, USA). The OCP experiments were conducted by monitoring the potential versus the SCE without applying any outside source for 3600 s, until equilibrium was reached at the corrosion potential E_{corr}. Once the equilibrium was achieved, EIS was started at the amplitude of 10 mV RMS and a frequency scan from 100 kHz to 1 Hz for all specimens. The interpretation of EIS results was made with the help of ZSimpWin software (Ametek Princeton Applied Research, Oak Ridge, TN, USA). The PDP experiments were done following the EIS, using a scan rate of 0.6 V/h, from −0.25 to +0.6 V vs. OCP.

The EN experiments were conducted on pure Mg, pure Zn, and Zn–0.5Al specimens only. Two identical working electrodes (mounted in resin) and a saturated silver chloride electrode (E = +0.197 V saturated) reference electrode were used and connected to a PC4/750 galvanostat/potentiostat (Gamry Instruments, Warminster, PA, USA). Data processing was done using an electrochemical signal analyzer (ESA400, Gamry Instruments, Warminster, PA, USA). The EN analysis was performed independently from the other electrochemical tests, as different equipment was used. The EN experiments were conducted for 48 h and at a frequency of acquisition of 10 Hz. The two identical coupled WEs were connected to a zero resistance ammeter (ZRA) to maintain the bias potential at 0 V. Each set of EN analysis, containing 1024 data points, was recorded with a data sampling rate of 0.1 s.

The artificial urine solution was made from analytical grade chemicals (Sigma-Aldrich, St. Louis, MO, USA) following the composition presented in Table 1. The pH of the artificial urine solution was adjusted to 6.0 using a solution of 1 N NH_4OH and monitored with a pH meter (Accumet pH meter 25, Fisher Scientific, Portsmouth NH, USA). To mimic the ureteral condition, the solution was kept at a normal human body temperature of 37 ± 1 °C using a water heater and a jacketed beaker. It was stirred with a magnetic stirrer at 80 rpm. The artificial urine was chosen over natural urine to allow more reproducible results. The natural variability of the chemical composition of urine is greatly inconsistent between individuals because of the different diet and physical activity level, and it also depends on the time of the day, with more concentrated urine in the morning [20].

Table 1. Artificial urine composition [21].

Chemical	NaCl	NaH_2PO_4	$Na_3C_6H_5O_7$	$MgSO_4$	Na_2SO_4	KCl	$CaCl_2$	$Na_2C_2O_4$
Mass (g)	6.17	4.59	0.944	0.463	2.408	4.75	0.638	0.043

Brought to 1 L by adding distilled water.

2.3. Surface and Corrosion Layer Characterization

The surface morphology observation and chemical composition analysis of the specimens after PDP were done using a scanning electron microscope (SEM, Quanta 250 FEI, Hillsboro, OR, USA) coupled with an energy dispersive X-ray spectroscope (EDS, SwiftED 3000, Oxford Instruments, Concord, MA, USA). The specimens were rinsed with distilled water and dried in desiccator for one day before being subjected to SEM observation. An X-ray photoelectron spectroscope (XPS, PHI 5600-ci spectrometer, Physical Electronics, Chanhassen, MN, USA) was then used to further determine the nature of the corrosion products at the surface of the corroded samples. A neutralizer was used to negate the charging effect at the surface of the metals. An XPS spectra survey was recorded from 1400 eV and ending at 0 eV using a standard aluminum anode (1486.6 eV) at 300 W. High-resolution spectra were then collected using a standard magnesium anode (1253.6 eV) at 300 W aimed at the following elements: carbon (300–280 eV), oxygen (545–525 eV), zinc (1037–1017 eV), phosphorus (147–127 eV), magnesium (64–44 eV), and, finally, aluminum (88–68 eV). The C1s peak was chosen as a reference at 285.0 eV to negate the effects of charging. Both survey and high-resolution scans were done in triplicate for each metal.

3. Results and Discussion

Once immersed in the artificial urine solution, all metals tended to stabilize after 1 h (Figure 1a). Pure Mg and Mg–2Zn–1Mn showed a more negative open circuit potential (−1.76 and −1.52 V, respectively) compared to all Zn groups (between −1.15 and −1.05 V with pure Zn having the highest value). Under potentiodynamic polarization, the Zn group shows a more noble behavior compared to the Mg group (Figure 1b). Corrosion rates vary from 2.16 mm/year for pure Mg to 0.87 mm/year for Zn (Table 2). The Mg–2Zn–1Mn corrodes slightly slower than pure Mg at 1.90 mm/year due to alloying effects. The corrosion of magnesium is highly dependent on the impurities that can act as active cathodic sites to accelerate corrosion. Zn is nobler than Mg and may act as an impurity, but the addition of Mn increases the tolerance limit of impurity's content by reducing the grain size of the alloys, thus lowering the corrosion rate of Mg–2Zn–1Mn compared to that of pure Mg [22–24]. In the Zn group, corrosion rate increases from pure Zn to Zn–0.5Al and Zn–Mg (Table 2). The addition of Mg into Zn produces an Mg_2Zn_{11} phase that creates microgalvanic action with the Zn matrix, which in turn increases the corrosion rates of Zn–Mg alloys [25]. Meanwhile, the high solubility of Al in Zn produces a single phase microstructure in the Zn–0.5Al, so its corrosion rate only slightly increases [19].

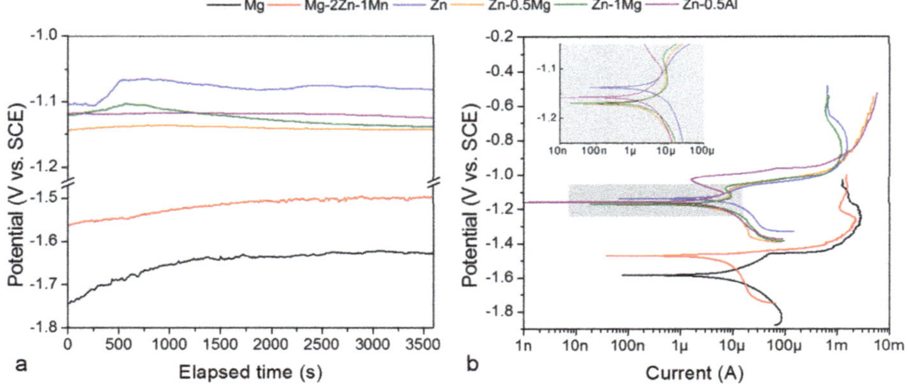

Figure 1. Typical open circuit potential (OCP) and potentiodynamic polarization (PDP) results: (**a**) potential vs. time, (**b**) potential vs. current.

Table 2. Corrosion parameters derived from PDP results.

Sample	OCP (V vs. SCE)	i_{corr} (μA/cm^2)	Corrosion Rate (mm/year)
Mg	-1.58 ± 0.02	94 ± 37	2.16 ± 0.84
Mg–2Zn–1Mn	-1.48 ± 0.07	84 ± 17	1.90 ± 0.38
Zn	-1.11 ± 0.06	58 ± 6	0.87 ± 0.09
Zn–0.5Mg	-1.18 ± 0.01	92 ± 5	1.39 ± 0.07
Zn–1Mg	-1.17 ± 0.01	99 ± 5	1.50 ± 0.08
Zn–0.5Al	-1.15 ± 0.01	77 ± 2	1.14 ± 0.03

After the polarization test, the surface of Mg group specimens is covered by a fairly uniform but not seemingly compact corrosion layer as indicated by the presence of cracks (Figure 2a,b). A pure Zn surface presents a uniform and dense corrosion layer (Figure 2c), as it tends to passivate in a near neutral aqueous solution, such as artificial urine [26]. Both the investigated Zn–Mg alloys show a non-uniform corrosion layer with the appearance of a flower-like crystal on the surface. The flower size is larger in Zn–0.5Mg (Figure 2d) with a seemingly compact underlying layer compared to that of Zn–1Mg (Figure 2e), which could explain the slightly faster corrosion rate of the latter. The flower is not observed on the Zn–0.5Al surface, but a rather uniform, smooth, and dense corrosion layer (Figure 2f).

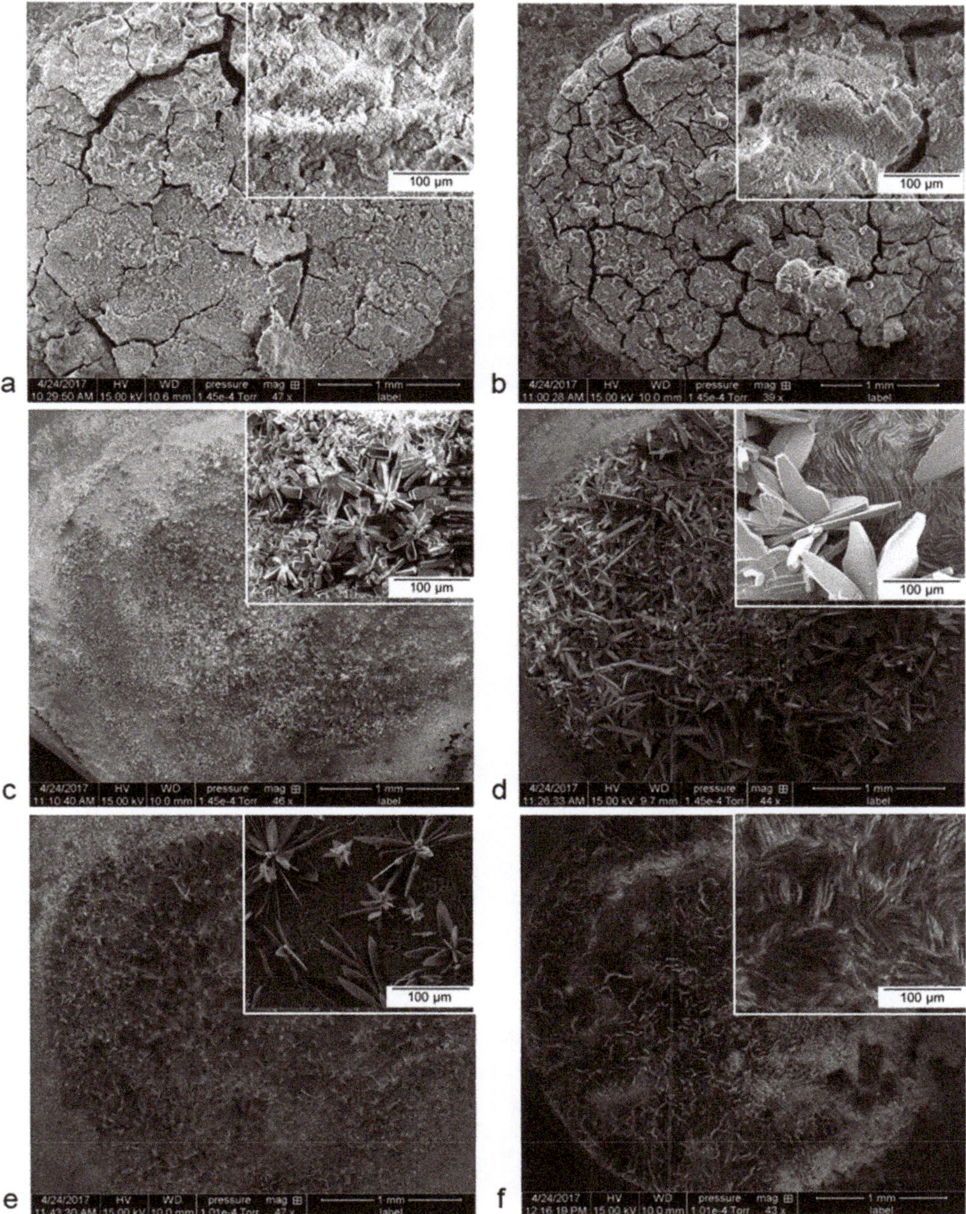

Figure 2. SEM images of the specimens' surfaces after polarization: (**a**) pure Mg, (**b**) Mg–2Zn–1Mn, (**c**) pure Zn, (**d**) Zn–0.5Mg, (**e**) Zn–1Mg, and (**f**) Zn–0.5Al.

The distinct difference on the corrosion layer's composition between the Mg group and the Zn group is the calcium content, which was only detected on that of the former group with a concentration range of 13–15 wt % (Table 3). The Mg group's composition has similarity to that of Mg alloys reported in many non-urinary in vitro and in vivo studies, and it generally forms hydroxide, phosphate, and carbonate [23,27–31]. Apart from the lack of calcium, the EDS spectra of the Zn group corrosion layer showed a lower content of oxygen and phosphorous, but a higher carbon content as compared to the Mg group. A focused EDS analysis on the flower-like formation on the surface of pure Zn and Zn–Mg alloys (Figure 2c–e) reveals a high content of zinc and oxygen, which supposedly forms ZnO, as later confirmed by XPS analysis. On the very top layer, as measured by the XPS, both groups have a similar content of carbon and oxygen, but a higher phosphorous content for the Zn group. Apart from the aluminum content, the surface of the Zn–0.5Al alloy has the lowest content of those elements, as measured by both techniques. Previous non-urinary in vitro and in vivo corrosion studies of pure Zn generally confirmed the formation of zinc oxide, zinc carbonate, zinc phosphate, and CaP on the very top layer [14,15,19,32,33]. The presence of calcium on the corrosion layer may initiate calcification and thus encrustation, a main current problem of ureteral stents [6], leading to the formation of urinary stones. These stones are composed mainly of calcium oxalate (CaO_x), struvite ($NH_4MgPO_4{\cdot}6H_2O$), calcium phosphate (CaP), and uric acid ($C_5H_4N_4O_3$) [34,35].

Table 3. Elemental composition of the corrosion layer detected by EDS and XPS.

Sample	Element						
EDS (wt %)	**C**	**O**	**Mg**	**P**	**Ca**	**Zn**	**Al**
Mg	4.9	50.6	11.9	19.7	13.0	-	-
Mg–2Zn–1Mn	2.3	47.2	13.3	22.1	15.3	-	-
Zn	16.4	17.5	-	8.8	-	57.3	-
Zn–0.5Mg	13.9	11.7	-	7.9	-	66.6	-
Zn–1Mg	22.3	14.2	-	5.6	-	57.9	-
Zn–0.5Al	9.2	6.7	-	4.0	-	65.5	14.6
XPS (at %)	**C1s**	**O1s**	**Mg2p**	**P2p**	**Ca2p**	**Zn3p2**	**Al2p**
Mg	31 ± 10	40 ± 8	15 ± 2	5 ± 4	9 ± 1	-	-
Mg–2Zn–1Mn	23 ± 3	46 ± 3	14 ± 4	9 ± 12	8 ± 3	-	-
Zn	25 ± 3	45 ± 2	-	18 ± 2	-	12 ± 2	-
Zn–0.5Mg	28 ± 5	42 ± 5	-	19 ± 1	-	11.1 ± 1	-
Zn–1Mg	31 ± 6	42 ± 2	-	17 ± 3	-	10 ± 2	-
Zn–0.5Al	26 ± 2	44 ± 2	-	9 ± 7	-	10 ± 2	5 ± 2

Further XPS analysis allowed for the observation of two Mg2p peaks at 51.15 and 49.66 eV on pure Mg and at 51.19 and 50.07 eV for Mg–2Zn–1Mn (Figure 3a), which correlate to $MgCO_3$ and $Mg(OH)_2$, respectively [36,37]. Three C1s peaks (Figure 3b) are also detected on these specimens at 285 eV for C–C, 286.6 eV for C–O(H) [38], and 289.7 eV for carbonate [37]. Three O1s peaks (Figure 3c) are present at 530.15 eV related to MgO or $Mg(OH)_2$, 531.7 eV for the double binding of P=O [39], and at 533.3 eV related to O–C=O of CO_3 species such as $MgCO_3$ or $CaCO_3$ [40]. Additionally, two P2p peaks were observed at 133.6 and 135 eV (Figure 3d), which correspond to $MgHPO_4$ and $Mg_3(PO_4)_2$, respectively [41].

Regarding the Zn–0.5Al alloy, the XPS analysis demonstrates the presence of three C1s peaks related to C–C at 285 eV, carbonate at 289.15 eV, and C–O(H) at 286.89 eV (Figure 4a). The first O1s peak was observed at 532 eV and corresponds to $Zn(OH)_2$ [42], the second at 529.6 eV is related to ZnO [42], and the third at 533 eV fits to carbonate (Figure 4b). Two P2p peaks were observed at 133.47 and 134.31 eV (Figure 4c), characterized as PO_4^{3-} and HPO_4^{2-} [41], which confirms the presence of a phosphate compound in the corrosion layer. One peak was observed for Zn2p3 at 1022.8 eV (Figure 4d), which corresponds to $Zn(OH)_2$ [42]. An additional peak of Al2p (not shown) was found at 74.76 eV for Al_2O_3 [43]. Overall, the XPS analysis confirms that no calcium-related compound was formed on the surface of Zn–0.5Al, which makes this alloy less susceptible to calcification compared with pure Mg and the Mg–2Zn–1Mn alloy.

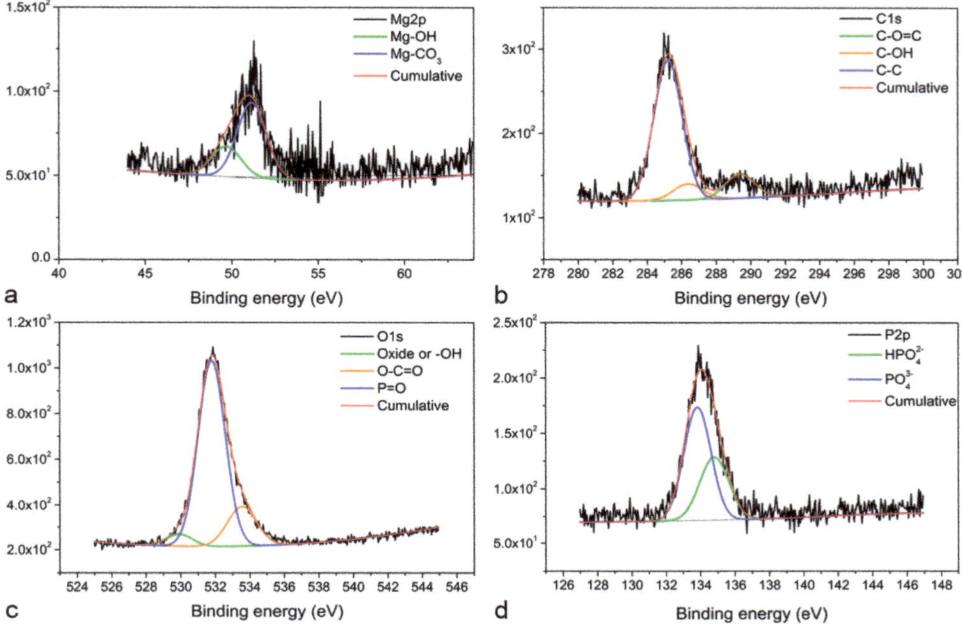

Figure 3. High resolution XPS spectra for pure Mg specimen: (**a**) Mg2p peaks, (**b**) C1s peaks, (**c**) O1s peaks, and (**d**) P2p peaks.

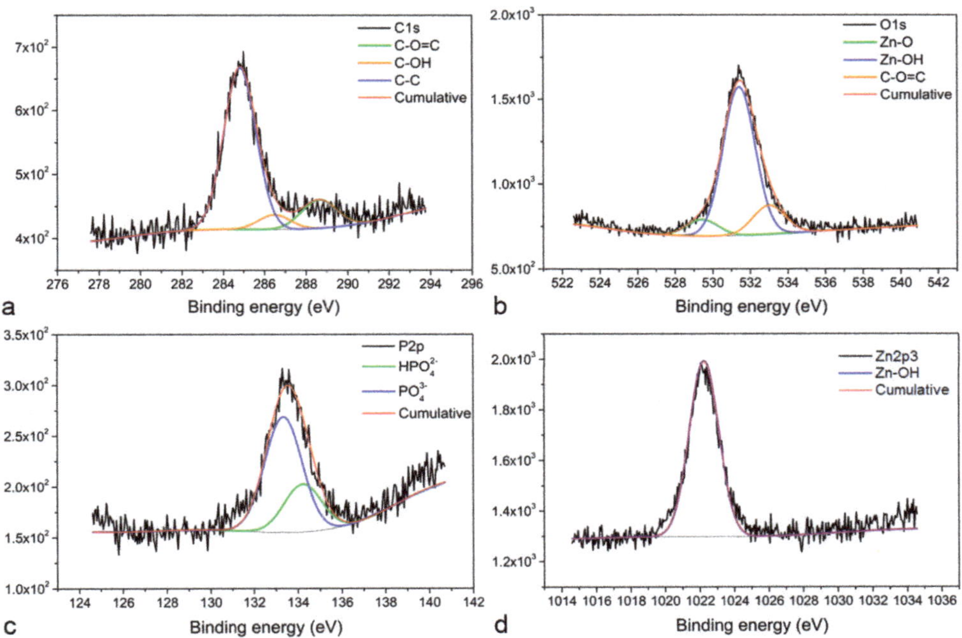

Figure 4. High resolution XPS spectra for Zn–0.5Al specimen: (**a**) C1s peaks, (**b**) O1s peaks, (**c**) P2p peaks, and (**d**) Zn2p3 peak.

The corrosion layers formed on the metals' surface exhibit a different impedance behavior, as indicated by their Nyquist plots (Figure 5a). The Mg group's plots form a semi-circle (one time constant), whilst those of the Zn group show two semi-circles (two time constant). Looking at the Bode plots, in the Mg group the phase angles increase from about 10° to 55° at a high frequency and then decrease to about 25° at a lower frequency (Figure 5b), indicating a capacitive loop [44]. Meanwhile, the plots for the Zn group show two peaks, one at a high frequency and another at a low frequency, except for the Zn–0.5Mg where the second peak is not that apparent. In the other Bode plots, all specimens show similar evolution from low impedance at high frequency to high impedance at low frequency, with the Mg group having the highest impedance (Figure 5c), indicating the corrosion layer exerts an electric effect between a resistor and a capacitor.

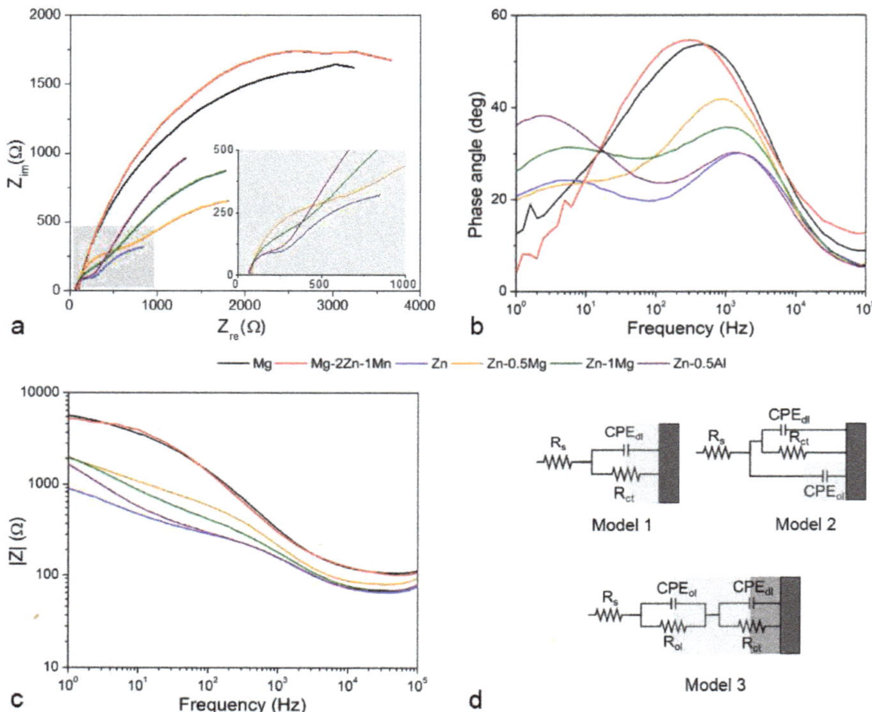

Figure 5. Typical electrochemical impedance spectroscopy (EIS) results: (**a**) Nyquist plots, (**b**) Bode phase angle plots, (**c**) Bode impedance plots, and (**d**) proposed equivalent electrical circuit models. Note that at very low frequency the Mg specimens show some negligible scatter.

By applying a single-frequency voltage to the interface, EIS allows a measurement of the electric properties of the corrosion layer, including a phase shift, a real and imaginary part of the impedance and the amplitude of the resulting current at that particular frequency [45,46]. All these properties can be simplified as an electrical circuit with each electrical component representing one of the elements of the electrochemical cell. For the Mg group, their semi-circle plots indicate the formation of a typical uniform oxide layer (Figure 2a,b). Its characteristic can be fitted with a simple electrical circuit (Model 1 (Figure 5d)) composed of the R_s (solution's resistance), R_{ct} (charge transfer resistance), and CPE_{dl} (constant phase element) of the double layer at the electrolyte–surface interface. The CPE represents a non-ideal electrical element in the circuit; in this case, it behaves as an imperfect capacitor and is very useful for assessing its dispersive effect at the surface of the electrode [47]. For a corroding metal, this effect is attributed to the microscopic increase in the surface roughness of the electrode [48]. When R_{ct}

is the only resistance in the circuit, it can often be substituted with R_p (polarization resistance) and it is inversely proportional with the corrosion rate. For the Zn group, its oxide layer characteristic is fitted with the Model 2 circuit that includes a CPE_{ol} for the capacitive properties of the oxide layer, considering that Zn is less reactive than Mg and formed a rather non-uniform corrosion layer (Figure 2c–f). A more complex electrical circuit can be fitted to all specimens, representing a bi-layered surface with outer CPE_{ol} and R_{ol} (Model 3). The calculated value of each electrical component from the equivalent circuits is presented in Table 4.

Table 4. EIS-derived parameters (R in Ω cm^2 and CPE in μF).

Specimen		Mg	Mg–2Zn–1Mn	Zn	Zn–Mg	Zn–0.5Al
Model 1 R(QR)	R_{ct}	634 ± 141	1357 ± 321	-	-	-
	CPE_{dl}	56 ± 8	277 ± 49	-	-	-
Model 2 R(Q(QR))	R_{ct}	-	-	9 ± 4	5 ± 5	13 ± 2
	CPE_{dl}	-	-	8480 ± 2770	1220 ± 1700	2740 ± 153
	CPE_{ol}	-	-	23 ± 15	2030 ± 1520	29 ± 10
Model 3 R(QR)(QR)	R_{ct}	45 ± 1	104 ± 47	20 ± 3	25 ± 4	495 ± 108
	R_{ol}	716 ± 178	583 ± 47	95 ± 27	242 ± 51	14 ± 2
	CPE_{dl}	85 ± 11	84 ± 11	2970 ± 1060	871 ± 119	74 ± 17
	CPE_{ol}	63 ± 19	317 ± 101	77 ± 48	82 ± 49	1850 ± 170

The solution's resistance, R_s, is in the constant range of 5–10 Ω throughout all the experiments. As shown in the Model 1 and Model 3 circuits, the Mg group exhibits a higher R_{ct} compared to the Zn group, conversely to their CPE_{dl}. The addition of Mn may play a role in improving the coating resistance of the Mg–2Zn–1Mn by increasing the passivity of the corrosion layer [28]. Both Zn–0.5Mg and Zn–1Mg exhibit similar electrical values, so in EIS analysis they are treated as one metal. The presence of a large value of CPE_{ol} (Model 2) indicates that a preferential corrosion occurred on the surface of the Zn–Mg alloys, leading to the creation of micropores on the corrosion layer (Figure 2e). The CPE_{dl} value of the Zn group is higher than those of the Mg group due to the higher capacitive properties of Zn, which produces fewer active ions than magnesium, and this could be the reason for the slower degradation rate of the former [49]. The value decreases from pure Zn to Zn–0.5Al and to Zn–Mg, meaning more charge should be accumulated at the electrode-electrolyte interface of pure Zn since no element acts as a captor of charged ions, i.e., Mg in Zn–Mg and Al in Zn–Al in the process of localize corrosion. In the Model 3 circuit, R_{ct} and CPE_{dl} are affected by the interaction of different layers, which in turn changes the behavior of both components. The R_{ol} value is much higher for the Mg group, which could be correlated to a thicker oxide layer (Figure 2a,b). The capacitive properties of the oxide layer are bound with CPE_{ol}. In this bi-layer model, the oxide layer of pure Zn and Zn–0.5Al has dielectric properties, which could be influenced by a thick and dense corrosion layer governed by the diffusion phenomenon (Figure 2c). However, this may not be the case for Zn–Mg alloys as the layer becomes less uniform due to the localized corrosion (Figure 2d,e).

The electrochemical noise data were analyzed in the time domain with the calculation of the noise resistance R_n. It is well established that R_n is inversely proportional to the corrosion rate [50]. This helps to monitor the corrosion rates after immersion of the specimens in the urine solution. Pure Mg, pure Zn, and Zn–0.5Al were used as samples for the EN investigation. Up to 12 h of immersion, pure Mg exhibited the highest corrosion rate followed by Zn–0.5Al and pure Zn (Figure 6), a similar trend was obtained by the accelerated PDP test. Pure Mg experienced a stabilization after 12 h, so, for longer periods, pure Mg possesses the lowest corrosion rate than the other two metals.

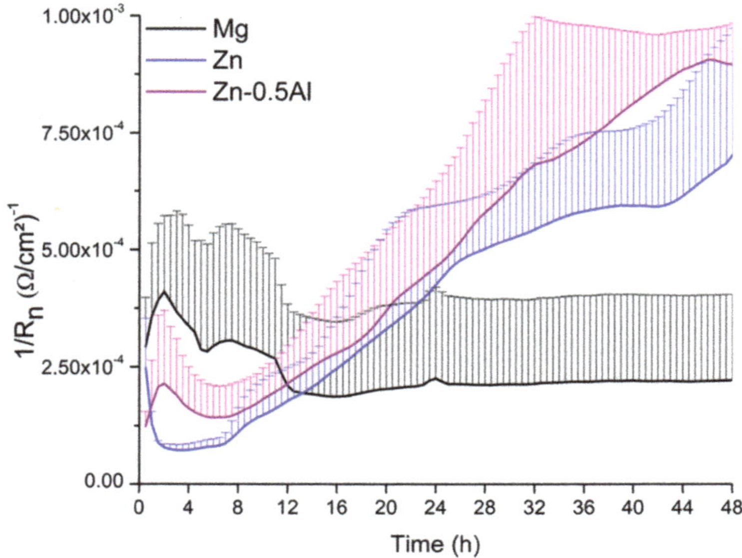

Figure 6. Electrochemical noise (EN)results presented as $1/R_n$ vs. time for pure Mg, pure Zn, and the Zn–0.5Al alloy over a period of 48 h.

In artificial urine, pure Mg dissolves quickly. Due to the high affinity of Mg ions with the ions in the solution, an insoluble corrosion layer (i.e., $Mg_3(PO_4)_3$) forms and protects the metals underneath from further corrosion [18]. This thick layer (Figure 7a) acts as a more effective barrier than $Mg(OH)_2$ or $Zn(OH)_2$ on pure Zn [17]. A rather smooth surface was observed on the latter (Figure 7b). The Zn–0.5Al is prone to microgalvanic corrosion due to the potential difference of the Zn and Al phases (Figure 7c).

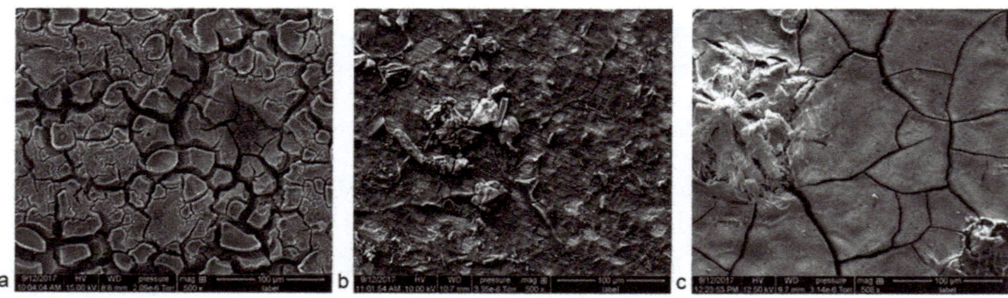

Figure 7. Surface morphology after 48 h of the EN test of (**a**) pure Mg, (**b**) pure Zn, and (**c**) Zn–0.5Al.

4. Conclusions

In artificial urine, the studied pure Mg and its alloy show a corrosion behavior distinct from that of pure Zn and its alloys. Based on the accelerated potentiodynamic polarization test, the Mg group exhibits a higher corrosion rate than does the Zn group. The corrosion layer of the Zn group has a more capacitive behavior than that of the Mg group. A flower-like structure of Zn oxide forms on the polarized surface of the Zn group, except for Zn–0.5Al, while precipitation of Ca and Mg was observed in the Mg group only. In the long-term electrochemical noise test, pure Mg exhibits stabilization after 12 h and thus has an overall lower corrosion rate than does pure Zn and Zn–0.5Al. The continuous metal dissolution of Zn–0.5Al and the low affinity to ions in artificial urine prevents the formation of a thick corrosion layer. In view of ureteral stent applications, a continuous dissolution of metal in urine, such as the case of the Zn–0.5Al alloy, is ideal to prevent calcification (stent encrustation) and bacterial adhesion on the surface.

Author Contributions: Conceptualization, H.H. and S.C.; Data Curation, S.C.; Formal Analysis, F.S., H.H. and S.C.; Methodology, F.S., H.H. and S.C.; Resources, E.M.; Supervision, H.H.; Writing—Original Draft Preparation, S.C.; Writing—Review & Editing, E.G., E.M., F.S., H.H., M.V. and S.C.

Funding: This research was funded by Natural Sciences and Engineering Research Council of Canada (NSERC) via Discovery grant.

Conflicts of Interest: The authors declare no conflict of interest.

References

1. ADAM Medical Encyclopedia. *Acute Bilateral Obstructive Uropathy*; ADAM Inc.: Atlanta, GA, USA, 2010.
2. Public Health Agency Canada. *Economic Burden of Illness in Canada 2005–2008*; Public Health Agency Canada: Ottawa, ON, USA, 2014.
3. Tseng, T.Y.; Stoller, M.L. Obstructive Uropathy. *Clin. Geriatr. Med.* **2009**, *25*, 437–443. [CrossRef] [PubMed]
4. Lange, D.; Bidnur, S.; Hoag, N.; Chew, B.H. Ureteral stent-associated complications—Where we are and where we are going. *Nat. Rev. Urol.* **2015**, *12*, 17–25. [CrossRef] [PubMed]
5. Auge, B.K.; Sarvis, J.A.; L'Esperance, J.O.; Preminger, G.M. Practice Patterns of Ureteral Stenting after Routine Ureteroscopic Stone Surgery: A Survey of Practicing Urologists. *J. Endourol.* **2007**, *21*, 1287–1292. [CrossRef] [PubMed]
6. Lange, D.; Chew, B.H. Update on ureteral stent technology. *Ther. Adv. Urol.* **2009**, *1*, 143–148. [CrossRef] [PubMed]
7. Talja, M.; Välimaa, T.; Tammela, T.; Petas, A.; Törmälä, P. Bioabsorbable and biodegradable stents in urology. *J. Endourol.* **1997**, *11*, 391–397. [CrossRef] [PubMed]
8. Wang, X.; Shan, H.; Wang, J.; Hou, Y.; Ding, J.; Chen, Q.; Guan, J.; Wang, C.; Chen, X. Characterization of nanostructured ureteral stent with gradient degradation in a porcine model. *Int. J. Nanomed.* **2015**, *10*, 3055–3064. [CrossRef] [PubMed]
9. Pedro, R.N.; Hendlin, K.; Kriedberg, C.; Monga, M. Wire-Based Ureteral Stents: Impact on Tensile Strength and Compression. *Urology* **2007**, *70*, 1057–1059. [CrossRef]
10. Hendlin, K.; Korman, E.; Monga, M. New Metallic Ureteral Stents: Improved Tensile Strength and Resistance to Extrinsic Compression. *J. Endourol.* **2011**, *26*, 271–274. [CrossRef]
11. Hermawan, H. Updates on the research and development of absorbable metals for biomedical applications. *Prog. Biomater.* **2018**, *7*, 93–110. [CrossRef]
12. Francis, A.; Yang, Y.; Virtanen, S.; Boccaccini, A.R. Iron and iron-based alloys for temporary cardiovascular applications. *J. Mater. Sci. Mater. Med.* **2015**, *26*, 138. [CrossRef]
13. Haude, M.; Ince, H.; Kische, S.; Abizaid, A.; Tölg, R.; Alves, L.P.; van Mieghem, N.; Verheye, S.; von Birgelen, C.; Christiansen, E. Sustained safety and clinical performance of a drug-eluting absorbable metal scaffold up to 24 months: Pooled outcomes of BIOSOLVE-II and BIOSOLVE-III. *EuroIntervention* **2017**, *20*, 432–439.
14. Drelich, A.J.; Zhao, S.; Guillory, R.J.; Drelich, J.W.; Goldman, J. Long-term surveillance of zinc implant in murine artery: Surprisingly steady biocorrosion rate. *Acta Biomater.* **2017**, *58*, 539–549. [CrossRef] [PubMed]
15. Yang, H.; Wang, C.; Liu, C.; Chen, H.; Wu, Y.; Han, J.; Jia, Z.; Lin, W.; Zhang, D.; Li, W.; et al. Evolution of the degradation mechanism of pure zinc stent in the one-year study of rabbit abdominal aorta model. *Biomaterials* **2017**, *145*, 92–105. [CrossRef] [PubMed]
16. Lock, J.Y.; Wyatt, E.; Upadhyayula, S.; Whall, A.; Nuñez, V.; Vullev, V.I.; Liu, H. Degradation and antibacterial properties of magnesium alloys in artificial urine for potential resorbable ureteral stent applications. *J. Biomed. Mater. Res. A* **2014**, *102*, 781–792. [CrossRef] [PubMed]
17. Zhang, S.; Zheng, Y.; Zhang, L.; Bi, Y.; Li, J.; Liu, J.; Yu, Y.; Guo, H.; Li, Y. In vitro and in vivo corrosion and histocompatibility of pure Mg and a Mg-6Zn alloy as urinary implants in rat model. *Mater. Sci. Eng. C* **2016**, *68*, 414–422. [CrossRef] [PubMed]

18. Zhang, S.; Bi, Y.; Li, J.; Wang, Z.; Yan, J.; Song, J.; Sheng, H.; Guo, H.; Li, Y. Biodegradation behavior of magnesium and ZK60 alloy in artificial urine and rat models. *Bioact. Mater.* **2017**, *2*, 53–62. [CrossRef] [PubMed]
19. Mostaed, E.; Sikora-Jasinska, M.; Mostaed, A.; Loffredo, S.; Demir, A.G.; Previtali, B.; Mantovani, D.; Beanland, R.; Vedani, M. Novel Zn-based alloys for biodegradable stent applications: Design, development and in vitro degradation. *J. Mech. Behav. Biomed. Mater.* **2016**, *60*, 581–602. [CrossRef] [PubMed]
20. Kirchmann, H.; Pettersson, S. Human urine—Chemical composition and fertilizer use efficiency. *Fertil. Res.* **1994**, *40*, 149–154. [CrossRef]
21. ASTM International. *ASTM F2129-17 Standard Test Method for Conducting Cyclic Potentiodynamic Polarization Measurements to Determine the Corrosion Susceptibility of Small Implant Devices*; ASTM International: West Conshohocken, PA, USA, 2017.
22. Ghali, E. *Corrosion Resistance of Aluminum and Magnesium Alloys: Understanding, Performance, and Testing*; John Wiley & Sons: Hoboken, NJ, USA, 2010.
23. Jamesh, M.; Kumar, S.; Narayanan, T.S.N.S. Corrosion behavior of commercially pure Mg and ZM21 Mg alloy in Ringer's solution—Long term evaluation by EIS. *Corros. Sci.* **2011**, *53*, 645–654. [CrossRef]
24. Song, G.L.; Atrens, A. Corrosion mechanisms of magnesium alloys. *Adv. Eng. Mater.* **1999**, *1*, 11–33. [CrossRef]
25. Dambatta, M.S.; Izman, S.; Kurniawan, D.; Farahany, S.; Yahaya, B.; Hermawan, H. Influence of thermal treatment on microstructure, mechanical and degradation properties of Zn–3Mg alloy as potential biodegradable implant material. *Mater. Des.* **2015**, *85*, 431–437. [CrossRef]
26. Zhang, X.G. *Corrosion and Electrochemistry of Zinc*; Springer Science & Business Media: Berlin, Germany, 2013.
27. Agarwal, S.; Curtin, J.; Duffy, B.; Jaiswal, S. Biodegradable magnesium alloys for orthopaedic applications: A review on corrosion, biocompatibility and surface modifications. *Mater. Sci. Eng. C* **2016**, *68*, 948–963. [CrossRef]
28. Rosalbino, F.; Negri, S.D.; Saccone, A.; Angelini, E.; Delfino, S. Bio-corrosion characterization of Mg–Zn–X (X = Ca, Mn, Si) alloys for biomedical applications. *J. Mater. Sci. Mater. Med.* **2010**, *21*, 1091–1098. [CrossRef] [PubMed]
29. Witte, F.; Kaese, V.; Haferkamp, H.; Switzer, E.; Meyer-Lindenberg, A.; Wirth, C.J.; Windhagen, H. In vivo corrosion of four magnesium alloys and the associated bone response. *Biomaterials* **2005**, *26*, 3557–3563. [CrossRef]
30. Xu, L.; Yu, G.; Zhang, E.; Pan, F.; Yang, K. In vivo corrosion behavior of Mg-Mn-Zn alloy for bone implant application. *J. Biomed. Mater. Res. A* **2007**, *83A*, 703–711. [CrossRef] [PubMed]
31. Abidin, N.I.Z.; Rolfe, B.; Owen, H.; Malisano, J.; Martin, D.; Hofstetter, J.; Uggowitzer, P.J.; Atrens, A. The in vivo and in vitro corrosion of high-purity magnesium and magnesium alloys WZ21 and AZ91. *Corros. Sci.* **2013**, *75*, 354–366. [CrossRef]
32. Törne, K.; Larsson, M.; Norlin, A.; Weissenrieder, J. Degradation of zinc in saline solutions, plasma, and whole blood. *J. Biomed. Mater. Res. B Appl. Biomater.* **2016**, *104*, 1141–1151. [CrossRef] [PubMed]
33. Kubásek, J.; Vojtěch, D.; Jablonská, E.; Pospíšilová, I.; Lipov, J.; Ruml, T. Structure, mechanical characteristics and in vitro degradation, cytotoxicity, genotoxicity and mutagenicity of novel biodegradable Zn–Mg alloys. *Mater. Sci. Eng. C* **2016**, *58*, 24–35. [CrossRef]
34. Kirejczyk, J.K.; Porowski, T.; Filonowicz, R.; Kazberuk, A.; Stefanowicz, M.; Wasilewska, A.; Debek, W. An association between kidney stone composition and urinary metabolic disturbances in children. *J. Pediatr. Urol.* **2014**, *10*, 130–135. [CrossRef]
35. Selvaraju, R.; Raja, A.; Thiruppathi, G. FT-IR spectroscopic, thermal analysis of human urinary stones and their characterization. *Spectrochim. Acta A Mol. Biomol. Spectrosc.* **2015**, *137*, 1397–1402. [CrossRef]
36. Hosking, N.C.; Ström, M.A.; Shipway, P.H.; Rudd, C.D. Corrosion resistance of zinc–magnesium coated steel. *Corros. Sci.* **2007**, *49*, 3669–3695. [CrossRef]
37. Liu, M.; Zanna, S.; Ardelean, H.; Frateur, I.; Schmutz, P.; Song, G.; Atrens, A.; Marcus, P. A first quantitative XPS study of the surface films formed, by exposure to water, on Mg and on the Mg–Al intermetallics: Al3Mg2 and Mg17Al12. *Corros. Sci.* **2009**, *51*, 1115–1127. [CrossRef]
38. Reid, G.; Davidson, R.; Denstedt, J.D. XPS, SEM and EDX analysis of conditioning film deposition onto ureteral stents. *Surf. Interface Anal.* **1994**, *21*, 581–586. [CrossRef]

39. Wu, L.; Zhao, L.; Dong, J.; Ke, W.; Chen, N. Potentiostatic Conversion of Phosphate Mineral Coating on AZ31 Magnesium Alloy in 0.1 M K_2HPO_4 Solution. *Electrochim. Acta* **2014**, *145*, 71–80. [CrossRef]
40. Santamaria, M.; di Quarto, F.; Zanna, S.; Marcus, P. Initial surface film on magnesium metal: A characterization by X-ray photoelectron spectroscopy (XPS) and photocurrent spectroscopy (PCS). *Electrochim. Acta* **2007**, *53*, 1314–1324. [CrossRef]
41. Song, Y.; Shan, D.; Chen, R.; Zhang, F.; Han, E.-H. A novel phosphate conversion film on Mg–8.8Li alloy. *Surf. Coat. Technol.* **2009**, *203*, 1107–1113. [CrossRef]
42. Biesinger, M.C.; Lau, L.W.M.; Gerson, A.R.; Smart, R.S.C. Resolving surface chemical states in XPS analysis of first row transition metals, oxides and hydroxides: Sc, Ti, V, Cu and Zn. *Appl. Surf. Sci.* **2010**, *257*, 887–898. [CrossRef]
43. Yan, Y.L.; Helfand, M.A.; Clayton, C.R. Evaluation of the effect of surface roughness on thin film thickness measurements using variable angle XPS. *Appl. Surf. Sci.* **1989**, *37*, 395–405. [CrossRef]
44. Tamilselvi, S.; Raman, V.; Rajendran, N. Corrosion behaviour of Ti–6Al–7Nb and Ti–6Al–4V ELI alloys in the simulated body fluid solution by electrochemical impedance spectroscopy. *Electrochim. Acta* **2006**, *52*, 839–846. [CrossRef]
45. Kirkland, N.T.; Birbilis, N.; Staiger, M.P. Assessing the corrosion of biodegradable magnesium implants: A critical review of current methodologies and their limitations. *Acta Biomater.* **2012**, *8*, 925–936. [CrossRef]
46. Macdonald, D.D. Reflections on the history of electrochemical impedance spectroscopy. *Electrochim. Acta* **2006**, *51*, 1376–1388. [CrossRef]
47. Jorcin, J.-B.; Orazem, M.E.; Pébère, N.; Tribollet, B. CPE analysis by local electrochemical impedance spectroscopy. *Electrochim. Acta* **2006**, *51*, 1473–1479. [CrossRef]
48. Scully, J.R.; Silverman, D.C. *Electrochemical Impedance: Analysis and Interpretation*; ASTM International: West Conshohocken, PA, USA, 1993.
49. Alves, M.M.; Prošek, T.; Santos, C.F.; Montemor, M.F. Evolution of the in vitro degradation of Zn–Mg alloys under simulated physiological conditions. *RSC Adv.* **2017**, *7*, 28224–28233. [CrossRef]
50. Gusmano, G.; Montesperelli, G.; Pacetti, S.; Petitti, A.; D'Amico, A. Electrochemical Noise Resistance as a Tool for Corrosion Rate Prediction. *Corrosion* **1997**, *53*, 860–868. [CrossRef]

Review

The Effect of Surface Treatments on the Degradation of Biomedical Mg Alloys—A Review Paper

Marcjanna Maria Gawlik [1,*], Björn Wiese [1], Valérie Desharnais [1,2], Thomas Ebel [1] and Regine Willumeit-Römer [1]

1 Helmholtz-Zentrum Geesthacht, Max-Planck-Straße 1, 21502 Geesthacht, Germany; bjoern.wiese@hzg.de (B.W.); valerie.desharnais@sympatico.ca (V.D.); thomas.ebel@hzg.de (T.E.); regine.willumeit@hzg.de (R.W.-R.)

2 School of Computer Science, McGill University, 845 Sherbrooke Street West, Montréal, QC H3A 2T5, Canada

* Correspondence: marcjanna.gawlik@hzg.de

Received: 19 November 2018; Accepted: 14 December 2018; Published: 16 December 2018

Abstract: This report reviews the effects of chemical, physical, and mechanical surface treatments on the degradation behavior of Mg alloys via their influence on the roughness and surface morphology. Many studies have been focused on technically-used AZ alloys and a few investigations regarding the surface treatment of biodegradable and Al-free Mg alloys, especially under physiological conditions. These treatments tailor the surface roughness, homogenize the morphology, and decrease the degradation rate of the alloys. Conversely, there have also been reports which showed that rough surfaces lead to less pitting and good cell adherence. Besides roughness, there are many other parameters which are much more important than roughness when regarding the degradation behavior of an alloy. These studies, which indicate the relationship between surface treatments, roughness and degradation, require further elaboration, particularly for biomedical Mg alloy applications.

Keywords: surface treatments; roughness; Mg-alloys; degradation behavior

---

1. Introduction

The study of Mg as degradable biomaterial for implants is an advanced research area. A second operation to remove the implant after bone healing can be avoided [1–6]. Mg is naturally available as trace element in the body, and is thus non-toxic and biocompatible [7–9]. Implant processing is feasible due to the ductility and workability of Mg [10]. Strength and toughness are higher than of polymer implants, which is beneficial for load-bearing implants [11,12]. Mg alloys are reported to show improved osseointegration and bone implant strength compared to permanent Ti alloys [13,14]. In particular, Mg alloys are suitable as biodegradable implant materials [1,15–19]. Mg is able to degrade in aqueous solutions with the formation of magnesium hydroxide and hydrogen [20–24]. In particular, aqueous salt solutions containing ions including chlorides or sulphates, with the exception of alkali metals or alkaline metal containing solutions, are able to dissolve the protective magnesium hydroxide layer, leading to enhanced degradation [24–28]. In order to improve the mechanical properties of Mg, elements are added to tailor, for example, its tensile strength and ductility. Thus, it is possible to produce implants that have tailored mechanical properties to use it as temporary bone fixation. However, when alloying and processing the material, impurities like Fe, Ni, and Cu or phases with a high electrochemical potential difference are found at or near to the surface of the material, which increases the degradation rate through galvanic corrosion [23,29]. For the application of biodegradable Mg implants to become feasible in the future, two different objectives must be met in order to achieve usable degradation behavior.

One objective is limiting the degradation rate of the initial state of the alloy, which, as explained later, is related to the amount of hydrogen evolution. The deeper and rougher the surface morphology, the more hydrogen gas will be produced [9]. An overly fast degradation with gas evolution in the initial state leads to degradation of the mechanical integrity. Excessive gas evolution can also modify the bone remodeling process and impair the consolidation of bones [30]. However, relatively strong hydrogen evolution is crucial for cell adherence and implant-bone integration [21,31,32]. Aqueous salt solutions including chloride ions, like those found in the human body fluids [33], increasing decomposing of $Mg(OH)_2$, release OH^- and raise the pH [34]. Besides hydrogen production, a local alkalization might provoke necrosis [9].

The second objective is to control the degradation rate of implants during the healing time. The required degradation rate depends on the application with lifetime and stability of the implant and the potential of the surrounding tissue to tolerate pH changes and high ion concentrations. It is reported that the properties of the material, e.g., crystallographic orientation [35–37], microstructure [21,38–48], grain size [41,49–53], secondary phases [51,54,55], contamination [38,40,56], and deformation [38,41,57–60], affect the degradation behavior, as well the environment, e.g., the immersion medium [61–64]. It is possible to control the degradation behavior of Mg alloys using chemical, physical, and mechanical surface treatments [27,32,65–69]. Additionally, surface uniformity has been shown to decelerate degradation [70]. Surface morphology can differ despite identical roughness parameters, and also affects the degradation process [39]. Studies have shown that surface roughness can affect the initial degradation [71], the degradation rate [38,71–77], degradation resistance [73,78–82], pitting behavior [38,71,72,83], bone integration [84–86], cell adherence [21,74,87,88], cell proliferation [88–91], and cell differentiation [92]. Besides roughness, surface unevenness can also influence the adhesion of cells [73]. In some cases, a smoother surface will reduce the degradation rate [72,74,93]. However, this behavior has been contradicted in other studies [73,79,81,82,94,95].

The aim of this review is to show the correlation between surface treatment, roughness, and the degradation behavior of Mg alloys in order to define meaningful roughness values and suitable surface treatments for biodegradable Mg implants. An overview of studies mentioning surface treatments, roughness, and degradation is given in Tables 1–5.

2. Mechanical Surface Treatments

2.1. Grinding and Polishing

The degradation behavior of sand-cast, ground, and polished AZ91 alloys were investigated by Walter and Kannan [96]. The use of a grinding paper with increased grit size decreased the surface roughness (Table 1, Ref. [96]). Three methods were used to evaluate the relationship between pitting and roughness: a 24 h immersion test in a 0.5 wt.% NaCl solution, 1 h of potentiodynamic polarization (PDP) and 1 h of Electrochemical Impedance Spectroscopy (EIS). For the polished samples, no inductive loop after EIS was observed. A low inductive loop is related to a low or negligible amount of surface pitting. This was found for all ground samples. Thus, it is suspected that no pitting will occur on polished surfaces due to a higher passivation. As a consequence, passivation is reduced for higher surface roughness values. The polarization curves in Ref. [96] show that a higher anodic current, as indicated by the current density i_{corr}, is produced with greater surface roughness (Table 1, Ref. [96] and Figure 1).

In particular, the polarization curve of material ground with 320 grit paper in Ref. [96] exhibits a strong increase in anodic current, which suggests a high number of pits being formed. The surface appearance was analyzed by scanning electron microscope (SEM) after immersion for 24 h and after galvanostatic testing. Numerous pits were observed after immersion when the 320 grit size paper was used, which confirms the results from the electrochemical testing. Less pitting was seen to occur when using a finer grinding paper. In the case of paper with a 1200 grit size, no localized pitting was found after testing. Surprisingly, more pits were observed on polished samples after galvanostatic testing.

This is likely to have been caused by a high anodic current which reduces passivation for all surface treatments. Walter and Kannan [96] concluded from these experiments that roughness does affect the passivation layer, but does not directly affect the likelihood of pitting. After removing the passivation layer, pitting occurred for all surface roughness values as seen by SEM [96].

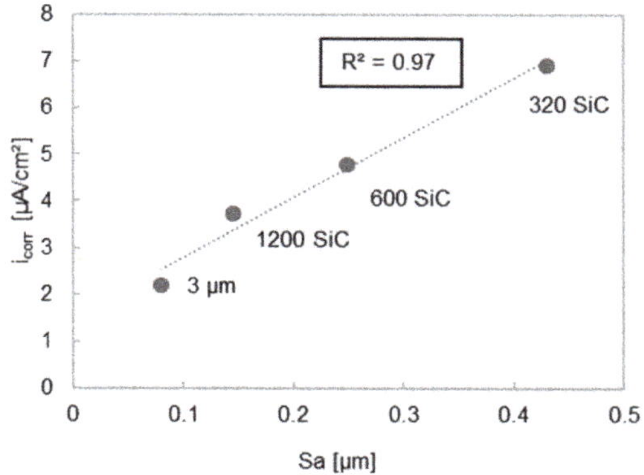

Figure 1. Graph shows a linear relationship between Sa and i_{corr} after grinding and polishing. R^2 is the coefficient of determination, which assess the linear mutual dependence of x and y. $R^2 = 1$ defines the highest linearity [97]. Sa and i_{corr} values were obtained from [96].

Walter et al. [71] also investigated the correlation between the degradation and surface roughness Sa (the arithmetic mean height within a sample area, three dimensionally determined roughness) [98] for samples which were ground using 120 SiC grit size paper (Sa = 973 nm) and samples that had been ground using 2500 grit size paper, followed by polishing with a 3 µm diamond paste (Sa = 22 nm). The samples were cleaned with acetone and ethanol. The degradation behavior was characterized under simulated body fluid (SBF) using EIS. The results for both surface finishes exhibited similar tendencies. The ground and the polished samples showed a mid-frequency capacitive loop at the beginning of testing [71]. A mid-frequency capacitive loop corresponds to a passivation layer [96]. For the ground sample, a mid-frequency capacitive loop was observed for the first 2 h and was then followed by an inductive loop at low frequencies. The polished samples had inductive loops at low frequencies after 4 h, which confirms that passivation layers on smoother surfaces last longer [71]. In general, an inductive loop implies surface pitting [99]. The polarization resistance was present for a maximum of 3 h for the polished sample in contrast to a maximum of 2 h for rougher ground surface. This observation agrees with the assumption that polished samples have a higher passivation. Thus polishing samples reduces the degradation behavior, especially at the beginning of immersion. The SEM results support these findings. After 2 h immersion, a general degradation for both finishes was observed. After 6 h, the rougher surface clearly showed more pitting, while the few pits seen on the polished surface indicated the start of pit formation. Pitting was studied for both surfaces after 12 h immersion. The initial pitting of the ground surface had progressed further compared to the polished finish. Walter et al. [71] explained this observation as a local pH drop caused by deep valleys in the rough surface. Additionally, the passivation layer of the rough surface broke up earlier than the smooth surface [71].

Alvarez et al. [94] found that polished AE44 samples encouraged more pitting compared to semi-polished samples in an immersion test. At the beginning of degradation, polished samples exhibited a higher pit volume compared to semi-polished samples. However, semi-polished samples had higher pit radii. The smoother surface of the polished sample and its related chlorine absorption capacity is given as a possible explanation. This behavior is distinct from reports that report that rougher surfaces on steel [75,76] and aluminum [77] lead to faster degradation and more pitting.

Walter and Kannan [96] suggested a change of the passivation layer provoked by shifting the local pH, initiated by aeration of the solution, as a reason for this behavior [96]. Pitting for both conditions leads to intergranular degradation after several hours. However, the start of intergranular degradation of the polished samples started earlier than for semi-polished samples [94].

Lorenz et al. [74] showed that surface roughness does not only influence the degradation resistance of pure magnesium; it also affects the cell (HeLa cells/GSP-C12 mouse fibroblasts) adhesion on the surface of Mg. For this study, discs were prepared with 600 paper grit size, a combination of 6 µm diamond paste, and an ethanol/glycerol solution. Sample cleaning was performed using an ultrasonic bath filled with ethanol for 3 min. In order to analyze the effects of surface morphology on cell adherence, one series of samples was immersed in 1 mol NaOH for 24 h and another series in modified simulated body fluid (M-SBF) at 37 °C for 5 d. Afterwards, the samples were flushed with ethanol and dried in air. The roughness increased after immersion in both solutions, but especially for the modified simulated body fluid (M-SBF) solution. pH measurements were also carried out on Mg samples degrading in a Minimum Essential Media (MEM) that included fetal bovine serum (FBS). A pH of 8.96 was observed for the M-SBF treated Mg samples and was higher compared to the other treatments after 2 h (Tables 1 and 4, Ref. [74]). The thicker Ca/Mg phosphate layer after M-SBF immersion does not protect the Mg sample due to its porosity, but the corrosion resistance increased by a factor of five compared to the untreated samples. In contrast, the cell density is higher compared to the polished and NaOH treated samples. The increase of the roughness by immersing in M-SBF improved the cell adhesion. The medium alkalization of the M-SBF samples is only suitable for short term applications. The smooth surface of the polished samples exhibits nearly no cell adherence and degrades very quickly. The passivation of Mg with NaOH reduces degradation, but cell adhesion is lower compared to M-SBF immersion [74].

Liu [61] compared the cell adherence of rolled pure Mg foils with an oxide layer and on ground foils without an oxide layer. He also studied the effects of roughness and degradation in Dulbecco's Modified Eagle's Medium (DMEM) and deionized (DI) water. The smoother ground samples varied by only 1.2% (oxide layer 13.6% and on ground foils 14.8%) in cell density from the rough oxidized samples (Tables 1 and 2, Ref. [61]). As such, it can be assumed, that surface roughness did not affect the cell density. No correlation between roughness and degradation rate was found, though ground samples in DMEM showed a slower degradation rate compared to oxidized samples. The opposite behavior was observed in DI water [61].

In contrast to the studies carried out by Liu [61] and Lorenz et al. [74], it was reported by Johnson et al. [21] that ground Mg-4Y samples demonstrated a better cell attachment than samples with an electrical discharged machined (EDM) surface. The roughness of the surfaces could be a possible explanation, as it was found that rough surfaces degrade faster than smooth surfaces [21,72,93]. Mg-4Y exhibits a contrary degradation behavior as pure Mg [61]. The ground surface leads to a lower mass loss in DI water which is opposite to the higher mass loss in DMEM (Tables 1 and 2, Ref. [21]). This effect is not thought to be due to roughness, but rather, from a different evolution of the pH under the different testing conditions.

Song and Xu [38] investigated the effect of tempering (HT), sandblasting, grinding, and etching on the degradation resistance of the alloy AZ31. Tempering and sandblasting reduced degradation resistance, while grinding or acid etching as a cleaning procedure decreased weight loss and hydrogen evolution. Heat treatment led to the precipitations of large Al-Mn-Fe particles which deteriorated the degradation resistance [38]. In addition to impurities, it is also known that roughness influences the degradation rate [100]. The roughness Ra (two-dimensionally determined roughness, arithmetic mean deviation of the roughness profile) and the hydrogen evolution of ground samples were very low compared to sandblasted samples (Tables 1 and 2, Ref. [38]). Sandblasting led to a very rough surface, accompanied by micro stresses in the surface layer. The Fe impurities rather than this surface roughening increased the degradation rate. Grinding the surface removes contaminations and leads to a slower degradation rate. Though ground surfaces are smoother than etched, the effect on the degradation is not as significant as removing a significant amount of Fe [38].

Zhao and Zhu [6] investigated, in addition to collagen monomer concentration, pH, and assembly time, the influence of ground surfaces on collagen fibril formation. They tested different surface finishes and the collagen formation with subsequent cell attachment. They ground Mg and AZ31 discs with 180 (Ra = 1.89 μm), 800 (Ra = 0.29 μm) and 1200 (Ra = 0.15 μm) SiC paper and apply 50 μL of 200 μg/mL D-phosphate-buffered solution (DPBS) diluted collagen solution for 2 h on the samples. By SEM they observed the morphology of collagen fibers for every surface finish and alloy. A clear difference of collagen formation was visible from roughest to smoothest surface for both alloys. While the collagen formation on both alloys for both smoothest surfaces was comparable, the roughest surface of AZ31 showed less dense structure in contrast to Mg. The roughest surface of both alloys adsorbed the highest amount of collagen after 2 h, while the smoothest surface showed the lowest adsorbed amount. This trend was more distinct for Mg compared to AZ31. Cell attachment observations after 2 h lead to the assumption that cells were more attached on collagen treated and smoother samples in contrast to the roughest surface finish. The roughest surface finish with a more fiber woven structure and highest collagen adsorption also showed in another Fluorescent live/dead cell analysis that, after one day, the collagen structure on a rough surface is more detrimental to cell density, independent of the alloy [6]. Nudelman et al. [101], reported a correlation between collagen and cell attachment [101]. In contrast to Nudelman et al. [101], Zhao and Zhu [6] evidenced a decrease in cell density with higher collagen adsorption. For this reason, it is assumed, that more collagen does not always result in a stronger cell attachment. In reference to roughness, this study shows an effect on the collagen formation which influences the cell density indirectly [6].

2.2. Burnishing

A comparison between ground and burnished sample degradation was performed using hydrogen evolution, PDP, and white light interferometry on the AZ31B alloy [39]. Ground and burnished (dry and cryogenic) samples had a very similar roughness before degradation. Burnishing was performed using a severe plasticity burnishing (SPB) process. Cryogenic burnishing is distinguished from dry burnishing by the use of liquid nitrogen. After degradation, the morphology of the ground samples differed from the burnished samples. Thus, roughness will not be the only factor to influence the degradation behavior. In addition, grain size and basal texture had an influence on the uniformity and amount of degradation. Dry and cryogenic burnishing decreased hydrogen evolution over a 7 h period with respect to ground surfaces. Both burnishing processes lead to a smoother finish with reduced pit depth and pit volume compared to the 4000 grit size paper treatment immersion test. The PDP analysis shows the same trend for both burnished surfaces with a higher degradation resistance, indicated by a broader capacitive loop [39]. The crystallographic orientation and grain refinement has to be considered, as well as the surface roughness [38]. In this report, the surface roughness did not affect the degradation, in disagreement with the prediction of Song and Xu [38]. Moreover, a small grain size and a strong basal texture led to a higher degradation resistance [39].

Table 1. Overview of different studies concerning grinding and polishing that consider roughness and degradation behavior. * Values were determined from the diagram with the corresponding reference.

Alloy	Sample	Experiment	Solution	Time	Grinding/ Polishing	Initial Roughness	Results	Ref
Mg	Disk	pH	MEM [1]	2 h	Polishing: 6 μm + lubricant	Ra = 0.10 μm	pH = 8.01	[74]
		Cell viability	MEM [1] +FBS [2]	24 h			* CD [3] = 10 cells/mm^2	
	Foil	Mass loss	DMEM [4] + 10% FBS [2]	up to 80 d	1200 grit	-	Max DR [5] = 0.09 mg cm^{-2} d^{-1}	[61]
			DI [6] Water				Max DR [5] = 0.28 mg/cmWütr^2/d	
		Cell adhesion	DMEM [4] + 10% FBS [2]	24 h	1200 grit	-	14.8% cell adhesion	
	Disk	Collagen quantification, cell attachment	-	2 h	180 grit	Ra = 1.89 μm	Trend of higher collagen adsorption with higher Ra, lowest CD [3] for highest Ra.	[6]
					800 grit	Ra = 0.29 μm		
					1200 grit	Ra = 0.15 μm		
AE44	Plate	Immersion	3.5 wt.% NaCl	4 and 12 h, 1.5 and 2.5 d	1400 grit	-	intergranular degradation started earlier after polishing	[94]
AZ31	-	PDP [7]	0.9 wt.% NaCl	-	P1000 emery Paper	Ra = 0.33 μm	i_{corr} = 3.64 μAcm^{-2}	[102]
		EIS [8]					Rp = 934 Ωcm^2	
	Sheet	Hydrogen	5 wt.% NaCl	24 h	1200 grit	Ra = 0.07 μm	1.11 mg/dcm^2	[38]
	Disk	PDP [7] (1 cm^2)	PBS [9]	-	1200 grit	Sa = 48.58 ± 23.45 nm	i_{corr} = 34.5 ± 3.5 μA cm^{-2}	[103]
							CR [10] = 0.76 ± 0.06 mm/y	
		Cytotoxicity	α-MEM [11]	21 d	1200 grit	Sa = 48.58 ± 23.45 nm	* Cell survival: 92 %	
	Disk	Immersion, Hydrogen, PDP [7]	5 wt.% NaCl	30, 200 h Immersion, 7 h Hydrogen	4000 grit	Ra = 0.2 μm	Burnishing lead to a better corrosion behavior	[39]
					Dry Burnishing			
					Cryogenic Burnishing			
	Disk	Collagen quantification, cell attachment	-	2 h	180 grit	Ra = 1.89 μm	Trend of higher collagen adsorption with higher Ra, lowest CD [3] for highest Ra.	[6]
					800 grit	Ra = 0.29 μm		
					1200 grit	Ra = 0.15 μm		

Table 1. *Cont.*

Alloy	Sample	Experiment	Solution	Time	Grinding/Polishing	Initial Roughness	Results	Ref
AZ91	-	EIS [8] (0.785 cm^2)	SBF [12]	12 h	120 grit	Sa = 0.022 μm	passivation layers on smoother surfaces last longer	[71]
					Polishing: 3 μm	Sa = 0.973 μm		
	-	PDP [7] (0.75 cm^2)	0.5 wt.% NaCl	-	320 grit	Sa = 0.430 μm	i_{corr} = 6.92 μA cm^{-2}	[96]
					600 grit	Sa = 0.248 μm	i_{corr} = 4.79 μA cm^{-2}	
					1200 grit	Sa = 0.145 μm	i_{corr} = 3.73 μA cm^{-2}	
					Polishing: 3 μm	Sa= 0.08 μm	i_{corr} = 2.19 μA cm^{-2}	
	Disk	PDP [7] (1 cm^2)	PBS [9]	-	1200 grit	Sa = 29.76 ± 12.69 nm	i_{corr} = 36.6 ± 3.2 μA cm^{-2}	[103]
							CR [10] = 0.78 ± 0.07 mm/y	
		Cytotoxicity	α-MEM [11]	21 d	1200 grit	Sa = 29.76 ± 12.69 nm	* Cell survival: 87 %	
M-4Y	Disk	Mass loss	DMEM [4] + 10% FBS [2]	9.04 d	1200 grit	Ra = 65 ± 31 nm	* 89.7 %	[21]
			DI [6] Water				* 0.33 %	
		pH	DMEM [4] + 10% FBS [2]	24 h	1200 grit	Ra = 65 ± 31 nm	* pH = 8.32	
			DI [6] Water				* pH = 9.00	
		Cell adhesion	DMEM [4] + 10% FBS [2]		1200 grit	Ra = 65 ± 31 nm	* 22.4 %	
ZK60A	Disk	PDP (1 cm^2)	PBS [9]	-	1200 grit	Sa = 78.30 ± 21.63 nm	i_{corr} = 32.3 ± 2.6 μA cm^{-2}	[103]
							CR [10] = 0.68 ± 0.01 mm/y	
		Cytotoxicity	α-MEM [11]	21 d	1200 grit	Sa = 78.30 ± 21.63 nm	* Cell survival: 32 %	
Mg-0.5Ca-6Zn	Rectangular prism	PDP [7] (1 cm^2)	Kokubo	-	2000 grit	Rq = 210 nm	i_{corr} = 365 μA cm^{-2}	[104]
							CR [10] = 8.34 mm/y	
		Hydrogen		10 d	2000 grit	Rq = 210 nm	4.92 mL/cm^2/d	
WE43	Plate		SBF [12]	-	Polishing: 1 μm	-	i_{corr} = 642 ± 125 μA cm^{-2}	[105]

Table 1. *Cont.*

Alloy	Sample	Experiment	Solution	Time	Grinding/ Polishing	Initial Roughness	Results	Ref
Mg-1.0Ca	Rectangular prism	Mass loss	SBF [12]	3 d	1200 grit	Sa = 4.67 nm	* 9.63 mg	[88]
		Cell viability	Extract DMEM [4] + 10% FBS [2]	3 d + 4 h	1200 grit	Sa = 4.67 nm	* 100%	
		EIS [8] (10 × 10 mm^2)	SBF [12]	-	1200 grit	Sa = 4.67 nm	$i_{corr} = 2.3 \times 10^2$ μA cm^{-2}	
Mg-0.5Sr		Mass loss	SBF [12]	3 d	1200 grit	Sa = 2.16 nm	* 14.3 mg	
		Cell viability	Extract DMEM [4] + 10% FBS [2]	3 d + 4 h	1200 grit	Sa = 2.16 nm	* 100 %	
		EIS [8] (10 × 10 mm^2)	SBF [12]	-	1200 grit	Sa = 2.16 nm	$i_{corr} = 1.0 \times 10^3$ μA cm^{-2}	

[1] MEM: Minimum Essential Media; [2] FBS: fetal bovine serum; [3] CD: Cell density; [4] DMEM: Dulbecco's Modified Eagle's Medium; [5] DR: degradation rate; [6] DI: deionized; [7] PDP: potentiodynamic polarization; [8] EIS: Electrochemical Impedance Spectroscopy; [9] PBS: phosphate buffered saline; [10] CR: corrosion rate; [11] α-MEM: MEM alpha modification Media; [12] SBF: simulated body fluid.

2.3. Machining

Turned, threaded and sandblasted Mg-0.8Ca samples were examined and tested in vivo [93]. Smooth (Ra = 3.65 μm) turned and threaded samples exhibited the best interlocking connection between the bone and implant. Rough (Ra = 32.7 μm) sandblasted rods degraded most rapidly with the highest number of visible gas bubbles. Turned surfaces led to the lowest gas evolution and decomposition in these studies [87,93]. Despite a similar integration of threaded and turned implants into the bone tissue, threaded implants showed a non-uniform bone resorption at the thread edges [93]. This is in agreement with the findings of Walter et al. [71], which may be explained by local variations in pH.

Mhaede et al. [102] reported a relationship between roughness and corrosion resistance. For the degradation test in 0.9 wt.% NaCl solution, eight different conditions of AZ31 alloy were prepared. Samples were either ground or shot-peened with 3 different Almen intensities (saturation value of residual arc height of an Almen strip, established by John Almen) [102,106,107], ground and coated, or shot-peened with 3 different Almen intensities and coated without prior grinding. Shot peening was performed with ceramic shot which had a diameter of 850 μm (Z850). The dicalcium phosphate dihydrate (DCPD) coating used was produced by electro-deposition of samples in a 0.1 mol $Ca(NO_3)_2$ $4H_2O$ + 0.06 mol NH_4HPO_4 solution. In Table 2, Ref. [102] it is shown that the current density i_{corr} for the shot-peened (SP) samples was increased compared to the other conditions in Table 4, Ref. [102]. It has been shown that having a rough surface after shot-peening affects i_{corr} (Table 2, Ref. [102] and Figure 2), as the resulting greater surface area increases the surface reactivity [102]. However, it is not possible to relay Ra linear to i_{corr} (Tables 1, 2 and 4, Ref. [102] and Figure 2) for all surface finishes due to the protective properties of the DCPD coating compared to only shot peened samples. The linear relationship between Ra and i_{corr} for the shot-peened and shot-peened/coated samples (Figure 2) agrees with the study of Walter and Kannan [96], whereas linear correlation was observed for only the ground samples. However, it should be noted that higher deformation and internal stress, arising from higher Almen intensities, could also affect the degradation behavior.

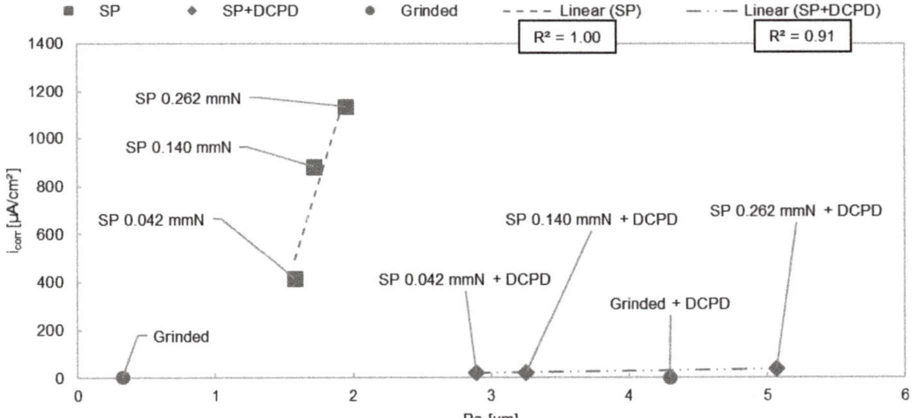

Figure 2. Diagram current density i_{corr} against roughness Ra for ground, shot peened, and shot peened + DCPD coated samples. A non-linear R^2 relationship between Ra and i_{corr} is shown by comparing all conditions (ground/SP/SP+DCPD) together. Linear R^2 is plotted for only shot peened or only shot peened and coated samples. A trend of linearity can be seen only for roughness values arising from same surface treatments. Ra and i_{corr} values were obtained from [102].

Denkena and Lucas [108] studied the surface and subsurface properties after turning and deep rolling a Mg-3Ca alloy. Three different conditions for each machining process were investigated. With regard to turning, the roughness Rz (distance from deepest valley to highest peak within sample length from a linear measurement) [109] decreased from around Rz ~ 4.48 μm to Rz ~ 3.75 μm after increasing the cutting speed from 10 m/min to 100 m/min at constant cutting depth and feed rate. By reducing the feed rate from 0.1 mm to 0.05 mm at a constant cutting speed (100 m/min) and cutting depth (0.5 mm), the roughness (Rz ~ 2.17 μm) was reduced (Table 2, Ref. [108]). After deep rolling with different rolling forces (F = 50 N, 200 N, 500 N) and constant rolling speed (25 m/min) and feed rate (0.1 mm), no significant change in roughness (Rz ~ 0.91–1.26 μm) occurred (Table 2, Ref. [108]). Degradation tests were performed in 0.9 wt.% NaCl solution and the hydrogen gas evolution was measured. The mass loss was calculated from the amount of hydrogen produced and a correction factor. The degradation rates for turning with higher roughness were greater compared to the degradation rates after rolling. However, for turning, the condition with the highest roughness showed the lowest mass loss compared to the smoother samples. For deep rolling, the condition with the lowest rolling force led to the highest, while not signifying mass loss (calculated from hydrogen generation) after around 240 h exposure time compared to conditions with higher rolling forces and comparable Rz values. No significant correlation between roughness and mass loss was found. High residual compressive stress was reported to reduce the degradation rate by about 100 times [108], and the degradation results were comparable with the results from high speed dry milled Mg-0.8Ca with the lowest roughness [110].

The influence of machining and deep rolling on Mg-3Ca and Mg-0.8Ca was analyzed by Denkena et al. [111], and the results were compared to those of Denkena and Lucas [108]. Only 3 conditions per alloy were tested. The turning was carried out with a cutting speed of 100 m/min, cutting depth of 200 μm, and a feed rate of 0.1 mm. Two deep rolling conditions were studied with rolling forces of 50 N and 200 N and the same cutting speed and feed as described by Denkena and Lucas [108]. The roughness Rz after turning and deep rolling was comparable for each alloy for every machining process. The roughness Rz after turning was about 4 μm, while Rz for deep rolling resulted in a lower Rz of between 0.44–0.76 μm (Table 2, Ref. [111],[108]) compared to Denkena and Lucas [108]. The corrosion behavior was tested by hydrogen evolution and performed in a 0.9 wt.% NaCl solution and μ-CT. It was shown that turned Mg-3Ca with the highest Rz resulted in the highest hydrogen evolution (~ 20.2 mL/cm^2 after 29 h) and greatest degradation in μ-CT compared to deep rolled samples (~0.76–1.27 mL/cm^2 after 29 h). For the Mg-0.8Ca alloy the hydrogen evolution (~5.42–6.22 mL/cm^2 after 29 h) showed no significant dependence on the method of machining. From these investigations, it is possible to say that roughness had no influence on the degradation behavior. Rather than roughness, a high compressive stress and the Mg_2Ca phase in the Mg-3Ca alloy was reported to affect the degradation behavior [111].

Table 2. Overview of different studies investigating SFF and machining and their influence on the degradation behavior. * Values were determined from the diagram with the corresponding reference.

Alloy	Sample	Experiment	Solution	Time	Machining	Initial Roughness	Results	Ref
Mg	-	PDP [1]	HBSS [2] + HEPES [3]	6 h	SFF [4]	Ra =0.59 ± 0.04 µm	i_{corr} = 94.52 µAcm^{-2}	[72]
						Ra = 2.68 ± 0.74 µm	i_{corr} ~ 189.04 µAcm^{-2}	
						Ra = 9.12 ± 0.44 µm	i_{corr} ~ 567.12 µAcm^{-2}	
		Mass loss				Ra =0.59 ± 0.04 µm	2.74 mg cm^{-2} d^{-1}	
						Ra = 2.68 ± 0.74 µm	28.43 mg cm^{-2} d^{-1}	
						Ra = 9.12 ± 0.44 µm	130.12 mg cm^{-2} d^{-1}	
	Foil	Mass loss	DMEM [5] + 10% FBS [6] + P/S [7]	80 d	Rolling	-	Max DR [8] = 1.2 mg cm^{-2} d^{-1}	[61]
			DI [9] Water				Max DR [8] = 0.14 mg cm^{-2} d^{-1}	
		Cell adhesion	DMEM [5] + 10% FBS [6] + P/S [7]	24 h	Rolling	-	13.6 % cell adhesion	
AZ31	Sheet	Hydrogen	5 wt.% NaCl	1.5 h	Milling	Ra = 2.02 µm	54.23 mg/dcm^2	[38]
				0.25 h	HT [10] + SB60 [11]		563.49 mg/dcm^2	
	-	PDP [1]	0.9 wt.% NaCl	-	SP [12] 0.042 mmN	Ra = 1.58 µm	i_{corr} = 416.17 µAcm^{-2}	[102]
					SP [12] 0.140 mmN	Ra = 1.72 µm	i_{corr} = 882.77 µAcm^{-2}	
					SP [12] 0.262 mmN	Ra = 1.95 µm	i_{corr} = 1136.5 µAcm^{-2}	
	Sheet	Hydrogen	5 wt.% NaCl	6.55 h	Rolling	-	* CR [13] = 7.17 mg cm^{-2} d^{-1}	[95]
	Plate	PDP [1]	3.5 wt.%	-	SB40 [14]	-	i_{corr} = 2.1 µA cm^{-2}	[80]
M-4Y	Disk			217 h	EDM [15]	Ra = 196 ±47 nm	* 75.2 %	[21]
			DI [9] Water				* 45.9 %	
				24 h	EDM [15]	Ra = 196 ±47 nm	* pH = 8.48	
			DI [9] Water				* pH = 8.98	
		Cell adhesion	DMEM [5] + 10% FBS [6] + P/S [7]		EDM [15]	Ra = 196 ±47 nm	* 7.82 %	

Table 2. *Cont.*

Alloy	Sample	Experiment	Solution	Time	Machining	Initial Roughness	Results	Ref
Mg-3.0Ca	Cylinder	Mass loss From hydrogen generation	0.9 wt.% NaCl	93 h	Turning: ap = 0.5 mm, vc = 10 m/min, f = 0.1 mm	* Rz = 4.48 µm	* 0.89 g/cm^2	[108]
					Turning: ap = 0.5 mm, vc = 100 m/min, f = 0.1 mm	* Rz = 3.75 µm	* 1.35 g/cm^2	
					Turning: ap = 0.5 mm, vc = 100 m/min, f = 0.05 mm	* Rz = 2.17 µm	* 1.29 g/cm^2	
				240 h	Deep Rolling [16]: Fr = 50 N	* Rz = 1.26 µm	* 0.07 g/cm^2	
					Deep Rolling [16]: Fr = 200 N	* Rz = 0.91 µm	* 0.02 g/cm^2	
					Deep Rolling [16]: Fr = 500 N	* Rz = 1.26 µm	* 0.02 g/cm^2	
	Cylinder	Hydrogen evolution	0.9 wt.% NaCl	29 h	Turning: ap = 200 µm, vc = 100 m/min, f = 0.1 mm	* Rz = 3.98 µm	* 20.2 mL/cm^2	[111]
					Deep Rolling [16]: Fr = 50 N	* Rz = 0.63 µm	* 1.27 mL/cm^2	
					Deep Rolling [16]: Fr = 200 N	* Rz = 0.47 µm	* 0.76 mL/cm^2	
		µ-CT	0.9 wt.% NaCl	29 h	Turning: ap = 200 µm, vc = 100 m/min, f = 0.1 mm	* Rz = 3.98 µm	* PV [17] = 19.6 mL	
					Deep Rolling [16]: Fr = 50 N	* Rz = 0.63 µm	* PV [17] = 1.44 mL	
					Deep Rolling [16]: Fr = 200 N	* Rz = 0.47 µm	* PV [17] = 1.05mL	

Table 2. *Cont.*

Alloy	Sample	Experiment	Solution	Time	Machining	Initial Roughness	Results	Ref
Mg-0.8Ca	Cylinder	Hydrogen evolution	0.9 wt.% NaCl	29 h	Turning: ap = 200 μm, vc = 100 m/min, f = 0.1 mm	* Rz = 4.00 μm	* 6.18 mL/cm^2	[111]
					Deep Rolling [16]: Fr = 50 N	* Rz = 0.44 μm	* 5.42 mL/cm^2	
					Deep Rolling [16]: Fr = 200 N	* Rz = 0.76 μm	* 6.22 mL/cm^2	
		μ-CT	0.9 wt.% NaCl	29 h	Turning: ap = 200 μm, vc = 100 m/min, f = 0.1 mm	* Rz = 4.00 μm	* PV [17] = 16.3 mL	
					Deep Rolling [16]: Fr = 50 N	* Rz = 0.44 μm	* PV [17] = 12.1 mL	
					Deep Rolling [16]: Fr = 200 N	* Rz = 0.76 μm	* PV [17] = 6.71 mL	
		Rabbit, μ-CT	-	3 and 6 months	Turning	Ra = 3.65 μm	Turning lead to the lowest gas evolution and decomposition	[93]
					Sand milling	Ra = 32.7 μm		
					Threading	-		
Mg-5Gd	Disk	Mass loss	DMEM [5] + 10% FBS [6] + P/S [7]	30 d	Milling	Sa =1.6 μm	CR [13] = 0.50 μm/d	[57]

[1] PDP: potentiodynamic polarization; [2] HBBS: Hank's Balanced Salt Solution; [3] HEPES: Biological buffer for cell culture media; [4] SFF: indirect solid free-form fabrication; [5] DMEM: Dulbecco's Modified Eagle's Medium; [6] FBS: fetal bovine serum; [7] P/S: Penicillin/Streptomycin; [8] DR: degradation rate; [9] DI: deionized; [10] HT: heat treated at 450 °C for 10 min (tempering); [11] SB60: sandblasting with glass bead type MS-6 at 60 psi; [12] SP: shot peening; [13] CR: corrosion rate; [14] SB40: sand blasting by #40 SiO_2; [15] EDM: wire electrical discharge machining; [16]: Deep Rolling: with vr = 25 m/min and Fr = 0.1 mm.

3. Chemical Surface Treatments and Coatings

3.1. Acid Etching

The treatments reported by Supplit et al. [95] indicated that it was possible to improve the degradation resistance of rolled AZ31 alloy by pickling with different acids like acetic acid, phosphoric acid, nitric acid, and hydrofluoric acid [95]. Acid pickling, especially with acetic acid decreased the degradation rate from 7.17 mg cm^{-2} d^{-1} (rolled condition) to 0.70 mg cm^{-2} d^{-1}. The second best etching method was found to be phosphoric acid. The degradation rates were determined by measuring hydrogen gas evolution in 5 % NaCl. A rougher surface after pickling with acetic acid was observed when compared to the other etching solutions. The samples with the lowest degradation rates had rougher surfaces, an observation that contradicts the findings of Nguyen et al. [72].

Organic acids like acetic, citric, or oxalic, and inorganic acids like phosphoric, nitric, and sulfuric acid were used to treat AZ31 alloy by Nwaogu et al. [40]. The aim was to remove contamination and impurities from resulting from rolling. After etching, 1–20 μm was removed from the surface. It was observed that more material was removed as the etching time increased. A roughness analysis showed that the roughness value Ra after etching is higher than Ra of rolled samples. Removing 5 μm of material generally reduced the number of Ni impurities. However, Fe impurities still remained at the surface even after material had been removed. To determine the degradation behavior, a 48 h salt spray test was used as a screening test. The lowest degradation rates were obtained from samples with the lowest impurity levels which had the greatest amount of material removed. Acetic acid-etched samples had the slowest degradation rates. EIS measurements supported the finding that acetic acid etching leads to the resistance, due to having the highest polarization resistance (Rp) [40]. This finding is in agreement with the results obtained by Supplit et al. [95] and Nwaogu et al. [40], who showed that a low impurity level and a 5 μm etching depth improved the degradation behavior. When more than 5 μm of material was removed, the surface became rougher [40].

The change in roughness after inorganic acid etching confirmed this finding [56]. For sulfuric acid etching, Ra (>2 μm) was much higher compared to other inorganic etching solutions when 7 μm of material had been removed. In addition, sulfuric acid etching leads to a lower degradation resistance in spite of the resulting low level of impurities. Degradation of sulfuric acid-etched material mostly results from galvanic degradation initiated by second phases. Though the effect of roughness were not the main focus in these investigations, nitric acid etching showed a high degradation resistance for a surface with an initially uniform roughness distribution and low roughness value [56]. Thus, the roughness of a sample after etching could also be a parameter which has to be considered in order to determine the full degradation behavior.

Gawlik et al. [57] measured the roughness after acetic acid etching with various combinations of acid concentration and immersion time. The surface roughness Sa increased after etching compared to the milled surface and varies with different conditions. After 30 days immersion, the same degradation rate was determined for all etched conditions (Table 3, Ref. [57]), in spite of different Sa and Sq (root mean square value of surface deviations) [98] values after etching. This leads to the conclusion that the initial roughness of the sample has no long-term effect on degradation. The varying surface morphology and near-surface deformation arising from milling also affected the degradation rate [57].

Similarly to Nwaogo et al. [40,56], Song and Xu [38] described that Fe impurities accelerate the degradation of the AZ31 alloy. As such, as-received samples and heat-treated samples have lower degradation resistance compared to ground and sulfuric acid etched samples, due to Fe particles remaining on the surface. Sulfuric acid etching roughens the surface much more than grinding, but both conditions lead to a similar degradation rate. Thus, the roughness of the etched samples itself does not contribute to the degradation rate. Acid cleaning removes contamination and the deformation zone arising from processing, and thus directly impacts the degradation behavior [38].

Table 3. List of different studies on acid etching being used as a surface treatment. * Values were determined from the diagram with the corresponding reference.

Alloy	Sample	Experiment	Solution	Time	Acid Etching	Initial Roughness	Results	Ref
AZ31	Sheet	SST [1]	5 wt.% NaCl	48 h	50 g/L H_2SO_4 [2], 15s	* Ra = 0.98 µm	CR [3] = 2.20 ± 0.18 mm/y	[56]
					80 g/L HNO_3 [4], 120 s	* Ra = 0.23 µm	CR [3] = 0.51 ± 0.10 mm/y	
					80 g/L H_3PO_4 [5], 60s	* Ra = 0.49 µm	CR [3] = 0.74 ± 0.31 mm/y	
		SST [1]	5 wt.% NaCl	48 h	300 g/L CH_3COOH [6], 120s	* Ra = 0.61 µm	CR [3] = 0.34 ± 0.08 mm/y	[40]
					80 g/L $C_2H_2O_4$ [7], 30s	* Ra = 0.48 µm	CR [3] = 0.59 ± 0.11 mm/y	
					80 g/L $C_6H_8O_7$ [8], 60s	* Ra = 0.34 µm	CR [3] = 0.72 ± 0.07 mm/y	
		Hydrogen	5 wt.% NaCl	~48 h	20% CH_3COOH [6], 30 s	-	* CR [3] = 0.70 mg cm^{-2} d^{-1}	[95]
				~30 h	50% H_3PO_4 [5], 30 s		* CR [3] = 1.58 mg cm^{-2} d^{-1}	
				~11 h	3.3% HNO_3 [4], 20 s		* CR [3] = 4.59 mg cm^{-2} d^{-1}	
				~23 h	12% HF [9], 1200 s		* CR [3] = 1.68 mg cm^{-2} d^{-1}	
		Hydrogen	5 wt.% NaCl	24 h	HT [10] + 10 % H_2SO_4 [2], 20 s	Ra = 2.50 µm	0.97 mg/dcm^2	[38]
	Foil	Immersion	SBF [11]	14 d	90% H_3PO_4 [5], 30 s	-	* CR [3] = 8.27 mg/d	[70]
Mg-5Gd	Disk	Mass loss	DMEM [12] + 10% FBS [13] + P/S [14]	30 d	150 g/L CH_3COOH [6], 150 s	Sa = 6.3 µm	CR [3] = 0.31 µm/d	[57]
					250 g/L CH_3COOH [6], 150s	Sa = 5.6 µm	CR [3] = 0.30 µm/d	
					300 g/L CH_3COOH [6], 90 s	Sa = 2.3 µm	CR [3] = 0.30 µm/d	

[1] SST: salt spray test; [2] H_2SO_4: sulfuric acid; [3] CR: corrosion rate; [4] HNO_3: nitric acid; [5] H_3PO_4: phosphoric acid; [6] CH_3COOH: acetic acid; [7] $C_2H_2O_4$: oxalic acid; [8] $C_6H_8O_7$: citric acid; [9] HF: hydrofluoric acid; [10] HT: heat treated at 450 °C for 10 min (tempering); [11] SBF: simulated body fluid; [12] DMEM: Dulbecco's Modified Eagle's Medium; [13] FBS: fetal bovine serum; [14] P/S: Penicillin/Streptomycin.

Gray-Munro et al. [70] also tested acetic treatments on AZ31 alloy. Gray and Luan's study [68] described that a strong passive oxide layer was formed during the etching process when compared to the as-received state. Gray-Munro et al. [70] found that as-received samples have a greater non-uniform morphology in comparison with phosphoric acid etched samples. Phosphoric acid treated samples showed a lower degradation rate of 8.27 mg/d compared to non-etched samples (~31 mg/d). Additionally, the modified surface after etching improves adhesion and minimizes the porosity of coatings [70].

3.2. Coatings

Gray-Munro et al. [70] discovered that biomimetic calcium phosphate coatings (Ca/P) and polymer coatings after phosphoric etching led to a uniform morphology, which in turn led to a uniform degradation over the surface of the AZ31 alloy. The degradation rates of 6.17 mg/d (Table 4, Ref. [70]) for a polymer poly(L-lactic acid) (PLA) coating and of 3.83 mg/d for a poly (desaminotyrosyl tyrosine hexyl) (DTH) carbonate coating are compared to etched samples with rates of 8.27 mg/d (Table 3, Ref. [70]) (sample size: 1 mm thick foil, 10 mm × 20 mm). The degradation rate (7.27 mg/d) resulting from the Ca/P coated samples did not strongly differ from the only etched alloys [70]. However, the Ca/P-coated Mg alloy exhibited non-toxic and biocompatible properties. Ca/P enhanced the osseointegration and bioresorption of the alloy in a physical environment [112–116], which is why a Ca/P coating is more favorable compared to phosphoric etching.

Bakhsheshi-Rad et al. [104] performed potentiodynamic polarization (PDP) tests and immersion tests in SBF (Kokubo solution) on polished Mg-0.5Ca-6Zn samples with and without a coating. The coatings tested were a fluoride conversion coating, a dicalcium phosphate dihydrate/magnesium fluoride (DCPD/MgF_2) coating, and a nano-hydroxyapatite/magnesium fluoride (nano-HA/MgF2) coating. A higher root mean square roughness (Rq) was measured for polished Mg-0.5Ca-6Zn samples with either a DCPD/MgF2 coating (Rq = 395 nm) or a nano-HA/MgF2 coating (Rq = 468 nm). The Rq for polished Mg-0.5Ca-6Zn (Rq = 210 nm) samples without coating and polished Mg-0.5Ca-6Zn samples with a fluoride coating (Rq = 280 nm) were somewhat lower. As seen in Table 4, Ref. [104], hydrogen evolution, i_{corr} and the degradation rate declined as the surface roughness increased, in contrast to studies of Walter and Kannan [96] and Mhaede et al. [102]. As shown in Table 4, Ref. [104], the degradation rate after coating compared to non-coated Mg alloy in Table 1, Ref. [104] is about a factor of 60 smaller, even though Rq only differs by 70 nm [104]. Thus, the protective coatings have a greater influence on corrosion than the Rq values.

Pompa et al. [103] investigated the morphology, surface roughness, cell viability, and degradation rate of ground and anodized AZ31B, AZ91E, and ZK60A alloys. Grinding was performed with a 1200 grit size grinding paper. Anodization was carried out using a mixture of alcohol and organic acid. The surface roughness increased dramatically from Sa = 29.76 nm to Sa = 204.81 nm after anodization for the AZ91E alloy. The anodization of AZ31B (Sa = 48.58 μm) and ZK60A (Sa = 78.30 μm) did not change the roughness significantly. It was shown that anodizing decreased the degradation rate compared to a ground surface. No correlation between surface roughness and degradation rate was found. This may be due to corrosion resistances being similar for all anodized surfaces despite their variation in roughness [103].

In the study of Chiu et al. [80], AZ31 plates were arc sprayed and hot pressed. Anodizing with oxalic acid was then performed [80]. The current density i_{corr} decreased after a combination of spraying and hot pressing or additional anodizing (Table 4, Ref. [80]) as compared to uncoated sandblasted samples (Table 2, Ref. [80]). Hot pressing decreased the surface roughness, which seems to improve the acid treatment afterwards. No correlation between roughness and degradation resistance was reported [80].

Table 4. Overview of different Coatings studies involving roughness and degradation behavior. * Values were determined from the diagram with the corresponding reference.

Alloy	Sample	Experiment	Solution	Time	Coatings	Initial Roughness	Results	Ref
Mg	Disk	pH	MEM [1]	2 h	Polished + NaOH	Ra = 0.23 μm	pH = 7.88	[74]
					Polished + M-SBF [2]	Ra = 1.12 μm	pH = 8.96	
		Cell viability	MEM [1] + FBS [3]	24 h	Polished + NaOH	Ra = 0.23 μm	* CD [4] = 177 cells/mm^2	
					Polishing + M-SBF [2]	Ra = 1.12 μm	* CD [4] = 838 cells/mm^2	
AZ31	Foil	Immersion	SBF [4]	2 weeks	90% H_3PO_4 [5], 30 s + Ca/P	-	* CR [6] = 7.27 mg/d	[70]
					90% H_3PO_4 [5], 30 s + PLA [7]		* CR [6] = 6.17 mg/d	
					90% H_3PO_4 [5], 30 s + poly (DTH [8] carbonate)		* CR [6] = 3.83 mg/d	
	-	PDP [9]	0.9 wt.% NaCl	-	P1000 Ground + DCPD [10]	Ra = 4.29 μm	i_{corr} = 1.57 μA cm^{-2}	[102]
					SP [11] 0.042 mmN + DCPD [10]	Ra = 2.89 μm	i_{corr} = 20.03 μA cm^{-2}	
					SP [11] 0.140 mmN + DCPD [10]	Ra = 3.25 μm	i_{corr} = 21.96 μA cm^{-2}	
					SP [11] 0.262 mmN + DCPD [10]	Ra = 5.07 μm	i_{corr} = 38.12 μA cm^{-2}	
	disk	PDP [9] (1 cm^2)	PBS [12]	-	1200 grit + anodizing	Sa = 49.0 ± 10.2 nm	i_{corr} = 2.72 ± 0.8 μA cm^{-2}	[103]
							CR [6] = 0.06 ± 0.01 mm/y	
		Cytotoxicity	MEM [1] alpha modification Media	21 d	1200 grit + anodizing	Sa = 49.0 ± 10.2 nm	* Cell survival: 67 %	
	Plate	PDP [9]	3.5 wt.%	-	SB [13] + Al ASC [14]	Ra = 11.6 μm	i_{corr} = 2.4 × 10^2 μA cm^{-2}	[80]
					SB [13] + Al ASC [14] + PHP [15] (800 MPa)	Ra = 4.89 μm	-	
					SB [13] + Al ASC [14] + PHP [15] (1600 MPa)	-	i_{corr} = 0.8 μA cm^{-2}	
					SB [13] + Al ASC [14] + PHP [15] (2000 MPa)	Ra = 1.12 μm	-	
					SB [13] + Al ASC [14] + PHP [15] + 7 wt.% oxalic acid anodizing	-	i_{corr} = 3.7 × 10^{-2} μA cm^{-2}	

Table 4. *Cont.*

Alloy	Sample	Experiment	Solution	Time	Coatings	Initial Roughness	Results	Ref
AZ91	Plate	PDP [9]	3.5 wt.% NaCl	-	PEO [16] without $K_4P_2O_7$	-	i_{corr} = 19.6 μA cm^{-2}	[73]
					PEO [16] + 0.03 mol/L $K_4P_2O_7$		i_{corr} = 1.22 × 10^{-2} μA cm^{-2}	
					PEO [16] + 0.06 mol/L $K_4P_2O_7$		i_{corr} = 2.27 μA cm^{-2}	
					PEO [16] + 0.15 mol/L $K_4P_2O_7$		i_{corr} = 4.77 μA cm^{-2}	
	Plate	PDP [9]	3.5 wt.% NaCl	-	Polishing 0.5 μm Al_2O_3 + PEO [16]	Ra = 0.5 μm	i_{corr} = 7.26 × 10^{-3} μA cm^{-2}	[81]
					1000 grit + PEO [16]	Ra = 1.0 μm	i_{corr} = 5.17 × 10^{-2} μA cm^{-2}	
					100 grit + PEO [16]	Ra = 2.5 μm	i_{corr} = 0.38 μA cm^{-2}	
	Plate				1000 grit + PEO [16] without KF [17]	-	Rp = 8.28 × 10^3 mΩm^2	[79]
					1000 grit + PEO + KF [17]		Rp = 4.67 × 10^3 mΩm^2	
	disk	PDP [9] (1 cm^2)	PBS [12]	-	1200 grit + anodized	Sa = 204.8 ± 62.7 nm	i_{corr} = 2.50 ± 0.5 μA cm^{-2}	[103]
							CR [6] = 0.05 ± 0.01 mm/y	
		Cytotoxicity	MEM [1] alpha modification Media	21 d	1200 grit + anodized	Sa = 204.8 ± 62.7 nm	* Cell survival: 102 %	
ZK60A	disk	PDP [9] (1 cm^2)	PBS [12]	-	1200 grit + anodizing	Sa = 75.88 ± 34.49 nm	i_{corr} = 1.86 ± 0.2 μA cm^{-2}	[103]
							CR [6] = 0.04 ± 0.01 mm/y	
		Cytotoxicity	MEM alpha modification Media	21 d	1200 grit + anodizing	Sa = 75.88 ± 34.49 nm	* Cell survival: 30 %	
Mg-0.5Ca-6Zn	Rectangular prism	PDP [9] (1 cm^2)	Kokubo	-	2000 grit + 40% HF [18]	Rq = 280 nm	i_{corr} = 6.20 μA cm^{-2}	[104]
							CR [6] = 0.14 mm/y	
					2000 grit + DCPD [10] /MgF_2	Rq = 395 nm	i_{corr} = 5.72 μA cm^{-2}	
							CR [6] = 0.13 mm/y	
					2000 grit + HA/MgF_2	Rq = 468 nm	i_{corr} = 5.23 μA cm^{-2}	
							CR [6] = 0.11 mm/y	
		Hydrogen	Kokubo	240 h	2000 grit + 40% HF [18]	Rq = 280 nm	1.31 mL/cm^2/d	
					2000 grit + DCPD [10] /MgF_2	Rq = 395 nm	1.12 mL/cm^2/d	
					2000 grit + nano-(HA [19] /MgF_2)	Rq = 468 nm	0.85 mL/cm^2/d	

[1] MEM: Minimum Essential Media; [2] M-SBF: modified simulated body fluid; [3] FBS: fetal bovine serum; [4] CD: Cell density; [5] H_3PO_4: phosphoric acid; [6] CR: corrosion rate; [7] PLA: polymer poly(L-lactic acid); [8] DTH: desaminotyrosyl tyrosine hexyl; [9] PDP: potentiodynamic polarization; [10] DCPD: dicalcium phosphate dihydrate; [11] SP: Shot peening; [12] PBS: phosphate buffered saline; [13] SB: Sand blasting; [14] ASC:arc-spray coating; [15] PHP: post hot pressing; [16] PEO: plasma electrolytic oxidation; [17] KF: potassium fluoride; [18] HF: hydrofluoric acid; [19] HA: hydroxyapatite.

Yoo et al. [81] studied the effect of roughness on the degradation resistance of a plasma electrolytic oxidation (PEO) layer on a AZ91 alloy. Surfaces with various Ra were prepared by grinding and polishing. Afterwards, all samples were coated using the same PEO process. Due to the differing roughness values of the ground and polished surfaces, the coating differentiated in pore size as well. This affects the degradation process. The current density increased with higher initial Ra (Table 4, Ref. [81]). A salt spray test also showed that the amount of pitting increased with higher Ra after 120 h, which indicates that the surface roughness before PEO indirectly influences the degradation resistance [81].

Cho et al. [73] compared the degradation resistance of PEO coatings on AZ91 alloy for different amounts of potassium pyrophosphate in the electrolyte. The size of the pores increased as the amount of potassium pyrophosphate was increased. It was also reported that the surface roughness increased with increasing pore size. There was a trend between pore size, surface roughness, and i_{corr} for additions of potassium pyrophosphate (Table 4, Ref. [73]). The PEO coating for the potassium pyrophosphate-free electrolyte exhibited the lowest degradation resistance compared to the rougher coatings [73].

In contrast to Cho et al. [73], Hwang et al. [79] compared PEO coatings on AZ91 alloy with and without potassium fluoride in the electrolyte. In addition to varying the surface roughness, the evolution of the degradation resistance with coating time was also examined. The surface roughness increased for longer coating times. The roughness of the coated samples, after dipping in the potassium fluoride containing electrolyte, was higher compared to coatings dipped into potassium fluoride free electrolyte. The roughness increases due to pore size enlargement as reported by Cho et al. [73]. The degradation resistance of the coatings when exposed to potassium fluoride-containing electrolyte was higher than for potassium fluoride-free electrolyte. Hwang et al. [79] explained that the oxide thickness is the reason for the improved degradation resistance, and did not assess the influence of roughness on the degradation resistance, as investigated in Hwang et al. [82]. The effect of the PEO coating roughness on the degradation behavior was examined in Hwang et al. [82] with PDP and three different coating surface roughness Ra values. The surface roughness increased with increased pore size, as was also seen in Hwang et al. [79] and by Cho et al. [73].

3.3. Ion Implantation

Jamesh et al. [105] implanted Si ions from a plasma on polished WE43 plates. Atomic force microscope (AFM) measurements after polishing and Si implantation showed that the surface became smoother after the ion implantation process. The smoother Si-implanted surfaces had improved degradation resistance. However, the roughness did not vary enough between the polished and Si-implanted surface types to obtain a correlation between roughness and degradation [105].

Zhao et al. [88] reported a slower degradation rate for Mg-Ca and Mg-Sr alloys after ion implantation (Zr and O ions) onto their surfaces. After the surfaces were implanted, measurements determined that the surfaces were uniformly rough. The roughness increased after implantation for both alloys (compare Tables 1 and 5, Ref. [88]). The current density i_{corr} decreased for surfaces with higher roughness, and the cell adherence and proliferation improved [88].

Table 5. Summary of different studies concerning the influence of ion implantation on the degradation behavior. * Values were determined from the diagram with the corresponding reference.

Alloy	Sample	Experiment	Solution	Time	Implantation	Initial Roughness	Results	Ref
WE43	Plate	PDP [1]	SBF [2]	-	Polishing: 1 μm + Si ion plasma	-	$i_{corr} = 27 \pm 32$ μA cm^{-2}	[105]
Mg-1.0Ca	Rectangular prism	Mass loss	SBF [2]	3 d	1200 grit + Zr	Sa = 5.34 nm	* 8.03 mg	[88]
					1200 grit + ZrO	Sa = 9.42 nm	* 6.77 mg	
		Cell viability	Extract assay (DMEM [3])	24 h + 72 h + 4 h	1200 grit + Zr	Sa = 5.34 nm	* 101 %	
					1200 grit + ZrO	Sa = 9.42 nm	* 103 %	
		EIS [4] (10 × 10 mm^2)	SBF [2]	-	1200 grit + Zr	Sa = 5.34 nm	$i_{corr} = 1.2 \times 10^2$ μA cm^{-2}	
					1200 grit + ZrO	Sa = 9.42 nm	$i_{corr} = 2.6 \times 10^1$ μA cm^{-2}	
Mg-0.5Sr	Rectangular prism	Mass loss	SBF [2]	3 d	1200 grit + Zr	Sa = 4.61 nm	* 13.7 mg	[88]
					1200 grit + ZrO	Sa = 7.29 nm	* 8.52 mg	
		Cell viability	Extract assay (DMEM [3])	24 h + 72 h + 4 h	1200 grit + Zr	Sa = 4.61 nm	* 110 %	
					1200 grit + ZrO	Sa = 7.29 nm	* 126 %	
		EIS [4] (10 × 10 mm^2)	SBF [2]	-	1200 grit + Zr	Sa = 4.61 nm	$i_{corr} = 2.5 \times 10^2$ μA cm^{-2}	
					1200 grit + ZrO	Sa = 7.29 nm	$i_{corr} = 1.7 \times 10^2$ μA cm^{-2}	

[1] PDP: potentiodynamic polarization; [2] SBF: simulated body fluid; [3] DMEM: Dulbecco's Modified Eagle's Medium; [4] EIS: Electrochemical Impedance Spectroscopy

4. Summary of the Influence of Roughness on Degradation

4.1. Mechanical Surface Treatments

Using grinding papers with higher grit size and/or polishing reduced surface roughness and reduced pitting during the degradation tests [71,96]. These results differ from those in studies by Alvarez et al. [94], where polished samples encouraged pitting compared to semi polished surfaces [94]. Some papers reported that roughness affects cell adherence to the surface [21,74]. However, one study showed that cell adherence was not influenced by surface roughness [61]. Some studies showed that roughness did not affect degradation [39,102,108,111]. In contrast, it was demonstrated that a linear relationship existed between roughness and degradation if the surface treatments were comparable (Figures 1 and 2) [72,96,102].

4.2. Chemical Surface Treatments and Coatings

Some studies investigated the correlation between etched AZ31 alloy samples and degradation behavior [38,40,56,70,80,95,103]. In studies by Chiu et al. [80], Supplit et al. [95], Nwaogu et al. [40,56], and Gray Munro et al. [70], it was found that acetic, nitric, and phosphoric acid surface treatment improved the degradation resistance. In a report by Song and Xu [38], sulfuric acid was shown to enhance degradation, contrasting the study by Nwaogu et al. [56]. The roughness after etching was not reported to affect degradation. Etching had positive effects on the surface as it removed contamination and manufacturing marks, resulting in a homogenous morphology [38,40,56,70]. Ca/P- and polymer coatings also led to a more uniform morphology which improved the overall degradation resistance compared to as-received samples [70]. Further investigations of the correlation between coatings and degradation behavior were performed by Yoo et al. [81], Cho et al. [73], and Hwang et al. [79,82]. They studied the influence of PEO on AZ91. They found that a PEO coating increased the surface roughness. In all reports except for that of Yoo et al. [81], the rougher surface resulted in a greater degradation resistance. Guo and An [78] reported, as did as Yoo et al. [81], Cho et al. [73], Hwang et al. [79,82], and Duan et al. [117], that coatings affect the surface roughness. Additionally, ion implantation is one technique that can be used to smooth the surface and increase the degradation resistance [105]. However, Zhao et al. [88] determined that the degradation rate slowed as the roughness increased after ion implantation.

5. Discussion

5.1. Suitable Roughness Values for Biodegradable Mg Implants

Nguyen et al. [72] investigated the influence of roughness on i_{corr} after 6 hours degradation in HBSS (Hank's Balanced Salt Solution) of pure Mg with indirect solid free form fabrication (SFF). After SFF, no postprocessing is necessary, which enables the production of different degrees of surface roughness with the same surface properties. They avoided the influence of different alloy compositions and surface treatments. It was shown that an increase in Ra led to an increase in i_{corr} and mass loss (Table 2, Ref. [72] and Figure 3a). Some reports concerning surface treatments established that there was no direct influence from the roughness on the degradation behavior. The roughness values from these reports are described using two-dimensional values such as Ra, Rq, and Rz, or three-dimensional parameters like Sa, which cannot be compared directly. The difference between macro-roughness and micro-roughness is also not defined. Macro roughness describes the height distribution which comes from a production process like sawing. As such, the macro-roughness of the sample is not going to influence the degradation process in the same way as micro-roughness. Macro-roughness is accompanied by the subsurface stress that results from production and machining, and which has also been reported to affect degradation [38]. Rougher surfaces influenced the pore enlargement of PEO coatings (Figure 3a) which indirectly controls the degradation rate [81].

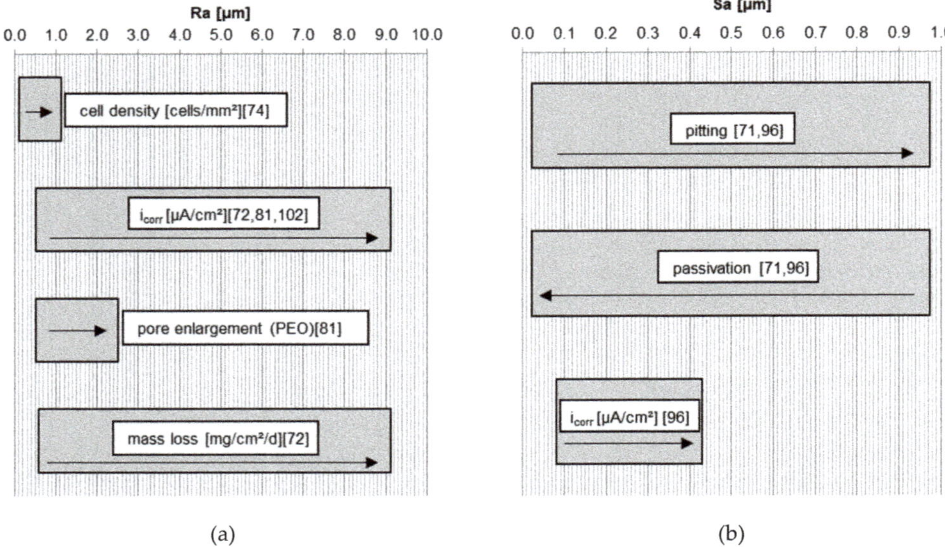

Figure 3. Trend of properties: (**a**) in relation to roughness Ra. The arrow shows the change of properties depending on the roughness according to [72,74,81,102]; (**b**) in relation to roughness Sa. The arrow shows the change of properties depending on the roughness according to [71,96].

There was also a trend of increasing current density when using different grinding papers as seen by Walter and Kannan [96] (Figures 1 and 3b) and using different Almen intensities during shot peening (Figure 2, Figures 1 and 3a) [96,102]. In general, higher roughness diminishes the passivation layer and raises the probability of pitting (Figure 3b). Initial pitting effects are only noticeable within the first six hours [71]. As such, it is suggested that roughness has no long-term effects on the degradation as the morphology changes during immersion in aqueous solutions. However, roughness can influence the initial degradation due to greater peak-to-valley height differences, which results in a higher anodizing surface area [102] with a lower pH solution inside the valleys [71]. This roughness effect fades after a short time as the higher surface peaks are eroded away. A more rapid degradation accelerates this process. Even if roughness has a noticeable effect at the start of degradation, as the surface flattens with time it will quickly become insignificant, as seen in the long-term experiments of Gawlik et al. [57]. A correlation of higher cell adherence with higher roughness for pure Mg was reported in [74] (Figure 3a). Cell toxicity and cell adherence have been mainly tested for Sa and Ra values around 1 µm and below. Some reports about osseointegration showed that the connection between dental implants and bone improves when using implants with rougher surfaces introduced by surface modifications [84,118]. One study reported that a roughness between 1–2 µm led to the best connection between a permanent Ti dental implant and bone [119]. Höh et al. [93] could not confirm a trend relating higher roughness to greater bone implant connectivity for biodegradable Mg-0.8Ca alloys. However, in vivo experiments in rabbits showed that sandblasted cylinders with a higher magnitude of roughness (Ra = 32.7 µm) led to strong gas evolution and material decomposition [93]. Mechanical integrity cannot be obtained if the initial degradation is too rapid. Significant hydrogen evolution hinders cell adherence and thus the formation of a good bond between the bone and implant. The required roughness for cell adherence depends on the kind of cells and the necessity of cell adherence. Stronger cell adherence resulting from higher roughness is needed for osseointegration for example, whereas smoother surfaces are preferred in stent applications where cell adhesion is less important. The influence of roughness on the degradation and cell adherence in in vitro and in vivo experiments is not possible either, due to different experimental set-ups and durations. The difficulty in comparing data of in vitro and in vivo testing is reviewed by Sanchez et al. [120]. From all of these studies, depending on the application, roughness values above Sa or Ra = 0.2 µm are suggested for Mg implants in relation to cell adherence, cell density, and cell survival. Generally, most surface roughness in the studies were in the nm range. Some of the roughness values investigated in the reviewed studies were within the range of 1–2 µm [38,56,71,74,102], or had roughness values below 10 µm [38,71,72,74,81,93,96]. Greatly higher roughness values should be avoided, due to strong initial degradation and gas evolution.

5.2. Suitable Treatments for Biodegradable Mg Implants

Etching is a chemical surface treatment which is highly suitable for biodegradable Mg implants. Depending on the etching solution and the alloy used, chemical etching can vary the surface properties of the alloy. Thus, it is possible to tailor the surface roughness depending on the etching conditions. Even a minor increase in roughness (nm) showed an influence on cell adhesion [74]. Smoother surfaces were also reported to minimize the porosity of coatings. A smaller pore size in a PEO coating decreases the degradation resistance [81]. As such, etching can be used as a pre-treatment for additional coatings or as a surface modification. In addition, etched material is reported to form a stronger passivation layer compared to non-etched material, thus slowing the degradation rate [70]. Etching enables a uniform treatment over the entire surface. It can possibly be used to homogenize the surface, change the surface roughness and morphology, and remove near surface material including contamination and impurities [38,40,56]. It is suspected that it increases the degradation resistance for specific implants such as stents, rods, tubes, and screws; this is very advantageous, as it is not possible to use mechanical surface treatments such as grinding, polishing, and burnishing on these geometries.

6. Conclusions

In general, this review shows that it is difficult to make reliable and clear comparisons between different studies, because several parameters and mechanisms influence the degradation behavior. One of these parameters is the amount and distribution of impurities, a factor that was not assessed in all of the investigations, although it is of critical importance. However, from this review, some rough rules can be derived:

- Considering different roughness values arising from the same type of surface treatment, especially mechanical surface treatments, a trend of increased degradation rate can be seen with higher surface roughness.
- Roughness values arising from different surface treatments are non-comparable, and thus, cannot be compared against a degradation result.
- The roughness of a Mg implant is thought to have a greater influence on initial degradation, compared to long-term degradation. The duration for implant acceptance by the body is negligibly affected by the implant's surface roughness.
- Implant surfaces with roughness values above Sa or Ra = 0.2 µm are unsuitable for initial cell adherence and cell viability. Higher roughness should be avoided, as increased degradation is expected, and consequently, greater local alkalization will occur.
- Ca/P coatings lead to a uniform surface morphology which results in a more uniform degradation over the surface, and decreases the degradation rate compared to uncoated material. Ca/P coated Mg alloys exhibited non-toxic and biocompatible properties.
- Differences in surface roughness and additions of $K_4P_2O_7$ or KF into the electrolyte varied the pore size of PEO coatings, which, in turn, affected the degradation rate of implant materials. A smaller pore size of the PEO coating resulted in higher degradation.
- Acid etching provides a treatment over the entire surface, removing contamination and impurities by removing surface material. In particular, acetic acid and phosphoric acid etching improved the

degradation behavior, i.e., by reducing the degradation rate. Etching allows the surface properties to be tailored in order to adjust the initial and long-term degradation.

Author Contributions: Conceptualization, M.M.G.; Data Curation, M.M.G, V.D., B.W.; Writing—Original Draft Preparation, M.M.G.; Writing—Review and Editing, M.M.G., V.D., B.W., T.E., R.W.R; Visualization, M.M.G., V.D., B.W.; Supervision, B.W., T.E., R.W.-R.

Funding: This research was funded by the Helmholtz Virtual Institute VH-VI-523 (In vivo studies of biodegradable magnesium-based implant materials).

Conflicts of Interest: The authors declare no conflict of interest. The funders had no role in the design of the study; in the collection, analyses, or interpretation of data; in the writing of the manuscript, or in the decision to publish the results.

References

1. Kraus, T.; Fischerauer, S.; Treichler, S.; Martinelli, E.; Eichler, J.; Myrissa, A.; Zötsch, S.; Uggowitzer, P.J.; Löffler, J.F.; Weinberg, A.M. The influence of biodegradable magnesium implants on the growth plate. *Acta Biomater.* **2018**, *66*, 109–117. [CrossRef] [PubMed]
2. Li, G.; Zhang, L.; Wang, L.; Yuan, G.; Dai, K.; Pei, J.; Hao, Y. Dual modulation of bone formation and resorption with zoledronic acid-loaded biodegradable magnesium alloy implants improves osteoporotic fracture healing: An in vitro and in vivo study. *Acta Biomater.* **2018**, *65*, 486–500. [CrossRef] [PubMed]
3. Rahim, M.; Ullah, S.; Mueller, P. Advances and Challenges of Biodegradable Implant Materials with a Focus on Magnesium-Alloys and Bacterial Infections. *Metals* **2018**, *8*, 532. [CrossRef]
4. Choo, J.T.; Lai, S.H.S.; Tang, C.Q.Y.; Thevendran, G. Magnesium-based bioabsorbable screw fixation for hallux valgus surgery—A suitable alternative to metallic implants. *Foot Ankle Surg.* **2018**. [CrossRef] [PubMed]
5. Höhn, S.; Virtanen, S.; Boccaccini, A.R. Protein adsorption on magnesium and its alloys: A review. *Appl. Surf. Sci.* **2019**, *464*, 212–219. [CrossRef]
6. Zhao, N.; Zhu, D. Collagen Self-Assembly on Orthopedic Magnesium Biomaterials Surface and Subsequent Bone Cell Attachment. *PLoS ONE* **2014**, *9*, e110420. [CrossRef] [PubMed]
7. Zeng, R.; Dietzel, W.; Witte, F.; Hort, N.; Blawert, C. Progress and Challenge for Magnesium Alloys as Biomaterials. *Adv. Eng. Mater.* **2008**, *10*, B3–B14. [CrossRef]
8. Vormann, J. Magnesium: nutrition and metabolism. *Mol. Aspects Med.* **2003**, *24*, 27–37. [CrossRef]
9. Song, G. Control of biodegradation of biocompatable magnesium alloys. *Corros. Sci.* **2007**, *49*, 1696–1701. [CrossRef]
10. Echeverry-Rendon, M.; Duque, V.; Quintero, D.; Robledo, S.M.; Harmsen, M.C.; Echeverria, F. Improved corrosion resistance of commercially pure magnesium after its modification by plasma electrolytic oxidation with organic additives. *J. Biomater. Appl.* **2018**, *33*, 725–740. [CrossRef]
11. Gao, Y.; Wang, L.; Li, L.; Gu, X.; Zhang, K.; Xia, J.; Fan, Y. Effect of stress on corrosion of high-purity magnesium in vitro and in vivo. *Acta Biomater.* **2018**, *83*, 477–486. [CrossRef] [PubMed]
12. Gu, X.-N.; Li, S.-S.; Li, X.-M.; Fan, Y.-B. Magnesium based degradable biomaterials: A review. *Front. Mater. Sci.* **2014**, *8*, 200–218. [CrossRef]
13. Castellani, C.; Lindtner, R.A.; Hausbrandt, P.; Tschegg, E.; Stanzl-Tschegg, S.E.; Zanoni, G.; Beck, S.; Weinberg, A.-M. Bone–implant interface strength and osseointegration: Biodegradable magnesium alloy versus standard titanium control. *Acta Biomater.* **2011**, *7*, 432–440. [CrossRef] [PubMed]
14. Li, Z.; Gu, X.; Lou, S.; Zheng, Y. The development of binary Mg-Ca alloys for use as biodegradable materials within bone. *Biomaterials* **2008**, *29*, 1329–1344. [CrossRef] [PubMed]
15. Ma, J.; Zhao, N.; Betts, L.; Zhu, D. Bio-Adaption between Magnesium Alloy Stent and the Blood Vessel: A Review. *J. Mater. Sci. Tech.* **2016**, *32*, 815–826. [CrossRef]
16. Song, G.; Song, S. A Possible Biodegradable Magnesium Implant Material. *Adv. Eng. Mater.* **2007**, *9*, 298–302. [CrossRef]
17. Lu, Y.; Huang, Y.; Feyerabend, F.; Willumeit-Römer, R.; Kainer, K.U.; Hort, N. Microstructure and Mechanical Properties of Mg-Gd Alloys as Biodegradable Implant Materials. In *TMS 2018 147th Annual Meeting & Exhibition Supplemental Proceedings*; Materials Society, T.M., Ed.; Springer International Publishing: Cham, Switzerland, 2018; ISBN 978-3-319-72525-3.

18. Li, P.; Zhou, N.; Qiu, H.; Maitz, M.F.; Wang, J.; Huang, N. In vitro and in vivo cytocompatibility evaluation of biodegradable magnesium-based stents: A review. *Sci. Chin. Mater.* **2018**, *61*, 501–515. [CrossRef]
19. Kim, B.J.; Piao, Y.; Wufuer, M.; Son, W.-C.; Choi, T.H. Biocompatibility and Efficiency of Biodegradable Magnesium-Based Plates and Screws in the Facial Fracture Model of Beagles. *J. Oral Maxillofac. Surg.* **2018**, *76*, 1055. [CrossRef]
20. Ferrando, W.A. Review of corrosion and corrosion control of magnesium alloys and composites. *J. Mater. Eng.* **1989**, *11*, 299–313. [CrossRef]
21. Johnson, I.; Perchy, D.; Liu, H. In vitro evaluation of the surface effects on magnesium-yttrium alloy degradation and mesenchymal stem cell adhesion. *J. Biomed. Mater. Res. Part A* **2012**, *100*, 477–485. [CrossRef]
22. Xin, Y.; Liu, C.; Zhang, X.; Tang, G.; Tian, X.; Chu, P.K. Corrosion behavior of biomedical AZ91 magnesium alloy in simulated body fluids. *J. Mater. Res.* **2007**, *22*, 2004–2011. [CrossRef]
23. Ghali, E. Corrosion and Protection of Magnesium Alloys. *Mater. Sci. Forum* **2000**, *350–351*, 261–272. [CrossRef]
24. Makar, G.L.; Kruger, J. Corrosion of magnesium. *Int. Mater. Rev.* **1993**, *38*, 138–153. [CrossRef]
25. Song, G.; Atrens, A.; Stjohn, D.; Nairn, J.; Li, Y. The electrochemical corrosion of pure magnesium in 1 N NaCl. *Corros. Sci.* **1997**, *39*, 855–875. [CrossRef]
26. *Corrosion*; ASM International; Korb, L.J. (Eds.) ASM handbook; [10. ed.], 7. print; ASM International: Materials Park, OH, USA, 2001; ISBN 978-0-87170-019-3.
27. Hu, H.; Nie, X.; Ma, Y. Corrosion and Surface Treatment of Magnesium Alloys. In *Magnesium Alloys—Properties in Solid and Liquid States*; Czerwinski, F., Ed.; Intech Open: London, UK, 2014; ISBN 978-953-51-1728-5.
28. Baboian, R.; Dean, S.; Hack, H.; Haynes, G.; Scully, J.; Sprowls, D. *Corrosion Tests and Standards: Application and Interpretation*; ASTM Manual Series: Philadelphia, PA, USA, 1995.
29. Song, G.L.; Atrens, A. Corrosion Mechanisms of Magnesium Alloys. *Adv. Eng. Mater.* **1999**, *1*, 11–33. [CrossRef]
30. Iglesias, C.; Bodelón, O.G.; Montoya, R.; Clemente, C.; Garcia-Alonso, M.C.; Rubio, J.C.; Escudero, M.L. Fracture bone healing and biodegradation of AZ31 implant in rats. *Biomed. Mater.* **2015**, *10*, 025008. [CrossRef]
31. Shalabi, M.M.; Gortemaker, A.; Hof, M.A.V.; Jansen, J.A.; Creugers, N.H.J. Implant Surface Roughness and Bone Healing: a Systematic Review. *J. Dent. Res.* **2006**, *85*, 496–500. [CrossRef]
32. Wang, J.; Tang, J.; Zhang, P.; Li, Y.; Wang, J.; Lai, Y.; Qin, L. Surface modification of magnesium alloys developed for bioabsorbable orthopedic implants: A general review. *J. Biomed. Mater. Res. B Appl. Biomater.* **2012**, *100*, 1691–1701. [CrossRef]
33. Kaesel, V.; Tai, P.-T.; Bach, F.-W.; Haferkamp, H.; Witte, F.; Windhagen, H. Approach to Control the Corrosion of Magnesium by Alloying. In *Magnesium*; Kainer, K.U., Ed.; Wiley-VCH Verlag GmbH & Co. KGaA: Weinheim, Germany, 2005; pp. 534–539. ISBN 978-3-527-60356-5.
34. Yun, Y.; Dong, Z.; Lee, N.; Liu, Y.; Xue, D.; Guo, X.; Kuhlmann, J.; Doepke, A.; Halsall, H.B.; Heineman, W.; et al. Revolutionizing biodegradable metals. *Mate. Today* **2009**, *12*, 22–32. [CrossRef]
35. Bland, L.G.; Gusieva, K.; Scully, J.R. Effect of Crystallographic Orientation on the Corrosion of Magnesium: Comparison of Film Forming and Bare Crystal Facets using Electrochemical Impedance and Raman Spectroscopy. *Electrochim. Acta* **2017**, *227*, 136–151. [CrossRef]
36. Zhao, Y.-C.; Huang, G.-S.; Wang, G.-G.; Han, T.-Z.; Pan, F.-S. Influence of Grain Orientation on the Corrosion Behavior of Rolled AZ31 Magnesium Alloy. *Acta Metall. Sin.* **2015**, *28*, 1387–1393. [CrossRef]
37. Liu, M.; Qiu, D.; Zhao, M.-C.; Song, G.; Atrens, A. The effect of crystallographic orientation on the active corrosion of pure magnesium. *Scr. Mater.* **2008**, *58*, 421–424. [CrossRef]
38. Song, G.-L.; Xu, Z. The surface, microstructure and corrosion of magnesium alloy AZ31 sheet. *Electrochim. Acta* **2010**, *55*, 4148–4161. [CrossRef]
39. Pu, Z.; Song, G.-L.; Yang, S.; Outeiro, J.C.; Dillon, O.W.; Puleo, D.A.; Jawahir, I.S. Grain refined and basal textured surface produced by burnishing for improved corrosion performance of AZ31B Mg alloy. *Corros. Sci.* **2012**, *57*, 192–201. [CrossRef]
40. Nwaogu, U.C.; Blawert, C.; Scharnagl, N.; Dietzel, W.; Kainer, K.U. Effects of organic acid pickling on the corrosion resistance of magnesium alloy AZ31 sheet. *Corros. Sci.* **2010**, *52*, 2143–2154. [CrossRef]

41. Aung, N.N.; Zhou, W. Effect of grain size and twins on corrosion behaviour of AZ31B magnesium alloy. *Corros. Sci.* **2010**, *52*, 589–594. [CrossRef]
42. Pardo, A.; Merino, M.C.; Coy, A.E.; Arrabal, R.; Viejo, F.; Matykina, E. Corrosion behaviour of magnesium/aluminium alloys in 3.5 wt.% NaCl. *Corros. Sci.* **2008**, *50*, 823–834. [CrossRef]
43. Ambat, R.; Aung, N.N.; Zhou, W. Evaluation of microstructural effects on corrosion behaviour of AZ91D magnesium alloy. *Corros. Sci.* **2000**, *42*, 1433–1455. [CrossRef]
44. Zheng, X.; Dong, J.; Xiang, Y.; Chang, J.; Wang, F.; Jin, L.; Wang, Y.; Ding, W. Formability, mechanical and corrosive properties of Mg-Nd-Zn-Zr magnesium alloy seamless tubes. *Mater. Des.* **2010**, *31*, 1417–1422. [CrossRef]
45. Liu, Z.; Schade, R.; Luthringer, B.; Hort, N.; Rothe, H.; Müller, S.; Liefeith, K.; Willumeit-Römer, R.; Feyerabend, F. Influence of the Microstructure and Silver Content on Degradation, Cytocompatibility, and Antibacterial Properties of Magnesium-Silver Alloys In Vitro. *Oxidat. Med. Cell. Longev.* **2017**, *2017*, 1–14. [CrossRef]
46. Stellwagen, E.; Babul, J. Stabilization of the globular structure of ferricytochrome c by chloride in acidic solvents. *Biochemistry* **1975**, *14*, 5135–5140. [CrossRef] [PubMed]
47. Jönsson, M.; Persson, D. The influence of the microstructure on the atmospheric corrosion behaviour of magnesium alloys AZ91D and AM50. *Corros. Sci.* **2010**, *52*, 1077–1085. [CrossRef]
48. Ben-Haroush, M.; Ben-Hamu, G.; Eliezer, D.; Wagner, L. The relation between microstructure and corrosion behavior of AZ80 Mg alloy following different extrusion temperatures. *Corros. Sci.* **2008**, *50*, 1766–1778. [CrossRef]
49. Ralston, K.D.; Birbilis, N. Effect of Grain Size on Corrosion: A Review. *Corrosion* **2010**, *66*, 075005. [CrossRef]
50. Liu, Y.; Liu, D.; You, C.; Chen, M. Effects of grain size on the corrosion resistance of pure magnesium by cooling rate-controlled solidification. *Front. Mater. Sci.* **2015**, *9*, 247–253. [CrossRef]
51. Lu, Y.; Bradshaw, A.R.; Chiu, Y.L.; Jones, I.P. Effects of secondary phase and grain size on the corrosion of biodegradable Mg-Zn-Ca alloys. *Mater. Sci. Eng. C* **2015**, *48*, 480–486. [CrossRef]
52. Kutniy, K.V.; Papirov, I.I.; Tikhonovsky, M.A.; Pikalov, A.I.; Sivtzov, S.V.; Pirozhenko, L.A.; Shokurov, V.S.; Shkuropatenko, V.A. Influence of grain size on mechanical and corrosion properties of magnesium alloy for medical implants. *Materialwiss. Werkstofftech.* **2009**, *40*, 242–246. [CrossRef]
53. Ullmann, B.; Reifenrath, J.; Seitz, J.-M.; Bormann, D.; Meyer-Lindenberg, A. Influence of the grain size on the in vivo degradation behaviour of the magnesium alloy LAE442. *Proc. Inst. Mech. Eng. H* **2013**, *227*, 317–326. [CrossRef]
54. Zeng, R.-C.; Chen, J.; Dietzel, W.; Zettler, R.; dos Santos, J.F.; Lucia Nascimento, M.; Kainer, K.U. Corrosion of friction stir welded magnesium alloy AM50. *Corros. Sci.* **2009**, *51*, 1738–1746. [CrossRef]
55. Zhang, T.; Li, Y.; Wang, F. Roles of β phase in the corrosion process of AZ91D magnesium alloy. *Corros. Sci.* **2006**, *48*, 1249–1264. [CrossRef]
56. Nwaogu, U.C.; Blawert, C.; Scharnagl, N.; Dietzel, W.; Kainer, K.U. Influence of inorganic acid pickling on the corrosion resistance of magnesium alloy AZ31 sheet. *Corros. Sci.* **2009**, *51*, 2544–2556. [CrossRef]
57. Gawlik, M.M.; Steiner, M.; Wiese, B.; González, J.; Feyerabend, F.; Dahms, M.; Ebel, T.; Willumeit-Römer, R. The Effects of HAc Etching on the Degradation Behavior of Mg-5Gd. *J. Med. Mater. Tech.* **2017**, *1*, 22–25.
58. Snir, Y.; Ben-Hamu, G.; Eliezer, D.; Abramov, E. Effect of compression deformation on the microstructure and corrosion behavior of magnesium alloys. *J. Alloys Compd.* **2012**, *528*, 84–90. [CrossRef]
59. Wang, B.J.; Xu, D.K.; Dong, J.H.; Ke, W. Effect of the crystallographic orientation and twinning on the corrosion resistance of an as-extruded Mg-3Al-1Zn (wt.%) bar. *Scr. Mater.* **2014**, *88*, 5–8. [CrossRef]
60. Zou, G.; Peng, Q.; Wang, Y.; Liu, B. The effect of extension twinning on the electrochemical corrosion properties of Mg-Y alloys. *J. Alloys Compd.* **2015**, *618*, 44–48. [CrossRef]
61. Liu, H. The effects of surface and biomolecules on magnesium degradation and mesenchymal stem cell adhesion. *J. Biomed. Mater. Res. Part A* **2011**, *99*, 249–260. [CrossRef]
62. Kieke, M.; Feyerabend, F.; Lemaitre, J.; Behrens, P.; Willumeit-Römer, R. Degradation rates and products of pure magnesium exposed to different aqueous media under physiological conditions. *BioNanoMaterials* **2016**, *17*, 131–143. [CrossRef]
63. Xin, Y.; Hu, T.; Chu, P.K. Influence of Test Solutions on In Vitro Studies of Biomedical Magnesium Alloys. *J. Electrochem. Soc.* **2010**, *157*, C238. [CrossRef]

64. Agha, N.A.; Feyerabend, F.; Mihailova, B.; Heidrich, S.; Bismayer, U.; Willumeit-Römer, R. Magnesium degradation influenced by buffering salts in concentrations typical of in vitro and in vivo models. *Mater. Sci. Eng. C* **2016**, *58*, 817–825. [CrossRef]
65. Uddin, M.S.; Hall, C.; Murphy, P. Surface treatments for controlling corrosion rate of biodegradable Mg and Mg-based alloy implants. *Sci. Tech. Adv. Mater.* **2015**, *16*, 053501. [CrossRef]
66. Yang, J.; Cui, F.; Lee, I.S. Surface Modifications of Magnesium Alloys for Biomedical Applications. *Ann. Biomed. Eng.* **2011**, *39*, 1857–1871. [CrossRef] [PubMed]
67. *Surface Modification of Magnesium and Its Alloys for Biomedical Applications*; Sankara Narayanan, T.S.N.; Park, I.-S.; Lee, M.-H. (Eds.) Woodhead Publishing series in biomaterials; Elsevier/Woodhead Publishing: Cambridge, UK; Waltham, MA, USA, 2015; ISBN 978-1-78242-077-4.
68. Gray, J.E.; Luan, B. Protective coatings on magnesium and its alloys—A critical review. *J. Alloys Compd.* **2002**, *336*, 88–113. [CrossRef]
69. Hornberger, H.; Virtanen, S.; Boccaccini, A.R. Biomedical coatings on magnesium alloys—A review. *Acta Biomater.* **2012**, *8*, 2442–2455. [CrossRef] [PubMed]
70. Gray-Munro, J.E.; Seguin, C.; Strong, M. Influence of surface modification on the in vitro corrosion rate of magnesium alloy AZ31. *J. Biomed. Mater. Res. Part A* **2009**, *91A*, 221–230. [CrossRef] [PubMed]
71. Walter, R.; Kannan, M.B.; He, Y.; Sandham, A. Effect of surface roughness on the in vitro degradation behaviour of a biodegradable magnesium-based alloy. *Appl. Surf. Sci.* **2013**, *279*, 343–348. [CrossRef]
72. Nguyen, T.L.; Blanquet, A.; Staiger, M.P.; Dias, G.J.; Woodfield, T.B.F. On the role of surface roughness in the corrosion of pure magnesium in vitro. *J. Biomed. Mater. Res. B Appl. Biomater.* **2012**, *100*, 1310–1318. [CrossRef]
73. Cho, J.-Y.; Hwang, D.-Y.; Lee, D.-H.; Yoo, B.; Shin, D.-H. Influence of potassium pyrophosphate in electrolyte on coated layer of AZ91 Mg alloy formed by plasma electrolytic oxidation. *Trans. Nonferr. Met. Soc. China* **2009**, *19*, 824–828. [CrossRef]
74. Lorenz, C.; Brunner, J.G.; Kollmannsberger, P.; Jaafar, L.; Fabry, B.; Virtanen, S. Effect of surface pre-treatments on biocompatibility of magnesium. *Acta Biomater.* **2009**, *5*, 2783–2789. [CrossRef]
75. Laycock, N.J.; Noh, J.S.; White, S.P.; Krouse, D.P. Computer simulation of pitting potential measurements. *Corros. Sci.* **2005**, *47*, 3140–3177. [CrossRef]
76. Burstein, G.T.; Vines, S.P. Repetitive Nucleation of Corrosion Pits on Stainless Steel and the Effects of Surface Roughness. *J. Electrochem. Soc.* **2001**, *148*, B504. [CrossRef]
77. Suter, T.; Müller, Y.; Schmutz, P.; von Trzebiatowski, O. Microelectrochemical Studies of Pit Initiation on High Purity and Ultra High Purity Aluminum. *Adv. Eng. Mater.* **2005**, *7*, 339–348. [CrossRef]
78. Guo, H.F.; An, M.Z. Growth of ceramic coatings on AZ91D magnesium alloys by micro-arc oxidation in aluminate–fluoride solutions and evaluation of corrosion resistance. *Appl. Surf. Sci.* **2005**, *246*, 229–238. [CrossRef]
79. Hwang, D.Y.; Kim, Y.M.; Shin, D.H. Corrosion Resistance of Plasma-Anodized AZ91 Mg Alloy in the Electrolyte with/without Potassium Fluoride. *Mater. Trans.* **2009**, *50*, 671–678. [CrossRef]
80. Chiu, L.-H.; Chen, C.-C.; Yang, C.-F. Improvement of corrosion properties in an aluminum-sprayed AZ31 magnesium alloy by a post-hot pressing and anodizing treatment. *Surf. Coat. Technol.* **2005**, *191*, 181–187. [CrossRef]
81. Yoo, B.; Shin, K.R.; Hwang, D.Y.; Lee, D.H.; Shin, D.H. Effect of surface roughness on leakage current and corrosion resistance of oxide layer on AZ91 Mg alloy prepared by plasma electrolytic oxidation. *Appl. Surf. Sci.* **2010**, *256*, 6667–6672. [CrossRef]
82. Hwang, D.K.; Yoo, B.Y.; Cho, J.Y.; Lee, D.H.; Shin, D.H. Effect of surface roughness on corrosion resistance of oxide layer on AZ91 Mg alloy prepared by plasma electrolytic oxidation. In Proceedings of the 214th ECS Meeting, Honolulu, HI, USA, 12–17 October 2008.
83. Burstein, G.T.; Pistorius, P.C. Surface Roughness and the Metastable Pitting of Stainless Steel in Chloride Solutions. *Corros. Sci.* **1995**, *51*, 380–385. [CrossRef]
84. Parekh, R.B.; Shetty, O.; Tabassum, R. Surface Modifications for Endosseous Dental Implants. *Int. J. Oral Implantol. Clin. Res.* **2012**, 3, 116–121. [CrossRef]
85. Lacefield, W.R. Materials Characteristics of Uncoated/Ceramic-Coated Implant Materials. *Adv. Dent. Res.* **1999**, *13*, 21–26. [CrossRef]

86. Suzuki, K.; Aoki, K.; Ohya, K. Effects of surface roughness of titanium implants on bone remodeling activity of femur in rabbits. *Bone* **1997**, *21*, 507–514. [CrossRef]
87. Von der Höh, N.; von Rechenberg, B.; Bormann, D.; Lucas, A.; Meyer-Lindenberg, A. Influence of different surface machining treatments of resorbable magnesium alloy implants on degradation—EDX-analysis and histology results. *Materialwiss. Werkstofftech.* **2009**, *40*, 88–93. [CrossRef]
88. Zhao, Y.; Jamesh, M.I.; Li, W.K.; Wu, G.; Wang, C.; Zheng, Y.; Yeung, K.W.K.; Chu, P.K. Enhanced antimicrobial properties, cytocompatibility, and corrosion resistance of plasma-modified biodegradable magnesium alloys. *Acta Biomater.* **2014**, *10*, 544–556. [CrossRef] [PubMed]
89. Mustafa, K.; Lopez, B.S.; Hultenby, K.; Wennerberg, A.; Arvidson, K. Attachment and proliferation of human oral fibroblasts to titanium surfaces blasted with TiO2 particles. A scanning electron microscopic and histomorphometric analysis. *Clin. Oral Implants Res.* **1998**, *9*, 195–207. [CrossRef]
90. Soskolne, W.A.; Cohen, S.; Shapira, L.; Sennerby, L.; Wennerberg, A. The effect of titanium surface roughness on the adhesion of monocytes and their secretion of TNF-alpha and PGE2. *Clin. Oral Implants Res.* **2002**, *13*, 86–93. [CrossRef] [PubMed]
91. Derhami, K.; Wolfaardt, J.F.; Wennerberg, A.; Scott, P.G. Quantifying the adherence of fibroblasts to titanium and its enhancement by substrate-attached material. *J. Biomed. Mater. Res.* **2000**, *52*, 315–322. [CrossRef]
92. Boyan, B.D.; Lohmann, C.H.; Dean, D.D.; Sylvia, V.L.; Cochran, D.L.; Schwartz, Z. Mechanisms Involved in Osteoblast Response to Implant Surface Morphology. *Ann. Rev. Mater. Res.* **2001**, *31*, 357–371. [CrossRef]
93. Höh, N.V.D.; Bormann, D.; Lucas, A.; Denkena, B.; Hackenbroich, C.; Meyer-Lindenberg, A. Influence of Different Surface Machining Treatments of Magnesium-based Resorbable Implants on the Degradation Behavior in Rabbits. *Adv. Eng. Mater.* **2009**, *11*, B47–B54. [CrossRef]
94. Alvarez, R.B.; Martin, H.J.; Horstemeyer, M.F.; Chandler, M.Q.; Williams, N.; Wang, P.T.; Ruiz, A. Corrosion relationships as a function of time and surface roughness on a structural AE44 magnesium alloy. *Corros. Sci.* **2010**, *52*, 1635–1648. [CrossRef]
95. Supplit, R.; Koch, T.; Schubert, U. Evaluation of the anti-corrosive effect of acid pickling and sol–gel coating on magnesium AZ31 alloy. *Corros. Sci.* **2007**, *49*, 3015–3023. [CrossRef]
96. Walter, R.; Kannan, M.B. Influence of surface roughness on the corrosion behaviour of magnesium alloy. *Mater. Des.* **2011**, *32*, 2350–2354. [CrossRef]
97. Sharma, A.K. *Text Book of Correlation and Regression*; Discovery Publishing House: New Delhi, Delhi, India, 2005; ISBN 978-81-7141-935-7.
98. *Characterisation of areal surface texture*; Leach, R. (Ed.) Springer: Berlin, Germany, 2013; ISBN 978-3-642-36458-7.
99. Jin, S.; Amira, S.; Ghali, E. Electrochemical Impedance Spectroscopy Evaluation of the Corrosion Behavior of Die Cast and Thixocast AXJ530 Magnesium Alloy in Chloride Solution. *Adv. Eng. Mater.* **2007**, *9*, 75–83. [CrossRef]
100. Song, G.L.; St John, D.H.; Abbott, T. Corrosion behaviour of a pressure die cast magnesium alloy. *Int. J. Cast Met. Res.* **2005**, *18*, 174–180. [CrossRef]
101. Nudelman, F.; Pieterse, K.; George, A.; Bomans, P.H.H.; Friedrich, H.; Brylka, L.J.; Hilbers, P.A.J.; de With, G.; Sommerdijk, N.A.J.M. The role of collagen in bone apatite formation in the presence of hydroxyapatite nucleation inhibitors. *Nat. Mater.* **2010**, *9*, 1004–1009. [CrossRef] [PubMed]
102. Mhaede, M.; Pastorek, F.; Hadzima, B. Influence of shot peening on corrosion properties of biocompatible magnesium alloy AZ31 coated by dicalcium phosphate dihydrate (DCPD). *Mater. Sci. Eng. C* **2014**, *39*, 330–335. [CrossRef] [PubMed]
103. Pompa, L.; Rahman, Z.U.; Munoz, E.; Haider, W. Surface characterization and cytotoxicity response of biodegradable magnesium alloys. *Mater. Sci. Eng. C* **2015**, *49*, 761–768. [CrossRef] [PubMed]
104. Bakhsheshi-Rad, H.R.; Idris, M.H.; Abdul-Kadir, M.R. Synthesis and in vitro degradation evaluation of the nano-HA/MgF2 and DCPD/MgF2 composite coating on biodegradable Mg-Ca-Zn alloy. *Surf. Coat. Technol.* **2013**, *222*, 79–89. [CrossRef]
105. Jamesh, M.; Wu, G.; Zhao, Y.; Chu, P.K. Effects of silicon plasma ion implantation on electrochemical corrosion behavior of biodegradable Mg-Y-RE Alloy. *Corros. Sci.* **2013**, *69*, 158–163. [CrossRef]
106. Almen, J.O. *Shot Blasting to Increase Fatigue Resistance*; SAE International: Warrendale, PA, USA, 1943.
107. Guagliano, M. Relating Almen intensity to residual stresses induced by shot peening: A numerical approach. *J. Mater. Process.Tech.* **2001**, *110*, 277–286. [CrossRef]

108. Denkena, B.; Lucas, A. Biocompatible Magnesium Alloys as Absorbable Implant Materials—Adjusted Surface and Subsurface Properties by Machining Processes. *CIRP Ann.-Manuf. Technol.* **2007**, *56*, 113–116. [CrossRef]
109. *American Society of Mechanical Engineers Surface Texture Symbols*; ASME: New York, 1996; ISBN 978-0-7918-2319-4.
110. Guo, Y.B.; Salahshoor, M. Process mechanics and surface integrity by high-speed dry milling of biodegradable magnesium–calcium implant alloys. *CIRP Ann.-Manuf. Technol.* **2010**, *59*, 151–154. [CrossRef]
111. Denkena, B.; Lucas, A.; Thorey, F.; Waizy, H.; Angrisani, N.; Meyer-Lindenberg, A. Biocompatible Magnesium Alloys as Degradable Implant Materials—Machining Induced Surface and Subsurface Properties and Implant Performance. In *Special Issues on Magnesium Alloys*; Monteiro, W.A., Ed.; Intech Open: London, UK, 2011; ISBN 978-953-307-391-0.
112. Dorozhkin, S.V. Calcium orthophosphate-based biocomposites and hybrid biomaterials. *J. Mater. Sci.* **2009**, *44*, 2343–2387. [CrossRef]
113. Dorozhkin, S.V. Calcium Orthophosphate Cements and Concretes. *Materials* **2009**, *2*, 221–291. [CrossRef]
114. Berzina-Cimdina, L.; Borodajenko, N. Research of Calcium Phosphates Using Fourier Transform Infrared Spectroscopy. In *Infrared Spectroscopy—Materials Science, Engineering and Technology*; Theophanides, T., Ed.; Intech Open: London, UK, 2012; ISBN 978-953-51-0537-4.
115. El Kady, A.M.; Mohamed, K.R.; El-Bassyouni, G.T. Fabrication, characterization and bioactivity evaluation of calcium pyrophosphate/polymeric biocomposites. *Ceram. Int.* **2009**, *35*, 2933–2942. [CrossRef]
116. *Introduction to Biomaterials*; Shi, D. (Ed.) Tsinghua University Press; World Scientific: Beijing, China; Singapore; Hackensack, NJ, USA, 2006; ISBN 978-7-302-10807-8.
117. Duan, H.; Yan, C.; Wang, F. Growth process of plasma electrolytic oxidation films formed on magnesium alloy AZ91D in silicate solution. *Electrochim. Acta* **2007**, *52*, 5002–5009. [CrossRef]
118. Novaes, A.B.; de Souza, S.L.S.; de Barros, R.R.M.; Pereira, K.K.Y.; Iezzi, G.; Piatelli, A. Influence of Implant Surfaces on Osseointegration. *Braz. Dent. J.* **2010**, *21*, 471–481. [CrossRef]
119. Wennerberg, A.; Albrektsson, T. Suggested guidelines for the topographic evaluation of implant surfaces. *Int. J. Oral Maxillofac. Implant.* **2000**, *15*, 331–344.
120. Sanchez, A.H.M.; Luthringer, B.J.C.; Feyerabend, F.; Willumeit, R. Mg and Mg alloys: How comparable are in vitro and in vivo corrosion rates? A review. *Acta Biomater.* **2015**, *13*, 16–31. [CrossRef]

MDPI
St. Alban-Anlage 66
4052 Basel
Switzerland
Tel. +41 61 683 77 34
Fax +41 61 302 89 18
www.mdpi.com

Materials Editorial Office
E-mail: materials@mdpi.com
www.mdpi.com/journal/materials

www.ingramcontent.com/pod-product-compliance
Lightning Source LLC
LaVergne TN
LVHW070121170726
843515LV00007B/1730